MICROBES IN THE SEA

ELLIS HORWOOD SERIES IN MARINE SCIENCE

Series Editor: T. D. ALLAN, Institute of Oceanograpic Sciences, Wormley, Surrey

A programme of authoritative books to keep professional scientists and students up-to-date with new advances in the marine disciplines — physics, chemistry, biology, geology and geophysics with particular attention to advanced techniques and including comprehensive studies of the role of remote sensing, amongst a wide range of important topics in the field.

SATELLITE MICROWAVE REMOTE SENSING
T. D. ALLAN, Institute of Oceanographic Sciences, Wormley, Surrey
PHYSICAL OCEANOGRAPHY OF COASTAL WATERS
K. F. BOWDEN, University of Liverpool
REMOTE SENSING IN METEOROLOGY, OCEANOGRAPHY AND HYDROLOGY
Edited by A. P. CRACKNELL, Carnegie Laboratory of Physics, University of Dundee
SATELLITE OCEANOGRAPHY
I. S. ROBINSON, University of Southampton
NEW PERSPECTIVES IN MARINE GEOLOGY
R. C. SEARLE and R. B. KIDD, Institute of Oceanographic Sciences, Surrey
MARINE CORROSION IN OFFSHORE STRUCTURES
Edited by J. R. MERCER, University of Aberdeen
MICROBES IN THE SEA
M. A. SLEIGH, University of Southampton

ELLIS HORWOOD SERIES IN MARINE SCIENCE

Series Editor: T. D. ALLAN, Institute of Oceanograpic Sciences, Wormley, Surrey

A programme of authoritative books to keep professional scientists and students up-to-date with new advances in the marine disciplines — physics, chemistry, biology, geology and geophysics with particular attention to advanced techniques and including comprehensive studies of the role of remote sensing, amongst a wide range of important topics in the field.

SATELLITE MICROWAVE REMOTE SENSING
T. D. ALLAN, Institute of Oceanographic Sciences, Wormley, Surrey
PHYSICAL OCEANOGRAPHY OF COASTAL WATERS
K. F. BOWDEN, University of Liverpool
REMOTE SENSING IN METEOROLOGY, OCEANOGRAPHY AND HYDROLOGY
Edited by A. P. CRACKNELL, Carnegie Laboratory of Physics, University of Dundee
SATELLITE OCEANOGRAPHY
I. S. ROBINSON, University of Southampton
NEW PERSPECTIVES IN MARINE GEOLOGY
R. C. SEARLE and R. B. KIDD, Institute of Oceanographic Sciences, Surrey
MARINE CORROSION IN OFFSHORE STRUCTURES
Edited by J. R. MERCER, University of Aberdeen
MICROBES IN THE SEA
M. A. SLEIGH, University of Southampton

MICROBES IN THE SEA

MICROBES IN THE SEA

M. A. SLEIGH, Ph.D.
Department of Biology
University of Southampton

ELLIS HORWOOD LIMITED
Publishers · Chichester

Halsted Press: a division of
JOHN WILEY & SONS
New York · Chichester · Brisbane · Toronto

First published in 1987 by
ELLIS HORWOOD LIMITED
Market Cross House, Cooper Street,
Chichester, West Sussex, PO19 1EB, England
The publisher's colophon is reproduced from James Gillison's drawing of the ancient Market Cross, Chichester.

Distributors:

Australia and New Zealand:
JACARANDA WILEY LIMITED
GPO Box 859, Brisbane, Queensland 4001, Australia

Canada:
JOHN WILEY & SONS CANADA LIMITED
22 Worcester Road, Rexdale, Ontario, Canada

Europe and Africa:
JOHN WILEY & SONS LIMITED
Baffins Lane, Chichester, West Sussex, England

North and South America and the rest of the world:
Halsted Press: a division of
JOHN WILEY & SONS
605 Third Avenue, New York, NY 10158, USA

© **1987 M.A. Sleigh/Ellis Horwood Limited**

British Library Cataloguing in Publication Data
Microbes in the sea.
1. Marine microbiology
I. Sleigh, M.A.
576'.192   QR106

Library of Congress CIP data available

ISBN 0–7458–0326–1 (Ellis Horwood Limited)
ISBN 0–470–20978–X (Halsted Press)

Printed in Great Britain by R.J. Acford, Chichester

Table of contents

Foreword

Professor P.J.LeB. Williams

In 1979, a group of French marine microbiologists organized a national symposium in Marseille. They followed this with another held two years later also in Marseille. These meetings attracted enough international attention to persuade the French to convene the first of the present series in 1982, again in Marseille. The second international meeting was held in Brest, again on French initiative, in October 1984, and was attended by 120 microbiologists representing 18 countries from 4 continents. At this meeting a decision was taken to continue the series on a pan-European basis, but to maintain its worldwide international character. The current meeting is therefore the third in the current series, but the first to be held outside France. The Federation of European Microbiological Societies (FEMS) has recognized the series. In line with the original pattern, that the meeting is organized by a national group from within the countries contributing to FEMS, the current meeting is being organized by British microbiologists and the next by scientists from West Germany in 1989.

In considering the general format of this third Symposium, the committee decided to deal with fundamental practical and methodological aspects of contemporary marine microbiology. Instead of publishing a selection of papers read at the meeting, keynote speakers have been invited to contribute to a Symposium volume to be published at the Symposium.

The committee wishes to thank warmly the distinguished panel of authors who have been kind enough to write chapters, and Michael Sleigh for his commitment to the volume and expert editorship. In his opening chapter John Sieburth of the University of Rhode Island has put forward some new ideas on methane generation, while Henry Blackburn, from Aarhus University, Denmark, explores food webs in the sediment. Julian Wimpenny of Cardiff has considered microbial ecology down at the microbe scale of size, while Eve Southward (MBA, Plymouth) reminds us that microbes do not live in isolation but are part of a living world including invertebrates. Energy considerations as seen from the microbe's point of view as it meets different levels of available nutrient are the theme of Steffan Kjelleberg from Gothenberg, Sweden.

The applied fields of marine microbiology are represented by a discussion of corrosion from Bill Hamilton of Aberdeen and of the related topic of adhesion from Brian Egan from Menai Bridge. Rita Colwell is the second United States author and her chapter peers into the future and considers some of the consequences of genetic engineering. The Symposium also deals with three developing techniques. The evaluation of biomarkers is undertaken by several workers from Scotland, Peter Burkhill (IMER, Plymouth) deals with flow cytometry, and Lena Gustafsson (Gothenburg) reviews the value of micro-calorimetry.

The present Symposium is very much due to the vision and insistance of one man: George Floodgate. George began his career in aquatic microbiology at the Torrey Research Station, working on the Adensonian classification of marine bacteria. Subsequently he moved to the University College of North Wales. George has been a great proponent of marine microbiology as a discipline and has crusaded and agitated for the subject with the energy and commitment of a zealot. He had the vision to see the need for an informal forum for the discussion of marine microbiology and that there was considerable common ground between freshwater and marine microbiology. He was part of the group that set up the Aquatic Microbiology Discussion Group which this year celebrates its 21st anniversary. George has been the major force ensuring its continuity and vigour over the years. The setting up of this group has had far-reaching effects: it brought together the UK aquatic microbiologists, gave them an opportunity to establish contacts, to develop their ideas and to lobby for the subject, and has without doubt been a major force giving rise to the present buoyant and healthy state of aquatic microbiology in the UK. George is a committed European and has taken every opportunity to set up and foster links between microbiologists from his own country and their colleagues elsewhere in Europe. Thus, directly or indirectly, the marine community in the UK and in Europe owes a great deal to George Floodgate's activities over the years; it is fitting and it is with great pleasure that we acknowledge this by dedicating the Symposium Volume to George.

1

Contrary habitats for redox-specific processes: methanogenesis in oxic waters and oxidation in anoxic waters

John McNeil Sieburth, Graduate School of Oceanography, The University of Rhode Island, Bay Campus, South Ferry Road, Narragansett, RI 02882–1197, USA

Process-specific bacteria, as part of consortia of bacteria and protists, can accomplish processes within the physical–chemical context of their microenvironment that at first appear to be unlikely for redox-specific habitats. Examples are fermentation including methanogenesis in oxic habitats, and oxidation of accumulating reduced compounds in anoxic habitats. A fuller understanding of these multi-species, multi-trophic and reciprocal processes may explain major pelagic paradoxes and challenge some of the dogmata of biogeochemistry and primary productivity.

INTRODUCTION

The sea is very dynamic and changeable. A place of extreme conditions and habitats. The physics of the sea modifies and controls its biology, and thus its chemistry. This major phenomenon is only starting to be recognized and has recently been termed dynamic biological oceanography (Legendre and Demers, 1984) or ecohydrodynamics (Nihoul, 1986). The concept has been applied to large scale processes, but it must also have far reaching implications on the microscale of microns that are important to the oxidation–reduction state of microenvironments (Alldredge and Cohen, 1987). To adapt to a changeable sea, the evolving associations of microorganisms have also had to be capricious. Their complex solutions often escape our logical and simplified approach to untangling the complicated web of life in the sea. We try to make them fit our mould instead of taking the necessary steps and time required really to observe and understand the composition and processes of microorganisms occurring in natural populations. Associations or consortia of a variety of trophic-specific bacteria and their protistan partners apparently occur in aggregations or associated with particles. The effects of these active redox sites, therefore, are exaggerated in interfaces such as seasonal and permanent pycnoclines, where the smaller microparticulates are trapped and accumulate (Garfield et al., 1983). Such crucial interfaces have been termed ergoclines (Legendre et al., 1986).

This brief narrative describes how I perceive the microbiota in the ergocline between the upper mixing layer and more static layer below, which can be one or more of the following: chemocline, pycnocline, thermocline or halocline. I am studying the processes in the pycnocline or ergocline with the personnel in my laboratory and with the help of colleagues cited in the acknowledgements who have had crucial inputs into my research. I have attempted to assemble enough of the puzzle so that we have the outlines of a picture of how microbiological processes could occur in stratified waters, both in the upper ocean and in shallow coastal waters.

BACKGROUND

The role of particulates accumulating in the pycnocline has been overlooked

Marine microbiologists do some funny things. When an aggregate becomes visible in a microscopic preparation, it is usually ignored so that it will not interfere with the count of free cells. When a ^{14}C incubation gives an extremely high number compared with the low values of most of the samples, owing presumably to the inclusion of an active particle, it is rejected as spurious. Are we counting and measuring worthless oceanic chaff, while throwing away the productive wheat of the sea? The use of only ^{14}C data misled Ryther (1969) to pronounce the open ocean a desert, an erroneous concept that is still cited. However, the long overdue work on particulates and their microbiology is now gaining momentum. Gordon Riley, who made classic studies on the waters of Long Island Sound, drew our attention to 'organic aggregates' over two decades ago (Riley, 1963). An early report downgraded the importance of bacteria on particles (Jannasch, 1973) because of the use of aerobic colony-forming units on C–C bonded substrates as an index of bacteria, despite a prior use of direct microscopy to demonstrate the occurrence of bacterial aggregations (Jones and Jannasch, 1959). A.E. Kriss and his Soviet colleagues described the pycnocline as a dynamic interface where microbial biomass and activity concentrated (Kriss, 1963), but this lead was not followed up by the marine microbiological community.

In hindsight, the stage was set by the mid-sixties to study the microorganisms and their processes in aggregates. What we lacked were qualified heretics to study phenomena that everyone knew did not exist. Today we are learning much about macroaggregates that rain out and can be caught and studied from sediment traps (Fowler and Knauer, 1986). However, what of the $< 50\,\mu m$ microparticulates that were shown to be so numerous, to be colonized by bacteria, and to fall out very slowly (Riley, 1963; Mullin, 1965; Lenz, 1972)? We microbiologists have virtually overlooked these important centres of microbiological activity. The bacteriological procedures applied to seawater at the time of Riley (Sieburth, 1967) and used by Altschuler in Long Island Sound (Altschuler and Riley, 1964) did show the sporadic activity of aggregations of bacteria (Sieburth, 1971), but were inadequate to detect the anaerobes and C_1-utilizing bacteria, among others, present on these particles.

The proceedings of the symposium on detritus held at Pallanza, Italy (Melchiorri-Santolini and Hopton, 1972) contain some five papers whose authors used direct microscopy including epifluorescence to document the intense bacterial colonization of aquatic particulates. The oxygen-consuming activities of these particles accumulating in the summer pycnocline of Long Island Sound were already being manifested by the summer hypoxia as noted by Hardy and Weyl (1971). This phenomenon has been ignored, until recently (Welch, 1987). It is amazing that the bacteriological processes in the upper water column responsible for creating the hypoxias and anoxias of coastal waters (Rossignol-Strick, 1985; Officer et al., 1984; Seliger et al., 1985), and the upper ocean methane peaks (Swinnerton et al., 1969; Brooks et al., 1973; Seiler and Schmidt, 1974; Scranton and Brewer, 1977) are virtually unstudied. Most marine biologists and chemists seem to ignore an upper ocean methane cycle, or regard it as minor or exotic. I suggest that methanogenesis occurs throughout the oxygenated ocean in microparticulates, but peaks in the false benthos of the upper ocean pycnocline where it plays a major role in carbon cycling.

Adequate knowledge of form and function is a prerequisite for marine microbial ecology

The modern marine microbiologist is not usually trained in, or conversant with, the ultrastructural and cultural aspects that are needed to study the taxonomic and trophic affinities of the microorganisms present in natural populations (Sieburth, 1976, 1979, 1983, 1984). This is because few ecologists feel that they should invest in the equipment and time to learn these procedures in addition to the other facets of their craft. Conversely, there are few taxonomists who can get past the intricacies of their craft to look at the bigger inter-taxonomic relationships. The poor state of taxonomy in marine microbiology is also due to the nature of the funding system that only pays lip service to systematics, and does not adequately fund this basic area of research required for meaningful ecological studies. The foreword to the *Annual Review of Ecology and Systematics* states the necessary dependence of these seemingly unrelated fields of study. The lack of adequate funding and training for microbiological taxonomists, especially protistologists, weakens the foundation of marine microbial ecology. Instead of a background in organismic microbiology, the modern marine microbiologists have a molecular bent. Elegant procedures have been developed for the incorporation of [^{3}H]thymidine (Fuhrman and Azam, 1982) and [^{3}H]leucine (Kirchman et al., 1985) that indicate rates of DNA and protein synthesis and thus bacterial cell division and growth, respectively. Such procedures should be correlated with those on trophic-specific bacteria and their processes, since the rates and the models built from such determinations are only as good as the knowledge of the form and function of the participants in the ecosystem.

The suspended microparticulates, with their associations of bacteria and the smaller protists, may be major loci of biochemical activity in the sea. Although their existence is well recognized (see papers in Melchiorri-Satolini and Hopton, 1972), the microbiology and biochemical activity of these microparticulates is relatively unstudied. Such microparticulates with their microbiota appear to accumulate in pycnoclines where they essentially form a false benthos (Sieburth, 1983, 1986) and carry out reducing processes thought to be exclusive to the benthos. The macroparticulates that rain down towards the sea floor are recognized for their dark CO_2 fixation which is assumed to be due to chemotrophy (Karl *et al.*, 1984), but could also be due in part to C_1 heterotrophy. The lack of ultrastructural and cultural observations in such studies prevents the trophic processes from being positively identified.

Paradoxes in the ergocline of the pycnocline

The potential role of the summer pycnocline in the ecohydrodynamics of the upper ocean, shallow coastal waters and basins is worthy of our attention. Key processes conducted by trophic-specific bacteria can apparently occur in illogical habitats, through associations of microorganisms in aggregations or consortia. Enigmatic chemical phenomena such as unbalanced cycles of CO_2 and O_2 (Johnson *et al.*, 1981a), nocturnal and pycnocline peaks in HCHO (Johnson *et al.*, 1983; Eberhardt and Sieburth, 1985), and the CH_4 peak in the oceanic pycnocline (Scranton and Brewer, 1977), among others, are probably just the tips of paradoxical icebergs screaming for attention.

My laboratory has spent the last four years studying the upper ocean methane cycle. In the process, we have stumbled upon what appears to be the key to *in situ* mineralization that might explain in part the mechanism for sustained upper ocean productivity and question the current thinking about old and new production (Dugdale and Goering, 1967; Pace *et al.*, 1987). If the hypotheses developed here and in Sieburth (1986, 1987a, b) can be substantiated, then we may have to rethink our pelagial paradigms. I will succinctly describe two of the major paradoxes associated with the pycnocline that I perceive at the moment, March 1987.

FERMENTATION IN OXIC HABITATS

The oceanic methane peak in the pycnocline

Methane-producing bacteria are obligate anaerobes whose metabolism and growth only proceed in the absence of oxygen. Therefore, methane production is thought to be restricted to anoxic habitats, where biopolymers are fermented to produce organic acids, C_1 compounds and the end products of fermentation, hydrogen and carbon dioxide, that are all used as substrates by methanogens (Large, 1983). Such anoxic habitats are the hypolimnion of lakes, peatlands, and the gastrointestinal tract of cellulose-utilizing mammals and insects such as ruminants and termites. The upper ocean is not regarded as a significant source of either oceanic or atmospheric methane (Holland, 1978; Rudd and Taylor, 1980; Harriss *et al.*, 1985).

It is curious that this popular notion persists, since a number of investigators have repeatedly shown that methane is produced in the upper ocean. Methane profiles have a pronounced maximum associated with the pycnocline at a depth between 50 and 150 m, where methane is 30–70% supersaturated relative to the atmospheric equilibrium concentration (Swinnerton *et al.*, 1969; Brooks and Sackett, 1973; Brooks *et al.*, 1973; Lamontagne *et al.*, 1973; Williams and Bainbridge, 1973; Seiler and Schmidt, 1974). Close to shore this might conceivably be ascribed to transport from

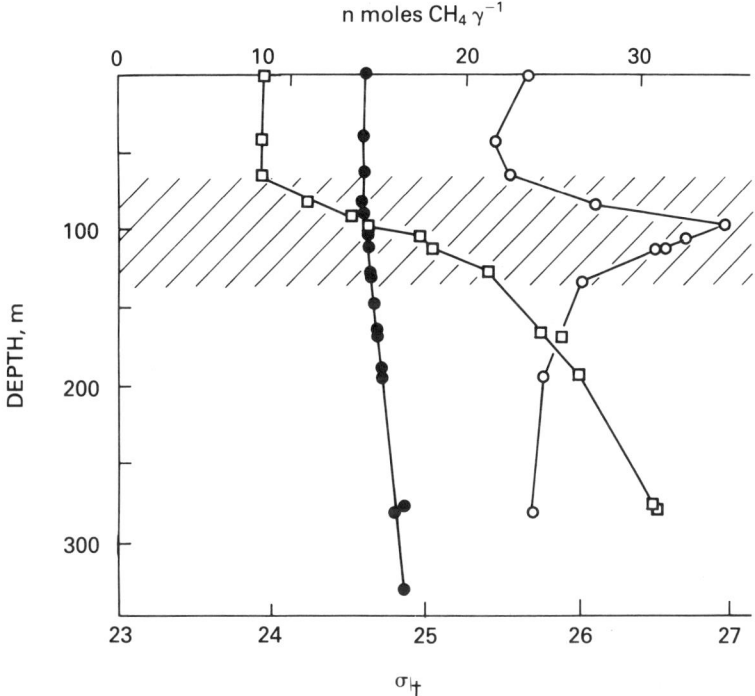

Fig. 1—The vertical profiles of (●) predicted and (○) observed methane concentrations plotted with (□) water density in the upper water column of the North Atlantic, east of the Caribbean Islands. Note the increased concentrations of methane within the stippled pycnocline to form a mid-pycnocline maximum. (Adapted from Scranton and Brewer, 1977.)

anoxic sediments or oil seeps, but in offshore waters as in Fig. 1, it is apparently due to *in situ* production (Lamontagne *et al.*, 1973; Scranton and Brewer, 1977; Scranton and Farrington, 1977).

Scranton and Brewer (1977), among others, have suggested that methane might arise from the by-products of algal metabolism or by methanogens living within reduced microenvironments. Traganza *et al.* (1979) observed a correlation of methane supersaturation with zooplankton ATP. Brooks *et al.* (1981) reported that chlorophyll or ATP maxima corresponded to methane maxima in the water column of the northwestern Gulf of Mexico. A very ambitious study that attempted to determine more accurately the source and site of methanogenesis in the eastern tropical North Pacific Ocean compared a variety of chemical and biological data (Burke *et al.*, 1983). They concluded that methane maxima are based on *in situ* biological production, controlled by physical processes, and that methanogenesis occurs within reducing microenvironments. The reducing environment of visible marine snow and fecal pellets has been well documented recently by Alldredge and Cohen (1987). There appears to be adequate indirect evidence that fermentation and methanogenesis, at least in the upper ocean, originates within particles that are probably largely from algal sources. What has been lacking is adequate direct evidence that bacteria involved in both the production and utilization of methane in the oxygenated upper ocean really do exist, and that the numbers and activities of methanogenic particles are adequate to account for a major decay of organic matter, through fermentation and the methane and sulphate reducing cycles, in the oxygenated water column. To paraphrase my friend, Professor Emeritus Nelson Marshall, the slight accumulation of methane in the pycnocline could be just the ashes of a very large fire.

Sielab's introduction to the paradoxical upper ocean methane cycle
During our studies on the nature and activities of *in situ* populations of the smaller marine microorganisms, we kept stumbling upon paradoxes that might be tied together. While making observations on the diel variation of populations of the picoplankton, nanoplankton, and concentrations of carbohydrates (Burney *et al.*, 1982), variations in TCO_2 and O_2 were also determined as indices of net metabolic activity. There was a consistent and enigmatically five-fold smaller variation in O_2 than CO_2 at the offshore stations (Johnson *et al.*, 1981a). Was there a tie-in with the dominant population of a bacterium in these waters with a distinctive ultrastructure: our type III cell that was described erroneously as a cyanobacterium (see Fig. 2A; Johnson and Sieburth, 1979)? As pointed out by Stanley Watson (personal communication), the distance between the cytomembranes in this bacterium is too close for thylakoids, and the distinctive cytomembranes are consistent with those found to be present in both ammonium- and methane-oxidizing bacteria. These facts we too began to realize. On the basis of ultrastructure and relative substrate availability, one must guess that they are methanotrophs. The populations and cell biovolume of this bacterium in upper oceanic waters of the Sargasso and Caribbean Seas range from 13–60% of the total bacteria present, with a mean of 28%. If this dominant bacterial cell is a planktonic methanotroph, it would indicate that a major methane cycle is operating in the upper mixed layer of the sea, and that unconsumed methane could contribute to atmospheric methane.

When methane or methanol produced by fermentation is oxidized, a key intermediate formed is formaldehyde (see Fig. 4; Anthony, 1982; Large, 1983). The possibility of a major oxidation of methane or methanol in the upper ocean was reinforced by estimating the daily variation in formaldehyde from the controls for our MBTH colorimetric procedure for carbohydrates (Johnson and Sieburth, 1977; Burney and Sieburth, 1977; Johnson *et al.*, 1981b; Burney *et al.*, 1981). There was a marked nocturnal peak (Johnson *et al.*, 1983). We refined this procedure for the determination of aldehydes in seawater and in seawater cultures of methylotrophs (Eberhardt and Sieburth, 1985). In a station with a pronounced oxygen minimum in the upwelling area off Peru, there was a background concentration of 'formaldehyde' equal to 1% of DOC, but there were also marked peaks correlating with pycnoclines and chemoclines that approached 15% of the DOC for these depths. These observations prompted us to speculate that our enigmatic discrepancies between the variations in TCO_2 and O_2 could possibly be due to a significant upper ocean methane cycle (Johnson *et al.*, 1983). If significant methane is being produced and oxidized by an upper ocean methane cycle, could we enrich and isolate the first oceanic methanotroph?

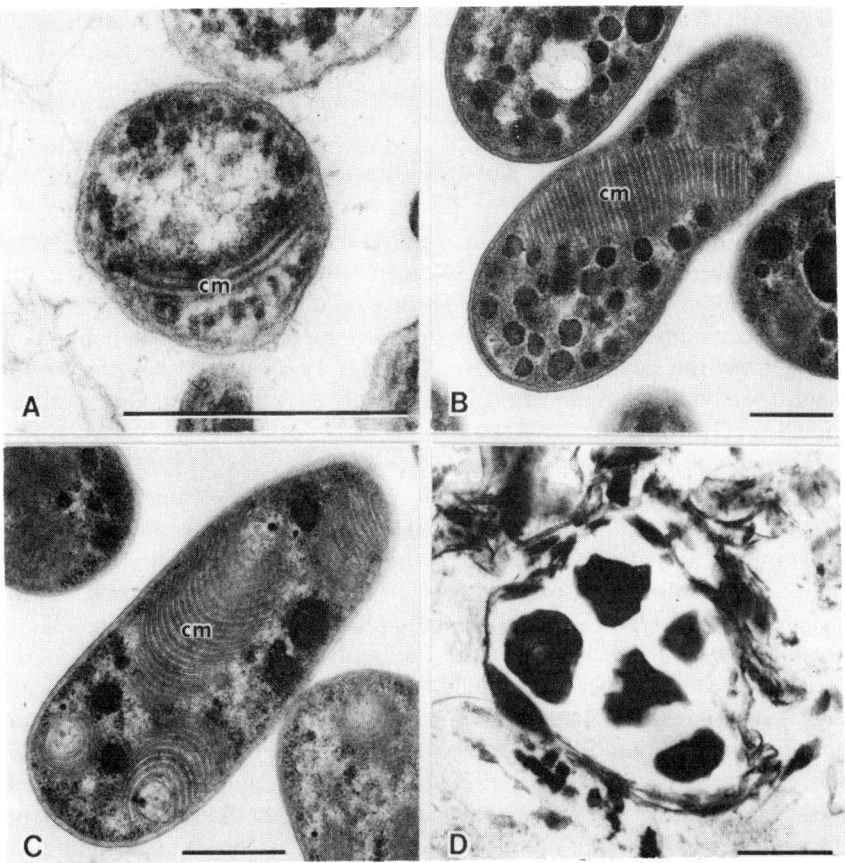

Fig. 2—Transmission electron micrographs of thin sections of bacteria in natural (A,D) and cultured (B,C) populations to show the ultrastructure of methane cycle bacteria. A, the type III cells of 'cyanobacteria' (Johnson and Sieburth, 1979), now thought to be a pelagic methanotroph with its central cytomembranes (CM). B, the first oceanic methanotroph *Methylomonas pelagica* with stacked cytomembranes (CM, Sieburth *et al.*, 1987). C, a type I methane-oxidizing bacterium, *Methylomonas methanica*, with disc-like cytomembranes (CM). D, an encapsulated microcolony with a methanogen-like ultrastructure in a fecal fragment. Marker bars = 0.5 μm.

The first oceanic methane-oxidizing bacterium

Methane-oxidizing bacteria were first isolated from methane enrichments of the leaves and stems of the aquatic macrophyte *Elodea* growing over methane-rich sediment (Söhngen, 1906). This pink bacterium, *Methylomonas methanica*, which is also associated with methane oxidation in soils, has been reisolated and shown to grow also on methanol which is further oxidized to formaldehyde and formate upon which it cannot grow (Dworkin and Foster, 1956). Only two other species were well described (Brown *et al.*, 1964; Foster and Davis, 1966) until the benchmark paper of Whittenbury *et al.* (1970) described the enrichment and isolation of more than 100 cultures of gram-negative, strictly aerobic, methane-utilizing bacteria whose only other growth substrate was methanol. About this time, Stanley Watson, who had done the benchwork studies on the ultrastructure of nitrifying bacteria that have similar cytomembranes, went to England to study the ultrastructure and culture requirements of the methanotrophs with Roger Whittenbury. A marvellous collection of methanotroph micrographs resulting from this collaboration are still largely unpublished. Progress on the study of terrestrial isolates has been summarized by Colby *et al.* (1979), Anthony (1982), Large (1983) and Crawford and Hanson (1984).

The first attempt to isolate marine forms was by Hutton and ZoBell (1949). They used a procedure similar to that of Söhngen (1906), and obtained methanotrophs from 54 of 62 samples (1 g) of coastal surface mud but failed to obtain enrichments from 100 ml volumes of seawater. Serious attempts to isolate methane-oxidizing bacteria from the nearshore waters of Woods Hole, Massachusetts, by Stanley Watson (personal communication) and the offshore waters of the southwestern Bering Sea (Griffiths *et al.*, 1982) have failed. No other attempts have apparently been reported.

My reason for making methanotroph enrichments in offshore waters, was an attempt to determine whether the type III 'cyanobacterium' so prevalent in natural populations of upper ocean seawater (Fig. 2A) could be cultured as either a methane- or an ammonium-oxidizing bacterium. This cell proved refractory to both enrichments, but a type I methanotroph that had an ultrastructure (Fig. 2B) similar to that of *M. methanica* (Fig. 2C) was repeatedly obtained in the Sargasso Sea and appears to be the first species obtained and described from the upper layer of deep pelagic seawater (Sieburth *et al.*, 1987). 23 portions (100 ml) of seawater from the pycnocline and half that depth at three stations yielded the same methanotroph that has been named *Methylomonas pelagica*. Unlike *M. methanica*, it is pigmented white, requires salt, grows well in seawater with either methane or methanol, but not on other C_1 compounds nor on C–C bonded organic matter. It uses either ammonia or nitrate but not dinitrogen as a nitrogen source. Formaldehyde is produced in considerable amounts from methanol, which is presumably a major fermentation product from algal polysaccharides and might be the source of much of the formaldehyde occurring nocturnally in the Sargasso Sea (Johnson *et al.*, 1983) and in pycnoclines (Eberhardt and Sieburth, 1985). *M. pelagica* grows well at 20 °C and 30 °C, but not at 10 °C or at high intensities of natural sunlight, perhaps explaining the nocturnal peak of formaldehyde in the upper mixing layer. This oceanic methanotroph assimilates one-carbon units via the ribulose monophosphate pathway for formaldehyde fixation. The DNA base composition of 49 mol% guanine plus cytosine, as well as the stacked cytomembranes, is consistent with this bacterium being a type I methanotroph. Since methanotrophs are present in the mixed layer and pycnocline of the upper ocean, then anoxic microenvironments required for methanogenesis by obligately anaerobic methanogens must exist just as Scranton and Brewer (1977) and Burke *et al.* (1983) have predicted. To persist in a hostile oxygenated environment, the methanogens must also be oxygen tolerant.

Methane-producing bacteria and their consorts

Methanogenic activity has been detected in anaerobic enrichments of plankton and fish guts using hydrogen and carbon dioxide as substrates for methanogenesis (Oremland, 1979). However, the methanogens present were not isolated or identified. The enrichment and isolation of methanogens from the particulates in seawater has apparently never been accomplished or published. The biochemistry and genetics of the methanogens have been reviewed by Balch *et al.* (1979). Up to that point, all methanogens examined were found to oxidize hydrogen and to reduce carbon dioxide to methane. The utilization of hydrogen by methanogens is essential to sustain fermentation, since a build-up of hydrogen inhibits the fermenters. Some species metabolize formate while members of the Methanosarcinae will also metabolize methanol, the methylated amines (MAs), and acetate as the sole electron donor for growth and methane formation. A number of methanogens that utilize hydrogen and formate for growth have been isolated from anoxic marine muds (references cited in Sowers *et al.*, 1984). Other species from anoxic marine habitats, however, failed to do this but utilized methanol and the MAs (König and Stetter, 1982; Sowers and Ferry, 1983) while another species from marine mud also utilized acetate. Methanogens from anoxic marine muds appear very facile in their substrate requirements.

Since obligately anaerobic methanogens must be surviving in anoxic microparticulates suspended in an otherwise oxygenated water column, how can they accomplish this? By persisting in association with other bacteria, methanogens can accomplish processes that they cannot do in isolation. Bryant *et al.* (1967) in a classic study showed that *Methanobacillus omelianskii* Barker maintained in an ethanol medium was actually a symbiotic association with the 'S-organism'. The latter fermented ethanol to hydrogen and acetate, thereby producing substrates for the methanogen. Ferry and Wolfe (1976) described a symbiotic bacterial consortium that degraded benzoate and produced methane. The former was degraded by non-methanogenic bacteria, while the two methanogens present were just the terminal bacteria in a multitrophic multicentre ecosystem. A methanogenic consortium of five bacterial species was constructed by Jones *et al.* (1984) by immobilizing the multicultures in a fixed bed under anaerobic conditions. Sucrose was fermented by the facultative anaerobe *Escherichia coli* to produce the methanogenic substrates acetate, formate, H_2, and CO_2, as well as the non-methanogenic substrates ethanol and lactate. The latter were oxidized to acetate by *Acetobacterium woodii* and *Desulfovibrio vulgaris* respectively using the combined oxygen present in sulphate. Methane was produced by the acetotrophic methanogen *Methanosarcina barkeri* and the hydrogenotrophic methanogen *Methanobacterium formicum*. An efficient 40% conversion of sucrose to methane was significantly decreased if any component of the five-membered consortium was omitted. It should be noted that both reductive and oxidative processes were occurring simultaneously in this process that originated with the fermentation of carbohydrate and terminated in the production of methane. Similar bacteriological processes must be happening in suspended and sedimenting particulates derived from the fecal pellets of herbivores, where the algal polysaccharides present must be undergoing fermentation during both suspension and sedimentation. Although some 75–95% of organic matter is decayed in the oxygenated zone of lakes and oceans, it is assumed to be due only to oxidative processes (Deuser, 1971; Rudd and Hamilton, 1978, 1979; Rudd and Taylor, 1980; Harrits and Hanson, 1980). Fermentation including methanogenesis, however, must also be a major pathway for decomposition in oxygenated waters and must occur in concert with oxidative processes.

Serendipitous aerobic enrichment of methanogens from suspended microparticulates

We initially enriched oceanic seawater samples in gas-permeable polycarbonate flasks with methane or ammonium supplements in an attempt to culture our type III bacterium (Sieburth et al., 1987). After repeated failures, in desperation we tried methylamine which combines these two substrates. There is good reason to do this as it is a C_1 compound used by many methylotrophs (Large, 1983), is produced by algae (Herrmann and Juttner, 1977; Kniefel, 1979), and is actively taken up by marine bacteria (Horstmann and Hoppe, 1981). Positive aerobic methylamine enrichments, indicated by marked turbidity, were obtained at virtually all depths sampled from seven diverse offshore and nearshore locations in both the Pacific and the Atlantic Oceans. Subcultures on MA agarose plates only yielded obligate methylotrophs, similar to those in the genus *Methylobacillus* (Urakami and Komagata, 1986), that grew on the MAs methylamine, dimethylamine and trimethylamine, as well as methanol and formaldehyde, but neither methane or C–C bonded organic matter. An apparent pellicle on the surface of some of these MA enrichments was similar to those present in the methane enrichments of *M. pelagica* (Sieburth et al., 1987). This suggested that perhaps some of the methylamine was being utilized paradoxically for methane formation by methanogens capable of anaerobic methylotrophy (Large, 1983). We detected the paradoxical accumulation of methane in the headspace of these originally aerobic enrichments, but they were very spotty and erratic. Several highly methanogenic anaerobic enrichments were obtained in the medium of Sowers and Ferry (1983).

We hypothesize that the paradoxical methanogenic activity in the oxygenated ocean is caused by oxygen-tolerant obligately anaerobic methanogenic bacteria present and active in microparticulates that are algal-derived fecal fragments suspended in oxygenated seawater, and which are kept anoxic through the exhaustion of oxygen by highly oxidative bacteria such as the methylotrophs consuming reduced C_1 compounds. In order to determine the presence and relative abundance of both types of bacteria, we made most probable number (MPN) estimates of both aerobic methylotrophs oxidizing methlyamine and anaerobic methylotrophs producing methane in the stratified waters of upper Chesapeake Bay. A single aerobic medium, identical to the one used previously in polycarbonate flasks, was used in glass serum bottles that were stoppered and sealed after inoculation. The observation of turbidity in the medium and the detection of methane accumulating in the headspace were taken as positive indications for methylamine-oxidizing particles (MOPs) and methane-producing particles (MPPs) respectively. The inocula consisted of whole water, particulates caught on 2.7 μm porosity, Whatman GF/D filters, and the filtrate passing through these filters.

The results from one station are shown in Table 1. The MOPs are apparently free bacteria since there were no significant differences between filtered and unfiltered water. Their MPNs were about $10^3 \, l^{-1}$ or, assuming the usual two order of magnitude underestimation (Sieburth, 1979), possibly $10^2 \, ml^{-1}$. Conversely, methane-producing bacteria were shown to be associated with particulates that were concentrated on filters and removed from the filtrate. The slightly higher MPN of MPPs in whole water than in filtrates may indicate that filtration may be damaging the delicate redox balance within the microparticulates. The MPNs of the MPPs

Depth	Methylamine oxidizing particles l^{-1}			Methane producing particles l^{-1}		
	Whole	Filter	Filtrates	Whole	Filter	Filtrates
2 m	3500	> 2400	> 2400	8	8	0
5.5 m	700	540	540	33	5	0
9.0 m	1100	180	350	33	13	0

Table 1   The distribution of methylamine oxidizing particles, and methane producing particles at an upper Chesapeake Bay station (904) as estimated by a most probable number procedure in whole water, filters (Whatman GF/D, 2.7 μm porosity), and filtrates

were about $30 l^{-1}$ or, again assuming the usual two order of magnitude underestimation in culture, some $3 ml^{-1}$. I have superimposed the cultured and extrapolated values for MOPs and MPPs on the hypothetical distribution of detrital particles in oceanic waters, constructed by Lenz (1972), which appears as Fig. 3. This suggests that the more numerous oxidative microparticulate MOPs are free cells and a minor part (0.1%) of the total bacterial population, while the anoxic microparticulate MPPSs are the larger suspended particulates about 40–45 μm in diameter.

Unlike the enrichments in polycarbonate flasks that remained aerobic while accumulating some methane, the enrichments in gas-tight serum bottles, although being initially aerobic, became reduced quickly, as indicated by the colour change in the resazurin redox indicator from pink to colourless, and went anoxic. Most of these methanogenic enrichments, which apparently consist of consortia of both oxidizing and reducing bacteria, lost their ability to produce methane under sustained anoxic conditions after three transfers (6 months). The few that persisted, when inoculated into prereduced methanogen medium, retained their methanogens and had an increased capacity to produce methane.

Our methanogenic MA enrichments made in aerobic media sometimes had the unmistakable odour of hydrogen sulphide. This is circumstantial evidence for the presence of sulphate-reducing bacteria in the genus *Desulfovibrio*. Although hydrogen sulphide can inhibit methanogenesis, sulphide is necessary for the formation of the essential coenzyme M in methanogens. If the bacterial aggregations suspended in oxygenated seawater are producing sulphide, this would explain the ubiquitous presence of *Thiobacillus*-like pseudomonads in seawater described by Tuttle and Jannasch (1972). Likewise, the methanogenic microparticulates may also house dinitrogen-fixing bacteria that require reduced oxygen conditions, and are associated with particulates < 50 μm (Guerniot and Colwell, 1985). The microbial consortia in suspended microparticulates may be quite complex in regard to bacterial species and their processes.

The algal connection

For methylotrophic and methanogenic microparticulates to be so numerous and widespread, they must be based upon primary production, as predicted by Scranton and Brewer (1977) and Burke *et al.* (1983), and be derived through animal feeding processes as indicated by Oremland (1979). The algae are known for their

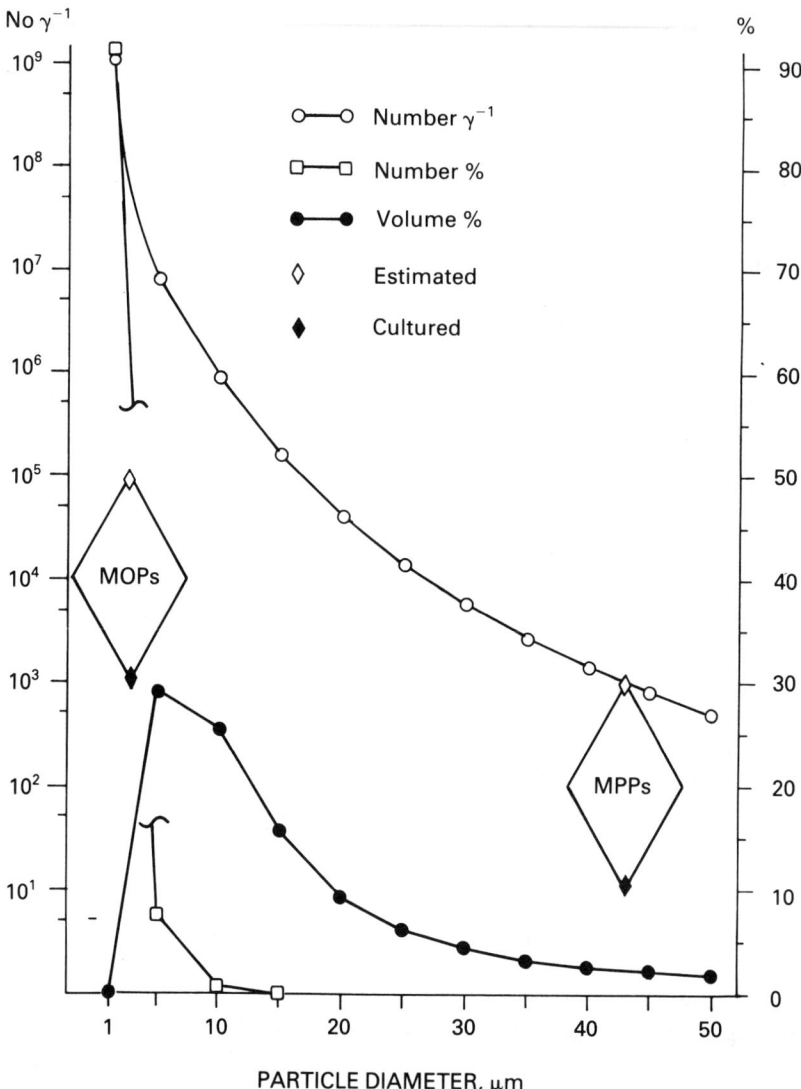

Fig. 3—The hypothetical exponential size distribution of detritus particles in oceanic surface waters adapted from Lenz (1972), with the cultured and extrapolated populations of MOPs and MPPs from Chesapeake Bay (Table 1) superimposed in regard to size and population, respectively. This indicates that MOPs are free cells that are a small fraction of the total population, while MPPs appear to be the dominant larger particulates.

production of polysaccharides as both structural and storage products. Besides these fermentable biopolymers that produce C_1 compounds and the precursors for methanogenesis, algae also directly produce C_1 compounds. These include dimethylsulphide, the other half of the sulphonium compound that hydrolyses to yield the antibacterial acrylic acid (Sieburth, 1960). Other C_1 compounds produced directly by algae are the MAs, as discussed previously, that are important in methanogenesis in sediments (King *et al.*, 1983). If methylated compounds and fermentable polysaccharides produced by algae in the sea are enriching methane cycle bacteria *in situ*, are they also doing it in captivity? Xenic cultures of algae, some maintained for decades, were tested for the presence of methylotrophic and methanotrophic bacteria. About 30% of the random cultures tested proved to be positive (Sieburth, 1987a). Included among the test cultures were nanoflagellates in the Prymnesiophyceae and Chrysophyceae that are dominant in oceanic waters (Estep *et al.*, 1984). In a life-long career of studying algal–bacterial interrelationships (Sieburth, 1968), I never thought that the algal class and its secondary compounds that got me into marine bacteriology in the first place (Sieburth, 1960) would three decades later reappear as the heart of the problem of the paradoxical methane cycle in the oxygenated ocean. The flip side, the role of bacteria on microparticulates in sustaining the thigmotactic nanoalgae where they peak in the dim pycnocline, is discussed by Sieburth (1987a).

Suggested mechanism for methanogenesis in oxygenated waters

The methane cycle that appears to be occurring in the oxygenated ocean and the bacteriological processes that paradoxically occur in our methanogenic but originally aerobic MA enrichments is presented diagramatically in Fig. 4. In enrichments of seawater with their suspended microparticulates, oxidative methylotrophs, among other bacteria on the particles, remove oxygen and permit the fermentative methylotrophs (methanogens) to produce methane. When this is produced in sufficient concentration in aggregating flocs that form in the pycnocline, then methanotrophs will also be enriched and be present. This explains our difficulty in obtaining methanotroph enrichments from seawater samples and the ease with which they can be enriched from sedimenting particles collected in sediment traps (see Fig. 2D). The vital role of algae and the zooplankton in the production and concentration of methanogenic particulates is also suggested.

OXIDATION IN ANOXIC HABITATS

Diverse anoxic basins

From an examination of the literature, there seems to be a reciprocal paradoxical phenomenon in anoxic waters, that of oxidation. Again the literature is small, somewhat overlooked, and has apparently not been synthesized. Bodies of water with an oxygenated mixing layer and anoxic water below a seasonal or permanent stratification are quite common. Examples are the Black Sea (Sorokin, 1970; Brewer and Murray, 1973; Sen Gupta and Jannasch, 1973), fjords such as those in Norway (Indrebø *et al.*, 1979a, b; Lidstrom, 1983) and British Columbia (Richards, 1965), stratified freshwater lakes (Harrits and Hanson, 1980), and stratified seawater lakes with a shallow sill like Lake Faro, Italy (Genovese, 1963) and the upper anoxic basins in the Pettaquamscutt River, Rhode Island (Gaines and Pilson, 1972).

The overall picture is one of the usual oxidative processes occurring in the oxygenated water above the pycnocline, with reduced products such as ammonia, sulphide and methane dramatically increasing with depth in the anoxic zone. In lakes, 75% of the organic matter is decomposed in the oxygenated water column above the anoxic hypolimnion (through unstudied processes assumed to be only

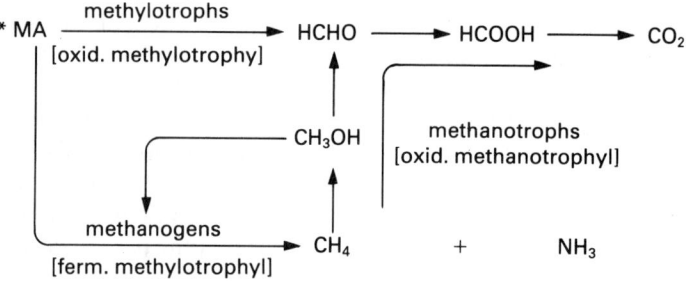

A. IN ENRICHMENT   * methylamine (MA) enrichment

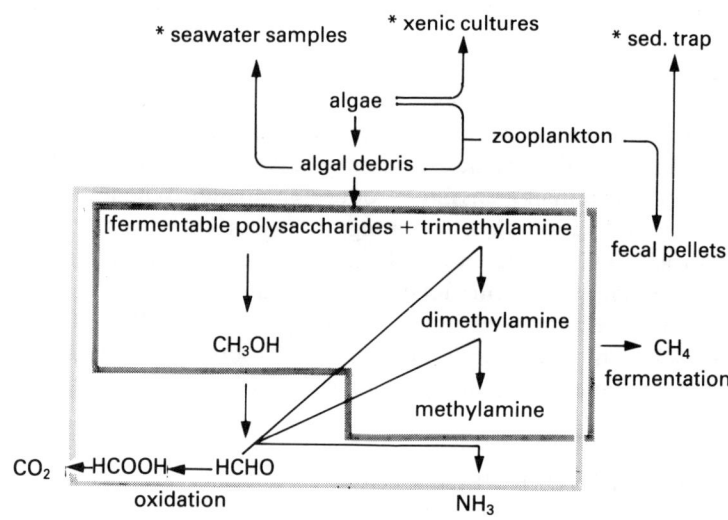

B. IN THE OCEAN

Fig. 4—Fermentative and oxidative processes apparently accomplished by methane cycle bacteria and their consorts in seawater particulates from the oxygenated ocean. A, the bacterial types and processes necessary to accomplish the paradoxical production of methane in originally aerobic methylamine enrichments. B, the role of algae and zooplankton in providing the organic debris and compounds required for the production of methane in suspended and sedimenting particulates in the oxygenated water column.

oxidative), the rest being decomposed mainly through fermentation, and the methane cycle (Rudd and Hamilton, 1978, 1979; Harrits and Hanson, 1980). In the saline waters of the Black Sea, Deuser (1971) found that 85–95% of the organic carbon input is recycled in the top oxygenated 200 m and at least 80% of the organic matter entering anoxic waters below 200 m was rendered soluble, presumably through the processes involved in methanogenesis and sulphate reduction, with only 4% of the input being permanently fixed in the sediment.

Evidence for oxidation in anoxic waters
Conventional wisdom excludes the anaerobic process of fermentation in the

oxygenated zone, and also excludes oxidative processes in the anoxic zone. There is a small body of literature which indicates that counter-intuitive processes are occurring, but that the observers tend to disbelieve their own observations. On the basis of high levels of CO_2 fixation in the dark, Sorokin (1970) postulated a high rate of sulphate reduction in the anoxic zone of the Black Sea that could not be accounted for by the organic matter produced by photosynthesis. High values for the dark assimilation of CO_2, found in and below the oxygen–sulphide interface in the Black Sea by Sen Gupta and Jannasch (1973), led to difficulties in explaining marked hydrogen sulphide oxidation in anoxic water. Jannasch *et al.* (1974) found a substantial population of hydrogen sulphide-oxidizing bacteria at the oxygen–hydrogen sulphide interface. By comparing estimates of productivity and rates of production and oxidation of sulphide they concluded that hydrogen acceptors required for further decomposition were not available and that excess organic matter must be deposited in the sediments. Tuttle and Jannasch (1979) report a maximum in the dark assimilation of CO_2 just below the oxygen–sulphide interface in the Cariaco trench, and then surprisingly present evidence for the oxidation of sulphide in anoxic waters. In studying the dark uptake of CO_2 in the Black Sea, Brewer and Murray (1973) obtained a value they thought was too high and attempted to find explanations to decrease this value. The disturbing feature of their proposed model was that sulphide oxidation appeared to be taking place in the absence of oxygen.

Indrebø *et al.* (1979a) studied a permanently stratified Norwegian estuary, the Saelenvann, with an anoxic basin and found that the discrepancies between primary production and sulphate reduction were seasonal with primary production exceeding sulphate reduction in summer, and the inverse during winter. They also found twice as much sulphate reduction in the water mass as in the sediment, with the maximum rate below the oxic–anoxic interface. This further indicates a marked oxidation within the anoxic water column. In companion work, Indrebø *et al.* (1979b) found that both particulate organic matter and bacterial populations peaked at the oxic–anoxic interface where the dark fixation of CO_2, due presumably to chemoautotrophy, was maximal. Because of the high value and partial inhibition by N-Serve they ascribed 10–50% of dark CO_2 fixation to ammonia oxidation (nitrification).

None of these papers considered the possibility that the oxidation of methane, in addition to that of ammonia and sulphide, could be accounting for the very high levels of dark CO_2 fixation. The uptake of methane in anoxic sediments has been ascribed to anaerobic oxidation (Reeburgh, 1980, 1982), although bacteria capable of this have not been demonstrated (Rudd and Taylor, 1980).

The similar fate of all reducing organic compounds
The consumption of methane in the anoxic water column of marine basins with shallow sills is also of interest. Lidstrom (1983) shows the distribution of methane in the Framvaren fjord and states that the profile indicates the depth where methane might be consumed. I disagree. The depths were obviously indicated by apparent methane consumption in bottle incubations of 35 ml samples. The methane profile actually suggests marked methane consumption in the upper anoxic zone. If oxidation is largely occurring on particulates, as suggested in a foregoing section, then larger volumes or filter-concentrated preparations may be needed to show the

actual distribution of oxidative activity. The controversy concerning anaerobic methane oxidation is nicely summarized by Rudd and Taylor (1980).

There are other reducing compounds in fjords, such as ammonia and hydrogen sulphide (Emerson *et al.*, 1979; Indrebø *et al.*, 1979a, b), whose profiles should also be compared to see whether methane is acting independently or similarly with the other reduced compounds. The upper anoxic basins of the Pettaquamscutt River (Gaines and Pilson, 1972; Gaines, 1975) and Lake Faro (Genovese, 1963) are in essence small fjords that are ideal experimental macrocosms close to laboratory facilities. The paradoxical utilization of all reduced substances in the anoxic water

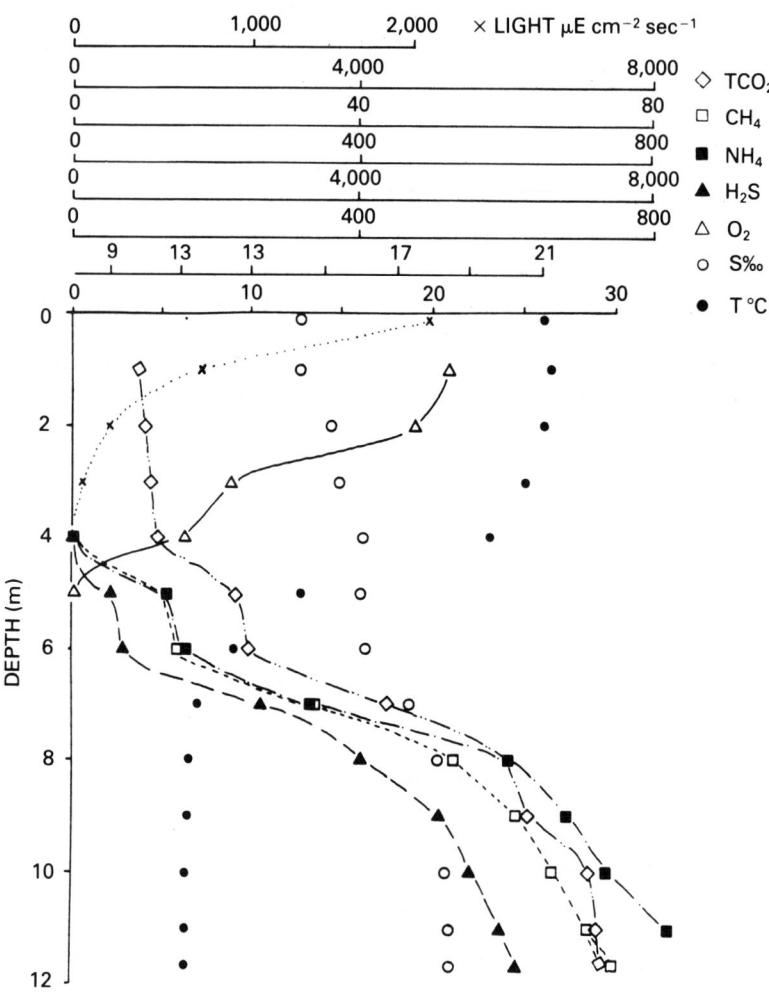

Fig. 5—The vertical profiles of oxygen and reduced compounds (μg At l^{-1}) with TCO_2, temperature, salinity and light (microeinsteins) in the upper basin of the Pettaquamscutt River on 18 August, 1980 (drawn from the data of Roques (1985)). Note the similar pattern of concentration of the reduced compounds with depth, indicating a common mechanism for consumption of reduced substances between 4 and 8 m.

column of the upper basin of the Pettaquamscutt River is well illustrated in Fig. 5, where data from Roques (1985) are plotted. The concentrations of sulphide, ammonia and methane decrease towards the oxic–anoxic interface in the water column. In the absence of mixing, as shown by salinity and temperature profiles, the concentrations of reduced substances should be uniform. They are not. When plotted as a function of salinity as in Fig. 6, the plots for reduced substances should be linear. They are not, indicating consumption through paradoxical oxidation. These vertical profiles may indicate that, below the oxic–anoxic interface, between a depth of 4 and 8 m, there could be significant oxidation that could account for the high rates of dark CO_2 fixation in this zone reported in the literature. An example of dark fixation from the pycnocline in upper Chesapeake Bay is given in Table 2. Marked dark fixation is apparently masked by the respiration of bacterivorous eucaryotes in the whole water. When the eucaryotes were removed by filtration or inhibited by the eucaryotic-specific cycloheximide, which is about 80% effective (Cynar *et al.*, 1985), there was marked dark CO_2 fixation.

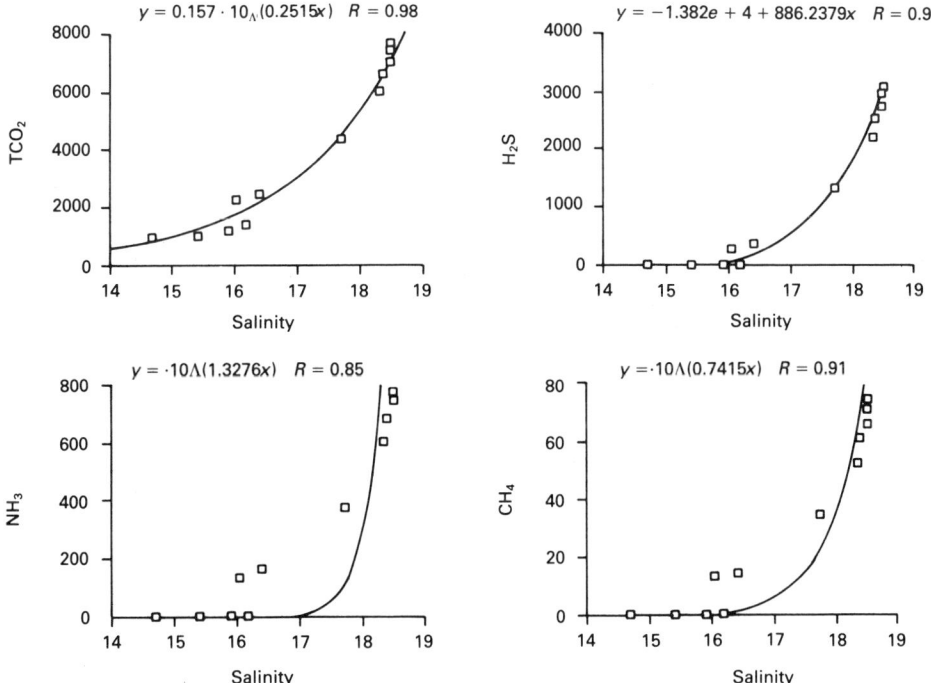

Fig. 6—The concentration of reduced compounds as a function of salinity for the data in Fig. 5. The curvilinear distributions indicate an active oxidative removal in the 'anoxic' waters below the oxygen–hydrogen sulphide interface (graphs prepared by Percy Donaghay).

Mechanisms for the imperceptible oxygenation of anoxic water

Several mechanisms for oxygenation of the upper hydrogen sulphide zone are possible. One is that it could come from the fixed oxygen in nitrate, phosphate and especially sulphate. However, this could be limiting. A second possibility, in shallow

Fraction–measurement	Incubation		Rate
	Period	Time (h)	(μmol kg^{-1} h^{-1})
Community			
Whole water — untreated	2200–0830	10.5	−0.2
Picoplankton			
<2.7 μm filtrate	2200–0900	11.0	−13.4
Whole water and cycloheximide	2200–1000	12.0	−8.9

Table 2   The dark fixation of coulometrically determined CO_2 in whole water, 2.7 μm filtrates and cycloheximide inhibited whole water from the pycnocline at 8 m at Chesapeake Bay station 904, May 1986

waters, might be the presence of high-velocity short-interval internal waves that are induced by tides and storm fronts, such as those that have been described for Chesapeake Bay (Sarabun *et al.*, 1986).

Although tidal dampening is substantial in the upper basin of the Pettaquamscutt River, appreciable tides still occur and storm fronts have a very strong effect up the estuary in the Pettaquamscutt River (Gaines, 1975). The heart of the problem is the quandary posed by the data in Fig. 5. There is evidence for both mixing and stagnation in the profiles of temperature and salinity. Some of the removal of reduced compounds must be by bacterial oxidation. A third, and very intriguing, possible source of oxygen in the anoxic zone could be that produced by *in situ* photosynthesis. Deep in the hydrogen sulphide zone, but still in the photic zone, a year-round population of the alga *Euglena proxima* (which could be mixotrophic, Bird and Kalff, 1986; Estep *et al.*, 1986a), persisted throughout the year and never occurred in the oxic zone (Miller, 1972). Although these populations do not appear to exceed $10^3 1^{-1}$, these are very large cells approximating cylinders 75 μm × 20 μm in size. A dinoflagellate, *Prorocentrum mariae-lebouriae* var. *minimum*, occurs in a chlorophyll maximum below the pycnocline in Chesapeake Bay and persists for long periods at suboptimal light conditions (Tyler, 1984). It is possible that this species may also be supplying oxygen to the pycnocline. The occurrence, nutrient effect, and photosynthetic contribution of algae within and below the pycnocline could have a profound effect in anoxic ecosystems. Whatever the various sources of oxygen are for the anoxic waters, the inability to detect measurable concentrations of oxygen does not mean that there is not a major input by flux and *in situ* production and an equal consumption that masks it to yield non-detectable concentrations of oxygen. Methane-oxidizing bacteria in fermentors fed with oxygen can achieve maximal oxidation rates at non-detectable concentrations of oxygen (Roger Whittenbury, personal communication, 1983).

Runaway fermentation in the ergocline and seasonal hypoxia and anoxia
In accounts of carbon inputs into upper estuaries, where freshwater mixes with seawater, such as in the upper Chesapeake Bay and the anoxic basin of the Pettaquamscutt River, a potentially significant source of utilizable organic matter, the input of flocculated freshwater DOC, has apparently been ignored. The average

concentration of DOC in river water is approximately $10\,mg\,Cl^{-1}$ (Sholkovitz, 1976). Terrestrial humic substances in the Nid River completely precipitated within the saline mixing plume of the Trondheim Fjord (Sieburth and Jensen, 1968). As the salinity of an estuary increases up to 15–20‰, some 3–11% of riverine DOC is flocculated at the halocline (Sholkovitz, 1976). These observations suggest that flocculated DOC could accumulate to significant values in the halocline and could be a significant source of particulate organic matter available for bacterial fermentation and oxidation. The influence of freshwater input into upper estuaries, such as Chesapeake Bay, where the flocculation and fermentation of allochthonous DOC must occur, could play a significant role in the summer anoxias (Officer *et al.*, 1984; Seliger *et al.*, 1985).

The concentration of particulate carbohydrate at the halocline in Chesapeake Bay has been reported by Biggs and Wetzel (1968). The interfaces between water masses such as the thermocline and pycnocline have been considered to form a barrier for the sedimentation of small particles (Garfield *et al.*, 1983). Sieburth (1983) referred to this barrier as a 'false benthos'. During a study of the highly stratified upper Chesapeake in May, 1986, Sieburth and Biggs (unpublished data) noted that the pycnocline, which coincided with a sound-scattering layer, contained much particulate organic matter rich in bacteria, and that the zooplankton which concentrated in this zone was dominated by larval forms of the estuarine polychaete *Polydora ligni* that exhibited delayed metamorphosis. On a subsequent cruise, a video camera with macro lens showed that this area was swarming with zooplankton activity, like the entrance of a beehive, and below it was like a snowstorm of particles up to 1 cm in diameter (Bob Biggs, personal communication). Harder (1968) has pointed out that zooplankton react strongly to water stratification. It is quite possible that the slowly sedimenting microparticulates that accumulate in the pycnocline are aggregating to macroflocs as described by Kranck and Milligan (1980), to produce the flakes of marine snow leaving the pycnocline. There is a trend in the data shown in Table 1 for the oxidative particulates to be more numerous in the upper water column and methanogenic particulates to be more numerous in the deeper layer. Such a trend was observed in the bacteriological activity of culturable bacteria with depth (Sieburth, 1971). We have much to learn about the qualitative differences in particulates with size and depth. Qualitative biochemical differences in suspended particles were observed along the Tamar estuary by Owens (1986). Owens (1985) had previously shown that nitrifying activity was associated with particles in suspension.

Gases, redox balance, carbon balance and GAIA

The role of bacteria in the production and utilization of gases is nicely summarized in the symposium volume edited by Schlegel *et al.* (1976). Marine biologists so conversant with oxygen liberated during photosynthesis and carbon dioxide released by respiration and taken up during photosynthesis tend to forget about the other oxidizing and reducing gases and the vital role that bacteria play in their production and utilization. Bacteriologists working with anaerobes are very conversant with the bacterial processes required for the decay of substrates like the polysaccharides that end up as methane. They determine the decrease of substrate, the production of by products, and calculate carbon recovery and cell efficiency. The oxidation–reduction state of the products of fermentation and oxidation gives a clue as to the nature of the compounds that are not analysed. Such work is described in the papers cited by Hungate (1975), who studied the microbial ecosystem of the rumen. Some two decades ago Bob Hungate, who was my advisor at one time, asked me why we marine bacteriologists did not try to balance the carbon budget in the sea. I believe that my reply was that the ocean is too big and the timescales are too large to do this. During the intervening years, a much clearer picture of production and transport has evolved. We can now treat the upper ocean, cut off from deeper water masses by the pycnocline and its false benthos, as a 100 m deep lake. Today I think we are ready for Bob Hungate's vital question. Not only are the organic inputs being characterized, but we now know that bacterial processes in the sea are the same as in any other ecosystem. We also have the tools to study them. Semi-automated image analysis of epifluorescent microscopy (Sieracki *et al.*, 1985; Estep *et al.*, 1986b) now takes the tedium out of cell counting and sizing. The development of very precise instruments for detecting environmental changes in oxygen (Williams and Jenkinson, 1982) and carbon dioxide (Johnson *et al.*, 1985, 1987) has permitted the determination of planktonic community metabolism (Bender *et al.*, 1987). Ken Johnson is developing a 'super bottle' for obtaining samples for determining volatile organic carbon and the component gases and C_1 compounds. By paddling our own canoes in a communal puddle with our colleagues in marine microbial ecology to form a formidable armada, we should be able to attack community composition and metabolism in a meaningful way. Only by studying how microbial consortia achieve their balance between oxidizing and reducing processes can we understand the basis for the GAIA hypothesis (Lovelock, 1982) that attempts to explain how our planet stays in equilibrium without going oxic like the other dead planets. It's a big job, let's do it!

SUMMARY

(1) The consortia of microorganisms that are associated with microparticulates carry out both reducing and oxidizing chemical processes throughout the water column.

(2) The microparticulates $< 50\,\mu m$ that slowly fall out are concentrated in the pycnocline (ergocline) to form a false benthos where anaerobic processes occur, such as methanogenesis, that are only normally considered to occur in the sediments of the sea floor.

(3) These processes may be very necessary for *in situ* nutrient regeneration in the photic zone, which is required for sustained oceanic productivity and to balance oxidation–reduction processes required by GAIA.

(4) When these processes are magnified in nearshore waters with terrestrial organic inputs, they may be responsible for hypoxic and anoxic events such as those occurring in the New York Bight, Chesapeake Bay and anoxic fjords.

(5) These are understudied problems, in part because the processes involved are contrary for their redox-specific habitats.

ACKNOWLEDGEMENTS

The topic of the chapter is the logical progression of three decades of observing marine microorganisms and their processes in various habitats. I acknowledge the late Sidney R. Galler for facilitating my switch from poultry pathology to marine bacteriology while he was at ONR and Professor Emeritus David M. Pratt who patiently introduced me to the basics of marine biology. During the latter half of this period I have been blessed with the competent and vital help of my labworkers Kenneth M. Johnson and Paul W. Johnson and many students whose names appear in the cited papers. Work in progress has been aided by recent visitors such as Nicholas J.P. Owens of IMER, Plymouth, Dag Klaveness of the University of Oslo, and Lars J. Tranvik of Lund University, and by colleagues such as Percy Donaghay of GSO, Bob Biggs of the University of Delaware, and Robert Guillard at the Bigelow Laboratory. I thank Michael L. Bender for a critical review of the draft and Michael A. Sleigh for improving the manuscript. The research described has been supported by the Biological Oceanography Program of the National Science Foundation through recent Grants OCE-8121881, -8316614 and -8511365.

REFERENCES

Alldredge, A.L. and Cohen, Y. 1987. Can microscale chemical patches persist in the sea? Microelectrode study of marine snow, fecal pellets. *Science*, **235**: 689–691.

Altschuler, S.J. and Riley, G.A. 1964. Microbiological studies in Long Island. *Bulletin of the Bingham Oceanographic Collection*, **19**: 81–88.

Anthony, C. 1982. *The biochemistry of methylotrophs*, 431 pp. Academic Press, London.

Balch, W.E., Fox, G.E., Magrum, L.J., Woese, C.R. and Wolfe, R.S. 1979. Methanogens: reevaluation of a unique biological group. *Microbiological Reviews*, **43**: 260–296.

Bender, M.L., Grande, K., Johnson, K., Marra, J., Williams, P.leB., Sieburth, J., Pilson, M., Langdon, C., Hitchcock, G., Orchardo, J., Donaghay, P. and Heinemann, C. 1987. A comparison of four methods for the determination of planktonic community metabolism. *Limnology and Oceanography*, in press.

Biggs, R.B. and Wetzel, C.D. 1968. Concentration of particulate carbohydrate at the halocline in Chesapeake Bay. *Limnology and Oceanography*, **13**: 169–171.

Bird, D.F. and Kalff, J. 1986. Bacterial grazing by planktonic lake algae. *Science*, **231**: 493–495.

Brewer, P.G. and Murray, J.W. 1973. Carbon, nitrogen and phosphorus in the Black Sea. *Deep-Sea Research*, **20**: 803–818.

Brooks, J.M. and Sackett, W.M. 1973. Sources, sinks, and concentrations of light hydrocarbons in the Gulf of Mexico. *Journal of Geophysical Research*, **78**: 5248–5258.

Brooks, J.M., Fredericks, A.D., Sackett, W.M. and Swinnerton, J.W. 1973. Baseline concentrations of light hydrocarbons in the Gulf of Mexico. *Environmental Science and Technology*, **7**: 639–642.

Brooks, J.M., Reid, D.F. and Bernard, B.B. 1981. Methane in the upper water column of the northwestern Gulf of Mexico. *Journal of Geophysical Research,* **86**: 11029–11040.

Brown, L.R., Strawinski, R.J. and McCleskey, C.S. 1964. The isolation and characterization of *Methanomonas methanooxidans* Brown and Strawinski. *Canadian Journal of Microbiology,* **10**: 791–799.

Bryant, M.P., Wolin, E.A., Wolin, M.J. and Wolfe, R.S. 1967. *Methanobacillus omelianskii,* a symbiotic association of two species of bacteria. *Archiv für Mikrobiologie,* **59**: 20–31.

Burke, R.A. Jr., Reid, D.F., Brooks, J.M. and Lavoie, D.M. 1983. Upper water column methane geochemistry in the eastern tropical Pacific Ocean. *Limnology and Oceanography,* **28**: 19–32.

Burney, C.M. and Sieburth, J.McN. 1977. Dissolved carbohydrates in seawater. II, A spectrophotometric procedure for total carbohydrate analysis and polysaccharide estimation. *Marine Chemistry,* **5**: 15–28.

Burney, C.M., Davis, P.G., Johnson, K.M. and Sieburth, J.McN. 1981. Dependence of dissolved carbohydrate concentrations upon small scale nanoplankton and bacterioplankton distributions in the western Sargasso Sea. *Marine Biology,* **65**: 289–296.

Burney, C.M., Davis, P.G., Johnson, K.M. and Sieburth, J.McN. 1982. Diel relationships of microbial trophic groups and *in situ* dissolved carbohydrate dynamics in the Caribbean Sea. *Marine Biology,* **67**: 311–322.

Colby, J., Dalton, H. and Whittenbury, R. 1979. Biological and biochemical aspects of microbial growth on C_1 compounds. *Annual Review of Microbiology,* **33**: 481–517.

Crawford, R.L. and Hanson, R.S. 1984. *Microbial growth on C_1 compounds,* 343 pp. American Society of Microbiology, Washington, DC.

Cynar, F.J., Estep, K.W. and Sieburth, J.McN. 1985. The detection and characterization of bacteria-sized protists in "protist-free" filtrates and their potential impact on experimental marine ecology. *Microbial Ecology,* **11**: 281–288.

Deuser, W.G. 1971. Organic carbon budget of the Black Sea. *Deep-Sea Research,* **18**: 995–1004.

Dugdale, R.C. and Goering, J.J. 1967. Uptake of new and regenerated forms of nitrogen in primary productivity. *Limnology and Oceanography,* **12**: 196–206.

Dworkin, M. and Foster, J.W. 1956. Studies on *Pseudomonas methanica* (Söhngen) nov. comb. *Journal of Bacteriology,* **72**: 646–659.

Eberhardt, M.A. and Sieburth, J.McN. 1985. A colorimetric procedure for the determination of aldehydes in seawater and in cultures of methylotrophic bacteria. *Marine Chemistry,* **17**: 199–212.

Emerson, S., Cranston, R.E. and Liss, P.S. 1979. Redox species in a reducing fjord: equilibrium and kinetic considerations. *Deep-Sea Research,* **26A**: 859–878.

Estep, K.W., Davis, P.G., Hargraves, P.E. and Sieburth, J.McN. 1984. Chloroplast containing microflagellate in natural populations of north Atlantic nanoplankton; including a description of five new species of *Chrysochromulina* (Prymnesiophyceae). *Protistologica,* **20**: 613–634.

Estep, K.W., Davis, P.G., Keller, M. and Sieburth, J.McN. 1986a. How important are oceanic algal nanoflagellates in bacterivory? *Limnology and Oceanography*, **31**: 646–650.

Estep, K.W., MacIntyre, F., Hjörleifsson, E. and Sieburth, J.McN. 1986b. MacImage: a user-friendly image-analysis system for the accurate mensuration of marine organisms. *Marine Ecology — Progress Series*, **33**: 243–253.

Ferry, J.G. and Wolfe, R.S. 1976. Anaerobic degradation of benzoate to methane by a microbial consortium. *Archives of Microbiology*, **107**: 33–40.

Foster, J.W. and Davis, R.H. 1966. A methane-dependent coccus, with notes on classification and nomenclature of obligate methane-utilizing bacteria. *Journal of Bacteriology*, **91**: 1924–1931.

Fowler, S.W. and Knauer, G.A. 1986. Role of large particles in the transport of elements and organic compounds through the oceanic water column. *Progress in Oceanography*, **16**: 147–194.

Fuhrman, J.A. and Azam, F. 1982. Thymidine incorporation as a measure of heterotrophic bacterioplankton production in marine surface waters: evaluation of field results. *Marine Biology*, **66**: 109–120.

Gaines, A.G. Jr. 1975. Papers on the geomorphology, hydrography and geochemistry of the Pettaquamscutt River Estuary. *PhD Thesis*, 278 pp. University of Rhode Island.

Gaines, A.G. Jr. and Pilson, M.E.Q. 1972. Anoxic water in the Pettaquamscutt River. *Limnology and Oceanography*, **17**: 42–49.

Garfield, P.C., Packard, T.T., Friederich, G.E. and Codispoti, L.A. 1983. A subsurface particle maximum layer and enhanced microbial activity in the secondary nitrite maximum of the northeastern tropical Pacific Ocean. *Journal of Marine Research*, **41**: 747–768.

Genovese, S. 1963. The distribution of the H_2S in the lake of Faro (Messina) with particular regard to the presence of "red water". In *Marine microbiology* (ed. C.H. Oppenheimer), pp. 194–204. Thomas, Springfield, Il.

Griffiths, R.P., Caldwell, B.A., Cline, J.D., Broich, W.A. and Morita, R.Y. 1982. Field observations of methane concentrations and oxidation rates in the southeastern Bering Sea. *Applied and Environmental Microbiology*, **44**: 435–446.

Guerniot, M.L. and Colwell, R.R. 1985. Enumeration, isolation and characterization of N_2-fixing bacteria from seawater. *Applied and Environmental Microbiology*, **50**: 350–355.

Harder, W. 1968. Reactions of plankton organisms to water stratification. *Limnology and Oceanography*, **13**: 156–168.

Hardy, C.D. and Weyl, P.K. 1971. Distribution of dissolved oxygen in the waters of western Long Island Sound. *Tech Report No. 11*, 37 pp. Marine Science Research Center, SUNY, Stony Brook, NY.

Harriss, R.C., Gorham, E., Sebacher, D.I., Bartlett, K.B. and Flebbe, P.A. 1985. Methane flux from northern peatlands. *Nature*, **315**: 652–654.

Harrits, S.M. and Hanson, R.S. 1980. Stratification of aerobic methane-oxidizing organisms in Lake Mendota, Madison, Wisconsin. *Limnology and Oceanography*, **25**: 412–421.

Herrmann, V. and Juttner, F. 1977. Excretion products of algae. Identification of biogenic amines by gas–liquid chromatography and mass spectrometry of their fluoroacetamides. *Analytical Biochemistry*, **78**: 365–373.

Holland, H.D. 1978. *The chemistry of the atmosphere and oceans*, 351 pp. Wiley, New York.

Hortsmann, W.E. and Hoppe, H.G. 1981. Competition in the uptake of methylamine/ammonium by phytoplankton and bacteria. *Kieler Meeresforschungen Sonderhefte*, **5**: 110–116.

Hungate, R.E. 1975. The rumen microbial ecosystem. *Annual Review of Ecology and Systematics*, **6**: 39–66.

Hutton, W.E. and ZoBell, C.E. 1949. The occurrence and charactistics of methane-oxidizing bacteria in marine sediments. *Journal of Bacteriology*, **58**: 463–474.

Indrebø, G., Pengerud, B. and Dundas, I. 1979a. Microbial activities in a permanently stratified estuary. I. Primary production and sulfate reduction. *Marine Biology*, **51**, 295–304.

Indrebø, G., Pengerud, B. and Dundas, I. 1979b. Microbial activities in a permanently stratified estuary. II. Microbial activities at the oxic–anoxic interface. *Marine Biology*, **51**: 305–309.

Jannasch, H.W. 1973. Bacterial content of particulate matter in offshore surface waters. *Limnology and Oceanography*, **18**: 340–342.

Jannasch, H.W., Truper, H.G. and Tuttle, J.H. 1974. Microbial sulfur cycle in the Black Sea. In *The Black Sea — geology, chemistry and biology* (eds. E.T. Degens and D.A. Ross), pp. 419–425. American Association of Petroleum Geologists, Tulsa, OK.

Johnson, K.M. and Sieburth, J.McN. 1977. Dissolved carbohydrates in seawater. I. A precise spectrophotometric analysis for monosaccharides. *Marine Chemistry*, **5**: 1–13.

Johnson, K.M., Burney, C.M. and Sieburth, J.McN. 1981a. Enigmatic marine ecosystem metabolism measured by direct diel Σ CO_2 and O_2 flux in conjunction with DOC release and uptake. *Marine Biology*, **65**: 49–60.

Johnson, K.M., Burney, C.M. and Sieburth, J.McN. 1981b. Doubling the production and precision of the MBTH spectrophotometric assay for dissolved carbohydrates in seawater. *Marine Chemistry*, **10**: 467–473.

Johnson, K.M., Davis, P.G. and Sieburth, J.McN. 1983. Diel variation of TCO_2 in the upper layer of oceanic waters reflects microbial composition, variation, and possibly methane cycling. *Marine Biology*, **77**: 1–10.

Johnson, K.M., King, A.E. and Sieburth, J.McN. 1985. Coulometric TCO_2 analyses for marine studies; an introduction. *Marine Chemistry*, **16**: 61–82.

Johnson, K.M., Williams, P.J.leB., Brändström, L. and Sieburth, J.McN. 1987. Coulometric total carbon dioxide analysis for marine studies: automation and calibration. *Marine Chemistry*, in press.

Johnson, P.W. and Sieburth, J.McN. 1979. Chroococcoid cyanobacteria in the sea: a ubiquitous and diverse phototrophic biomass. *Limnology and Oceanography*, **24**: 928–935.

Jones, G.E. and Jannasch, H.W. 1959. Aggregates of bacteria in seawater as determined by treatment with surface active agents. *Limnology and Oceanography*, **4**: 269–276.

Jones, W.J., Guyot, J.-P. and Wolfe, R.S. 1984. Methanogenesis from sucrose by defined immobilized consortia. *Applied and Environmental Microbiology*, **47**: 1–6.

Karl, D.M., Knauer, G.A., Martin, J.H. and Ward, B.B. 1984. Bacterial chemolithotrophy in association with sinking particles. *Nature*, **309**: 54–56.

King, G.M., Klug, M.J. and Lovely, D.R. 1983. Metabolism of acetate, methanol, and methylated amines in intertidal sediments of Lowes Cove, Maine. *Applied and Environmental Microbiology*, **45**: 1848–1853.

Kirchman, D., K'nees, E. and Hodson, R. 1985. Leucine incorporation and its potential as a measure of protein synthesis by bacteria in natural aquatic systems. *Applied and Environmental Microbiology*, **49**: 599–607.

Kniefel, H. 1979. Amines in algae. In *Marine algae in pharmaceutical science* (eds. H.A. Hoppe, T. Levring and Y. Tanaka), pp. 365–401. de Gruyter, Berlin.

König, H. and Stetter, K.O. 1982. Isolation and characterization of *Methanolobus tindarius* sp. nov., a coccoid methanogen growing only on methanol and methylamines. *Zentralblatt für Bakteriologie, Parasitenkunde, Infektionskranksheiten und Hygiene, Abt Orignale*, **3**: 478–490.

Kranck, K. and Milligan, T. 1980. Macroflocs: production of marine snow in the laboratory. *Marine Ecology Progress Series*, **3**: 19–24.

Kriss, A.E. 1963. *Marine microbiology*, 536 pp. Oliver and Boyd, Edinburgh and London.

Lamontagne, R.A., Swinnerton, J.W., Linnenbom, V.J. and Smith, W.D. 1973. Methane concentrations in various marine environments. *Journal of Geophysical Research*, **78**: 5317–5324.

Large, P. 1983. Methylotrophy and methanogenesis. In *Aspects of microbiology* Vol. 8 (eds. J.A. Cole, C.J. Knowles and D. Schlessinger), pp. 1–88. American Society of Microbiology, Washington DC.

Legendre, L. and Demers, S. 1984. Towards dynamic biological oceanography and limnology. *Canadian Journal of Fisheries and Aquatic Sciences*, **41**: 2–19.

Legendre, L., Demers, S. and Lefaivre, D. 1986. Biological production at marine ergoclines. In *Marine interfaces ecohydrodynamics* (ed. J.C.J. Nihoul), pp. 1–29. Elsevier, Amsterdam.

Lenz, J. 1972. The size distribution of particles in marine detritus. *Memorie Dell'Istituto Italiano Di Idrobiologia*, **29(S)**: 17–35.

Lidstrom, M.E. 1983. Methane consumption in Framvaren, an anoxic marine fjord. *Limnology and Oceanography*, **28**: 1247–1251.

Lovelock, J.E. 1982. *GAIA, a new look at life on earth*, 157 pp. Oxford University Press, Oxford.

McCave, I.N. 1975. Vertical flux of particles in the ocean. *Deep-Sea Research*, **22**: 491–502.

Melchiorri-Santolini, U. and Hopton, J.W. 1972. *Detritus and its role in aquatic ecosystems. Memorie Dell'Istituto Italiano Di Idrobiologia*, **29(S)**: 540 pp.

Miller, B.T. 1972. The phytoplankton and related hydrography in the south basin of the Pettaquamscutt River. *MS Thesis*, 119 pp. University of Rhode Island.

Mullin, M.M. 1965. Size fractionation of particulate organic carbon in surface waters of the Western Indian Ocean. *Limnology and Oceanography*, **10**: 459–462.

Nihoul, J.C.J. 1986. *Marine interfaces ecohydrodynamics*, 670 pp. Elsevier, Amsterdam.

Officer, C.B., Biggs, R.B., Taft, J.L., Cronin, L.E., Tyler, M.A. and Boynton, W.R. 1984. Chesapeake Bay anoxia: origin, development, and significance.

Science, **223**: 22–27.

Oremland, R.S. 1979. Methanogenic activity in plankton samples and fish intestines: a mechanism for *in situ* methanogenesis in oceanic surface waters. *Limnology and Oceanography*, **24**: 1136–1141.

Owens, N.J.P. 1985. Variations in the natural abundance of ^{15}N in estuarine suspended particulate matter: a specific indicator of biological processing. *Estuarine Coastal Shelf Science*, **20**: 505–510.

Owens, N.J.P. 1986. Estuarine nitrification: a naturally occurring fluidized bed reaction? *Estuarine Coastal Shelf Science*, **22**: 31–44.

Pace, M.L., Knauer, G.A., Karl, D.M. and Martin, J.H. 1987. Primary production, new production and vertical flux in the eastern Pacific Ocean. *Nature*, **325**: 803–804.

Reeburgh, W.S. 1980. Anaerobic methane oxidation: rate depth distributions in Skan Bay sediments. *Earth and Planetary Science Letters*, **47**: 345–352.

Reeburgh, W.S. 1982. A major sink and flux control for methane in marine sediments: anaerobic consumption. In *The dynamic environment of the ocean floor* (eds. K.A. Fanning and F.T. Manheim), pp. 203–217. Lexington Books, Lexington, MA.

Richards, F.A. 1965. Anoxic basins and fjords. In *Chemical oceanography* (eds. J.P. Riley and G. Skirrow), pp. 611–645. Academic Press, New York.

Riley, G.A. 1963. Organic aggregates in seawater and the dynamics of their formation and utilization. *Limnology and Oceanography*, **8**: 372–381.

Roques, P.F. 1985. Rate and stoichiometry of nutrient mineralization in an anoxic estuary, the Pettaquamscutt River (Rhode Island, USA). *PhD Thesis*, 330 pp. University of Rhode Island.

Rossignol-Strick. M. 1985. A marine anoxic event on the Brittany Coast, July 1982. *Journal of Coastal Research*, **1**: 11–20.

Rudd, J.W.M. and Hamilton, R.D. 1978. Methane cycling in an eutrophic shield lake and its effects on whole lake metabolism. *Limnology and Oceanography*, **23**: 337–348.

Rudd, J.W.M. and Hamilton, R.D. 1979. Methane cycling in Lake 227 in perspective with some components of the carbon and oxygen cycles. *Archiv für Hydrobiologie ergenbnisse der Limnologie*, **12**: 115–122.

Rudd, J.W.M. and Taylor, C.D. 1980. Methane cycling in aquatic environments. *Advances in Aquatic Microbiology*, **2**: 77–150.

Ryther, J.H. 1969. Photosynthesis and fish production in the sea. *Science*, **166**: 72–76.

Sarabun, C.C., Brandt, A., Tyler, M.A. and Smith, G.D. 1986. Biological transport, internal waves, and mixing in the Chesapeake Bay. *Johns Hopkins API Technical Digest*, **6**: 227–235.

Schlegel, H.G., Gottschalk, G. and Pfennig, N. 1976. *Microbial production and utilization of gases (H$_2$, CH$_4$, CO)*, 425 pp. Goltze KG, Göttingen.

Scranton, M.I. and Brewer, P.G. 1977. Occurrence of methane in the near-surface waters of the western subtropical North Atlantic. *Deep-Sea Research*, **24**: 127–138.

Scranton, M.I. and Farrington, J.W. 1977. Methane production in waters off Walvis Bay. *Journal of Geophysical Research*, **82**: 4947–4953.

Seiler, W. and Schmidt, V. 1974. Dissolved nonconservative gases in seawater. In *The sea* (ed. E.D. Goldberg), Vol. 5, pp. 219–243. Wiley, New York.

Seliger, H.H., Boggs, J.A. and Biggley, W.H. 1985. Catastrophic anoxia in the Chesapeake Bay in 1984. *Science*, **228**: 70–73.

Sen Gupta, R. and Jannasch, H.W. 1973. Photosynthetic production and dark-assimilation of CO_2 in the Black Sea. *International Revue Der Gesamten Hydrobiologie*, **58**: 625–632.

Sholkovitz, E.R. 1976. Flocculation of dissolved organic and inorganic matter during the mixing of river water and seawater. *Geochimica et Cosmochimica Acta*, **40**: 831–845.

Sieburth, J.McN. 1960. Acrylic acid, an "antibiotic" principle in *Phaeocystis* blooms in Antarctic waters. *Science*, **132**: 676–677.

Sieburth, J.McN. 1967. Seasonal selection of estuarine bacteria by water temperature. *Journal of Experimental Marine Biology and Ecology*, **1**: 98–121.

Sieburth, J.McN. 1968. The influence of algal antibiosis on the ecology of marine microorganisms. *Advances in Microbiology of the Sea*, **1**: 63–94.

Sieburth, J.McN. 1971. Distribution and activity of oceanic bacteria. *Deep-Sea Research*, **18**: 1111–1121.

Sieburth, J.McN. 1976. Bacterial substrates and productivity in marine ecosystems. *Annual Review of Ecology and Systematics*, **7**: 259–285.

Sieburth, J.McN. 1979. *Sea microbes*, 491 pp. Oxford University Press, New York. .

Sieburth, J.McN. 1983. Microbiological and organic-chemical processes in the surface and mixed layers. In *Air sea exchange of gases and particles* (eds. P.S. Liss and W.G.N. Slinn), pp. 121–172. NATO Advanced Study Institute, Reidel, Dordrecht, The Netherlands.

Sieburth, J.McN. 1984. Protozoan bacterivory in pelagic marine waters. In *Heterotrophic activity in the sea*, (eds. J.E. Hobbie and P.J.leB. Williams), pp. 405–444. Plenum Press, New York.

Sieburth, J.McN. 1986. Dominant microorganisms of the upper ocean: form and function, spatial distribution and photoregulation of biochemical processes. In *Dynamic processes in the chemistry of the upper ocean* (eds. J.D. Burton, P.G. Brewer and R. Chesselet), pp. 173–195. Plenum Press, New York.

Sieburth, J.McN. 1987a. The nanoalgal peak in the dim pycnocline: is it sustained by microparticulates and their bacterial consortia? In *Biogeochemical cycling and fluxes between the deep euphotic zone and other oceanic realms* (ed. C.R. Agegian), Vol. 3, Part 2, in preparation. Symposia Series for Undersea Research, NOAA Undersea Research Program.

Sieburth, J.McN. 1987b. The role of microorganisms in marine ecosystems: the unique orchestration of mandatory processes by versatile microbial consortia. In *Proc. 1st International Symposium on Microbial Ecology of the Mediterranean Sea, Sorrento, Italy, 25–30 May 1987*.

Sieburth, J.McN. and Jensen, A. 1968. Studies on algal substances in the sea. I. Gelbstoff (humic material) in terrestrial and marine waters. *Journal of Experimental Marine Biology and Ecology*, **2**: 174–189.

Sieburth, J.McN., Johnson, P.W., Eberhardt, M.A., Sieracki, M.E., Lidstrom, M. and Laux, D. 1987. The first methane-oxidizing bacterium from the upper mixing layer of the deep ocean: *Methylomonas pelagica* sp. nov. *Current Microbiology*,

14: 285–293.

Sieracki, M.E., Johnson, P.W. and Sieburth, J.McN. 1985. Detection, enumeration, and sizing of planktonic bacteria by image-analyzed epifluorescence microscopy. *Applied and Environmental Microbiology*, **49**: 799–810.

Söhngen, N.L. 1906. Über Bakterien, welche Methan als Kohlenstoffnahrung und Energiequelle gebrauchen. *Zentrablatt für Bakteriologie Parasitenkenkunde Infektions Krankheiten Abstract 2*, **15**: 513–517.

Sorokin, Y.I. 1970. Experimental investigation of the rate and mechanism of oxidation of hydrogen sulfide in the Black Sea using S-35. *Oceanology*, **10**: 37–46.

Sowers, K.R. and Ferry, J.G. 1983. Isolation and characterization of a methylotrophic marine methanogen, *Methanococcoides methylutens* gen. nov., sp. nov. *Applied and Environmental Microbiology*, **45**: 684–690.

Sowers, K.R., Baron, S.F. and Ferry, J.G. 1984. *Methanosarcina acetivorans* sp. nov., an acetotrophic methane producing bacterium isolated from marine sediments. *Applied and Environmental Microbiology*, **47**: 971–978.

Stockner, J.G. and Antia, N.J. 1986. Algal picoplankton from marine and freshwater ecosystems: a multidisciplinary perspective. *Canadian Journal of Fisheries and Aquatic Sciences*, **43**: 2472–2503.

Swinnerton, J.W., Linnenbom, V.J. and Cheek, C.H. 1969. Distribution of methane and carbon monoxide between the atmosphere and natural waters. *Environmental Science and Technology*, **3**: 836–838.

Traganza, E.D., Swinnerton, J.W. and Cheek, C.H. 1979. Methane supersaturation and ATP-zooplankton blooms in near surface waters of the Western Mediterranean and the subtropical North Atlantic Ocean. *Deep-Sea Research*, **26**: 1237–1245.

Tuttle, J.H. and Jannasch, H.W. 1972. Occurrence and types of thiobacillus-like bacteria in the sea. *Limnology and Oceanography*, **17**: 532–543.

Tuttle, J.H. and Jannasch, H.W. 1979. Microbial dark assimilation of CO_2 in the Cariaco Trench. *Limnology and Oceanography*, **24**: 746–753.

Tyler, M.A. 1984. Dye tracing of a subsurface chlorophyll maximum of a red-tide dinoflagellate to surface frontal regions. *Marine Biology*, **78**: 285–300.

Urakami, T. and Komagata, K. 1986. Emendation of *Methylobacillus* Yordy and Weaver 1977, a genus for methanol-utilizing bacteria. *International Journal of Systematic Bacteriology*, **36**: 502–511.

Welch, B.L. 1987. Deepwater hypoxia in western Long Island Sound, 1986 joint cruise preliminary results. *Proc. 1st science–management workshop, Long Island Sound Study, University of Connecticut, Avery Point, October 1986*, 71 pp.

Whittenbury, R., Phillips, K.C. and Wilkinson, 1970. Enrichment, isolation and some properties of methane-utilizing bacteria. *Journal of General Microbiology*, **61**: 205–218.

Williams, P.J.leB. and Jenkinson, N.W. 1982. A transportable microprocessor-controlled precise Winkler titration suitable for field station and shipboard use. *Limnology and Oceanography*, **27**: 576–584.

Williams, R.T. and Bainbridge, A.E. 1973. Dissolved CO, CH_4 and H_2 in the southern ocean. *Journal of Geophysical Research*, **78**: 2691–2694.

2

MICROBIAL FOOD WEBS IN SEDIMENTS

T. Henry Blackburn, Department of Genetics and Ecology, Aarhus University, DK 8000 Aarhus C, Denmark

The production of cells, which can serve as substrate in benthic food chains, can be calculated from the rate of microbial carbon oxidation, or more indirectly from the rate of electron acceptor reduction or from the rate of organic-N mineralization. In measuring the sum of the reduction rates of the various electron acceptors, it is necessary to determine whether sulphide oxidation contributes to oxygen consumption. Bioturbation and other physical disturbances of the sediment can influence sulphide reoxidation. Two examples of sulphide reoxidation are discussed, one in a highly eutrophic fish pond and the other in a Bering Sea sediment. Detrital degradability in sediment is very important as it determines microbial production; this is discussed in relation to the Bering Sea sediments and to an artificial microcosm system. Models are presented to describe the mineralization processes and the importance of C:N ratios for both situations. The contribution of the benthic macrofauna in increasing the rate of detrital mineralization and in directly competing for substrate is examined.

INTRODUCTION

The main purpose of this presentation is to examine the factors regulating the growth and productivity of bacteria in sediment. It is this production of bacterial biomass that determines their importance as contributors to a food chain. This aspect of microbial metabolism has largely been overlooked in a consideration of the role of bacteria in sediments. The emphasis has been on the important part played by bacteria in the mineralization of sediment detritus, and in returning mineralized nutrients to the overlying water, to be utilized by the primary producers. The researches by Moriarty (Moriarty and Pollard, 1981, 1982) and Karl (Karl and Winn, 1984) are exceptions to this general trend, and some aspects of bacterial production in sediments have been reviewed by Blackburn (1987b). Much more attention has been devoted to research into bacterial productivity in the pelagic environment, with less emphasis on mineralization (Williams, 1984).

In most situations, algal cells and organic detritus reach the sediment surface and are mineralized by bacteria and other organisms in the sediment. There is little accumulation of organic matter due to burial. Maximum rates of bacterial productivity could, theoretically, be calculated from the input of organic matter multiplied by the efficiency of incorporation of carbon into the bacterial cells. This would set an upper limit to the bacterial biomass that would be available as a resource to higher trophic levels. Unfortunately, in most situations the input of organic material to the sediment is unknown. The calculation cannot be made, although upper limits on the bacterial biomass could be set by the primary productivity in the water column multiplied by an efficiency factor. More definitive measurements can be made by ^{3}H-thymidine (Moriarty and Pollard, 1981) and adenine (Karl and Winn, 1984) incorporations, or more crudely from mineralization rates in the sediment. The latter include the rate of CO_2 production, the rate of electron acceptor reduction and the rate of NH_4^+ production (Blackburn, 1987a, b).

In this presentation, the relationship between C- and N-mineralizations is formulated, depending on whether bacteria are themselves mineralized in the sediment. It is usually difficult to determine whether O_2 consumption rates include the reoxidation of sulphide, Fe and Mn, or whether they give an exclusive measurement of carbon oxidation. Some examples are given where sulphide plays an important role in this type of calculation and where SO_4^{2-} reduction has been dominant in carbon oxidation. One example is in a fish pond, where fish bioturbation played a major role on a short time interval of days; another example is that of a sediment in the Bering Sea where sulphide build-up took place over the winter and reoxidation occurred during the early summer. The processes in the latter are illustrated by a conceptual model, to explain the interrelationship between pools and rates on a yearly basis. The partitioning of the incoming food resource (algal cells) between microorganisms (bacteria + other microbes) and the benthic macrofauna is assessed, although no attempt is made to estimate the actual rates of production of bacterial or macrofaunal biomass. In general, there is considerable uncertainty regarding micro-, meo- and macrofaunal production rates and the extent to which they are dependent on bacterial biomass as a food resource (Tenore, 1987).

The quality and quantity of organic detritus in the Bering Sea sediments is examined and correlations made with sediment mineral composition and sediment porosity. The Bering sediments are of considerable interest, since grey whales and walruses act as both bioturbators and predators on the benthic macrofauna.

The quality and quantity of degradable organic detritus in sediment of aquaria is also considered, where the sediments were aerobic, anaerobic or contained *Nereis*. A simulation model is presented which describes the relationship between rates of organic-C and organic-N mineralization, the production of mineral products, and the C:N ratio of the material undergoing degradation at different times. It is concluded that differential mineralization of organic-N occurs first, but is followed by mineralization of detritus having a very high C:N ratio. This has implications for the efficiency of bacterial biomass production, and thus for food webs.

| Substrates | | | Zone | Products | | | C |
C	N	Oxidant		C	N	Reduced oxidant	incorporated (%)
Organic-C	Organic-N	O_2	Oxygen	CO_2+cells	NH_4^+	H_2O	50
Organic-C	Organic-N	NO_3^-	Nitrate	CO_2+cells	NH_4^+	N_2	40
Organic-C	Organic-N	Fe^{3+}	Oxides	CO_2+cells	NH_4^+	Fe^{2+}	40
Organic-C	Organic-N	SO_4^{2-}	Sulphate	CO_2+cells	NH_4^+	S^{2-}	30
Organic-C	Organic-N	Fermentation	Methane	CO_2+cells +CH_4	NH_4^+		20

Table 1   Substrates and products of bacterial metabolism in different sediment zones

BACTERIAL PROCESSES

The types of bacteria that are responsible for the mineralization of organic detritus in sediments vary with sediment depth, which in turn corresponds to increasing *Eh*. This zonation is summarized in Table 1. The depth of any zone depends largely on the organic loading of the system, but typically the oxygen zone would extend from 0 to 7 mm, nitrate from 7 to 20 mm, oxides of iron and manganese immediately below the nitrate zone to an unknown depth, sulphate extending downward for some metres, and the methane zone to below the point of sulphate disappearance. It should be emphasized that all these mineralizing bacteria are heterotrophic, and are limited by energy (carbon) availability, rather than by other nutrients, e.g. N or P. They utilize the reduced carbon substrates with different degrees of efficiency, i.e. they vary in the proportion of substrate that must be oxidized in order to generate the necessary energy to build new biomass. The amount of energy that is available from a substrate decreases with increasing *Eh*, as is seen in the decreasing efficiency of the percentage incorporation of carbon (Table 1). The products of bacterial mineraliza-tion in sediments are, therefore, mineralized substrate (CO_2 and NH_4^+), reduced oxidant (H_2O, N_2, Fe^{3+} and S^{2-}) and bacterial cells. The mineralized CO_2 and NH_4^+ diffuse to the sediment surface where NH_4^+ can be oxidized to NO_2^- and NO_3^-. All these species may leave the sediment to the overlying water, and be utilized by the pelagic primary producers. Some increase in the biomass of the nitrifying population occurs as a result of autotrophic NH_4^+ oxidation, but this is not quantitatively significant in the context of bacterial cells being components of a food web (Fenchel and Blackburn, 1979). Similarly, N_2 fixation is not significant in most sediments.

Denitrification, which is extremely important in the N-cycle, is generally of little importance in the carbon–cell generation cycle, except under conditions where high NO_3^- concentrations occur in the overlying water (Jørgensen and Sørensen, 1985). As with autotrophic NH_4^+ oxidation, the bacterial oxidation of reduced Fe and Mn and of S^{2-} results in cell production. There is little information available on the quantitative significance of Fe and Mn oxidations, but in many situations in marine sediment SO_4^{2-} reduction rates are high and result in the formation of large amounts of S^{2-}, which is mostly bound to Fe as an insoluble but oxidizable precipitate (acid-volatile sulphide). There is much energy to be released from the oxidation of S^{2-}, which could be used by bacteria to generate biomass. Fenchel and Blackburn (1979) calculated that 5.1 mmol cell-C m^{-2} day^{-1} could be assimilated by sulphide-oxidizing cells in Limfjorden sediments, compared with 36.6 mmol cell-C m^{-2} day^{-1} that could be generated from heterotrophic processes. The limitation of O_2 to the top 1 cm of sediments precludes the direct interaction of O_2 with insoluble S^{2-} in the lower sediment layers, except when O_2 is directly mixed downward by physical disturbance of the sediment, induced by bioturbation or wave action.

The oxidation of S^{2-} is considered in more detail in the next section but, before moving on to that topic, it should be emphasized that much carbon that is metabolized in the sediment is incorporated into bacterial cells. As these cells are probably themselves mineralized in the sediment, a high proportion of the organic detritus may originate from degraded bacterial cells, rather than from non-degraded residues of the algal cells etc. which settled to the sediment surface. This is sometimes ignored in diagenetic models of sediment mineralizations (Berner, 1980).

In practical terms, the rate of cell biomass production can itself be calculated from the rate of carbon oxidation, which can be measured in three ways. The rate of CO_2 production can be measured directly, or it can be deduced from the sum of the rates of electron acceptor reduction, or from the rates of N-cycling (Table 2). All these methods have their individual problems, which will be examined in subsequent sections. One of the greatest problems in summing the reduction rates of the electron acceptors is to determine whether O_2 is being used to oxidize reduced products of anaerobic mineralization (e.g. S^{2-}) during the measurement of O_2 consumption, or whether all O_2 is used to oxidize reduced carbon.

Nitrogen rate	Nitrogen rate defined in terms of carbon oxidation: cells not mineralized *in situ*[a]	Nitrogen rate defined in terms of carbon oxidation: cells mineralized *in situ*[a]
i (incorporation)	$C_oN_cE/(1-E)$	$C_oN_cE/(1-E)$
d−i (net mineralization)	C_oN_s	$C_o(N_s-N_cE)/(1-E)$

[a] *i* is the rate of ammonium incorporation, C_o is the rate of carbon oxidation. *E*, the efficiency of carbon incorporation, is defined as the rate of carbon incorporation divided by the sum of the rates of carbon oxidation and carbon incorporation. N_c and N_s are the N:C ratios in cells and substrate respectively (based on Blackburn, 1987a).

Table 2   C- and N-cycling in sediment depending on site of biomass mineralization

SULPHIDE OXIDATION

An interesting example of the effect of bioturbation on S^{2-} oxidation and on sediment mineralizations in a eutrophic fish pond is presented in Fig. 1. Rates of exchange of CO_2, O_2 and NH_4^+ between the surface of sediment cores and the overlying water were measured. Rates of SO_4^{2-} reduction were measured in cores injected with $^{35}SO_4^{2-}$ and are expressed in mmol m^{-2} day^{-1} from 0 to 12 cm depth (Jørgensen, 1979). Net rates of NH_4^+ production were

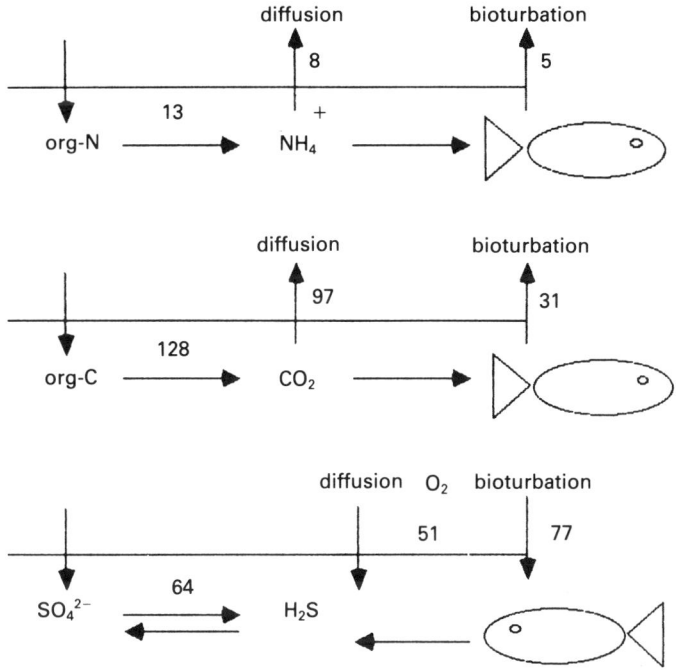

Fig. 1—Role of fish bioturbation in sediment mineralization. All rates are in mmol m^{-2} day^{-1}. Measurements were made in sediment cores from pond number 6, Eilat, Israel, in March 1986. The influence of fish was calculated from mass balance considerations, as explained in the text, (Blackburn *et al.*, unpublished data).

measured in sediment slices (Blackburn, 1979). There was no accumulation of NH_4^+ or of S^{2-} in the sediments, and it was deduced that these species left, or were oxidized in the sediment. As there was no evidence of nitriification and there was no free sulphide, it is probable that all NH_4^+ left the sediment as NH_2^+ and that all S^{2-} was oxidized in the sediment. The rate of net mineralization of organic-N was 13 mmol m^{-2} day^{-1}, the diffusional flux to the overlying water was 8 mmol m^{-2} day^{-1}, and it was deduced that 5 mmol m^{-2} day^{-1} left the sediment as a result of fish bioturbation. Rates of SO_4^{2-} reduction were very high, 64 mmol m^{-2} day^{-1}, and it was assumed that all organic-C was oxidized by SO_4^{2-} reduction and that all O_2 was used in the reoxidation of S^{2-}. On the basis of these assumptions, 128 mmol CO_2 m^{-2} day^{-1} was produced, but only 76% left the sediment by diffusion, indicating that 24% left by bioturbation. The diffusional O_2 flux into the sediment was sufficient to reoxidize only 40% of the S^{2-}, indicating that 60% sulphide oxidation was due to O_2 introduced by bioturbation. It is not known whether bacteria grew at the expense of S^{2-} oxidation, or indeed what the fate was of bacterial biomass produced in the sediment. It is very clear from this example that it is very dangerous to assume that all S^{2-} is oxidized by O_2 during the course of the experiment.

Fig. 2 presents some information on S^{2-} oxidation, but on a very different

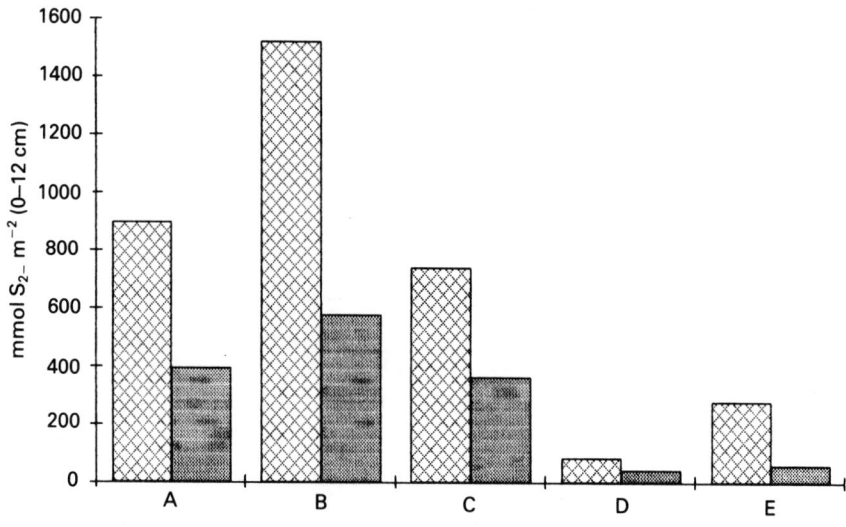

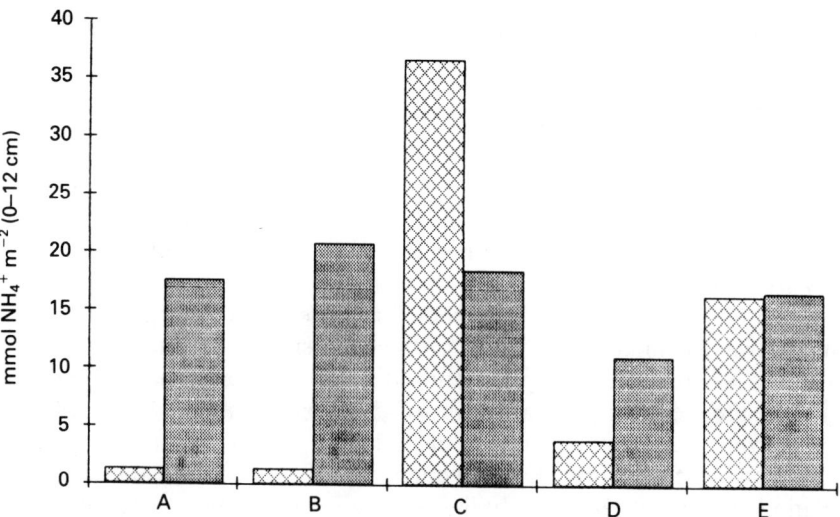

Fig. 2—Sulphide and ammonium pools in sediments at different sites and times. The location of sites A, B, C, D and E are indicated on the map of the North Bering and Chukchi Seas in Fig. 3. The first columns (crosshatched) were sampled in 1986, in July for A, B, C and D, and in August for E. The sites were sampled a second time (stippled) one month later for A, B, C and D, and two weeks later for E. The ammonium is the total KCl extractable pool integrated from 0 to 12 cm depth. The sulphide is the total acid-volatile sulphide pool integrated from 0 to 12 cm depth.

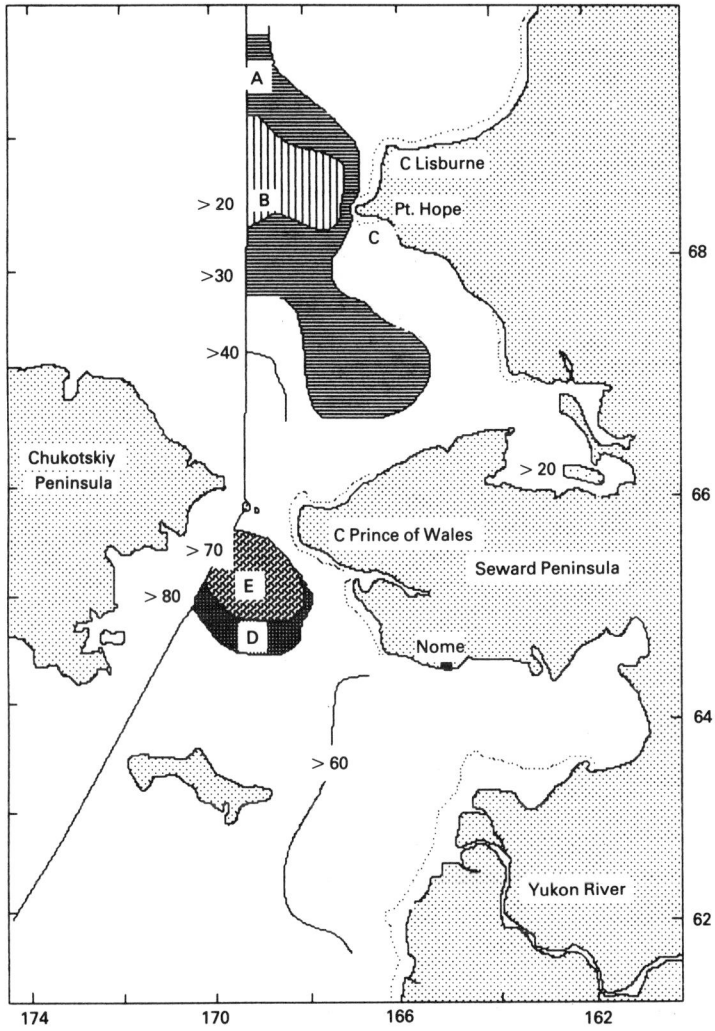

Fig. 3—Location of sites and distribution of sand (125 to 250 μm particle size) in sediments in the North Bering and Chukchi Seas. Isopleths show the content of sand as a percentage of the dry weight (Blackburn *et al.*, unpublished data).

time scale. Measurements were made of S^{2-} and NH_4^+ pools at five sites (A through E, Fig. 3) in North Bering and Chukchi sediments, at times separated by 4 or 2 week intervals. At all sites, the S^{2-} pool was higher at the first sampling, whereas there was a tendency for NH_4^+ pools to be lower at this time. The implication is that S^{2-} increased during the wintertime, but that NH_4^+ may have decreased. A plot of measured CO_2 flux from the sediment versus a sum of the rates of reduction of the electron acceptors (Fig. 4) showed a difference between the sediments in July and in August, one month later. The slope of the plot for July suggests that more O_2 diffused in than CO_2 diffused out (SO_4^{2-} reduction was small), whereas in August both rates were equal.

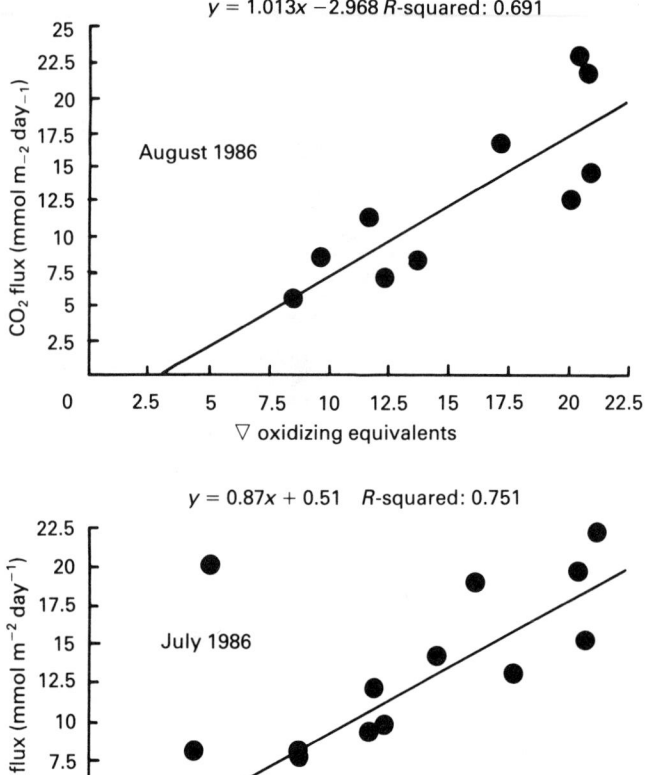

Fig. 4—Relationship between CO_2 produced and oxidants consumed in sediment cores from stations in the North Bering and Chukchi Seas. The CO_2 flux from the sediment to the overlying water was measured directly. The oxidizing equivalents were calculated from the rates of reduction of O_2, NO_3^- and SO_4^{2-}. Some of the stations (sites A, B, C, D and E in Fig. 3) were sampled twice. One station (cross) was not included in the regression (Blackburn *et al.*, unpublished data).

(Incidentally, Fig. 4 shows that CO_2 production rates correlate well with electron acceptor reduction rates.)

These data can be explained as follows (Fig. 5). In early and mid-summer, algal cells with a low C:N ratio settle to the sediment surface. By mid-summer these cells are being mixed downward, and in the process the C:N ratio increases because of preferential N-mineralization. By the onset of winter much of the degradable content of the algal cells has been mineralized, and that which remains is deficient in N. During the winter this detritus is further mineralized in the sediment, principally by SO_4^{2-} reduction, leading to the production of sulphide.

The difference between the sulphide pools at the two different times indicates that

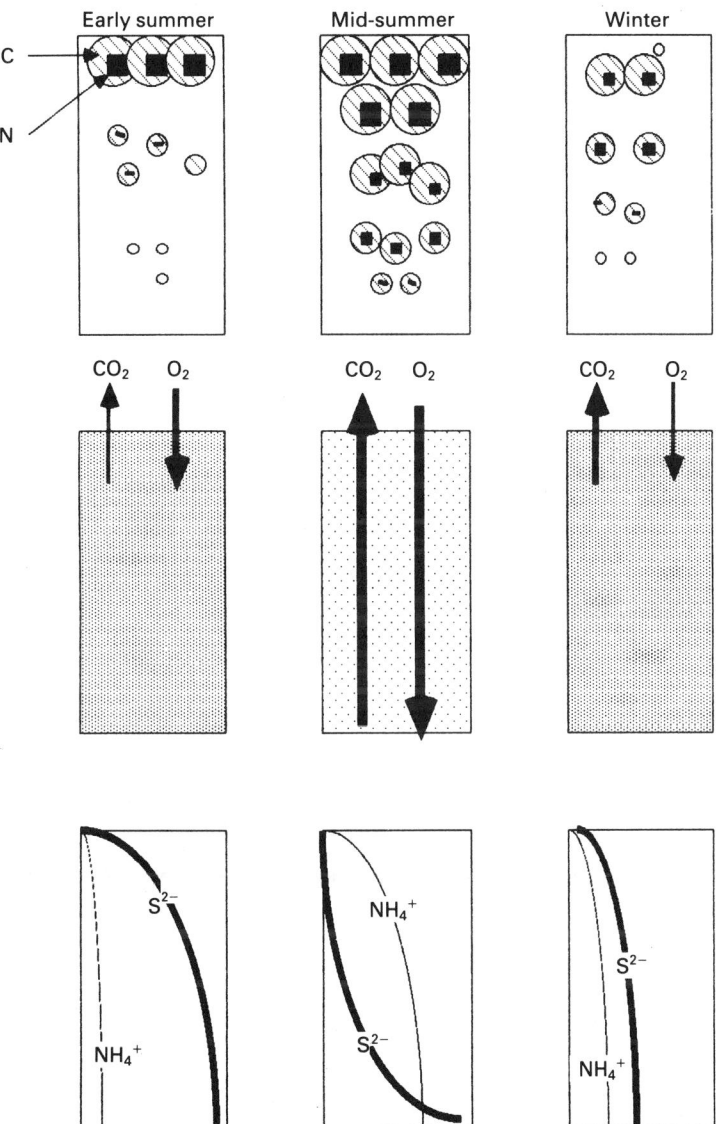

Fig. 5—Model of seasonal mineralizations in sediment from the North Bering and Chukchi Seas (see text for details).

up to 2000 mmol $C\,m^{-2}$ (24 g C) must have been oxidized to produce this S^{2-}. A useful indication is obtained of the minimum rate of carbon oxidation during the winter. Winter sulphide is not oxidized, possibly because of a lack of macrofaunal bioturbation in the winter, which would also explain the high rate of SO_4^{2-} reduction. Because the detritus being degraded has a low N-content, little NH_4^+ is produced. Thus, in the early summer, the sediments contain high S^{2-}, low NH_4^+ and extra O_2 is used to oxidize the S^{2-} reservoir. By mid-summer the S^{2-} has gone, NH_4^+ has increased and the O_2 and CO_2

fluxes are equal. Again, it is not known whether an increase in bacterial biomass is associated with the oxidation of S^{2-}, but it seems quite likely that bacteria benefit from the process.

It has been suggested that the activities of the benthic macrofauna can result in the controlled entry of O_2 to the S^{2-}-rich anoxic sediment layers, where bacterial growth occurs with energy supplied by S^{2-} oxidation (Dobbs and Scholly, 1986). The bacterial cells can then be used by the animal as food, and the process is a type of resource cultivation or 'gardening'. In a similar way the fauna associated with thermal and other oceanic vent systems use S^{2-}-oxidizing bacteria, grown internally, to supply reduced carbon for their nutrition (Karl et al., 1984). It seems likely that the pogonophore *Siboglinum fiordicum* and the bivalve *Lucinoma borealis*, which live in deep anoxic sediment, also obtain energy from S^{2-} indirectly through symbiotic bacterial growth (Dando et al., 1986).

DIRECT MACROFAUNAL UTILIZATION OF DETRITUS

There is considerable controversy regarding the role of sediment bacteria in the nutrition of the benthic infauna (Tenore, 1987). If the food source is of a high nutritional value, such as algal cells, it is an advantage for the fauna to use it directly without bacterial modification. This is likely to result in a preferential loss of nitrogen and nutritive value, such as occurs in ruminants (Fenchel and Blackburn, 1979). In the Bering–Chukchi sediments, the macrofauna can obtain a major portion of their food directly (Table 3). Sites A and B in the Chukchi are dominated by bivalves, sites D and E in the North Bering Sea have mostly amphipods, while the coastal site C contains little fauna. The very high amphipod populations at sites D and E, sustained as a result of direct utilization of the abundant algal input, have made these sites a major feeding resource for the grey whale (Nelson and Johnson, 1987). At sites A, B, D and E quite a high proportion of the sediment respiration is due to the benthic macrofauna. A smaller role is played by macrofauna in the sparsely inhabited coastal sediments.

Site	C-oxidation total (mmol m^{-2} day^{-1})	Mineralization by macrofauna (%)
A,B	26	5–50
D,E	26	60–70
C	9	10–20

[a]The location of the sites is indicated in Fig. 3. The data are from Henriksen and Blackburn (unpublished).

Table 3 Partitioning of mineralization at different sites[a]

We have just begun to examine the turnover of urea in these sediments, and have been surprised to discover that urea plays a very significant role in nitrogen cycling.

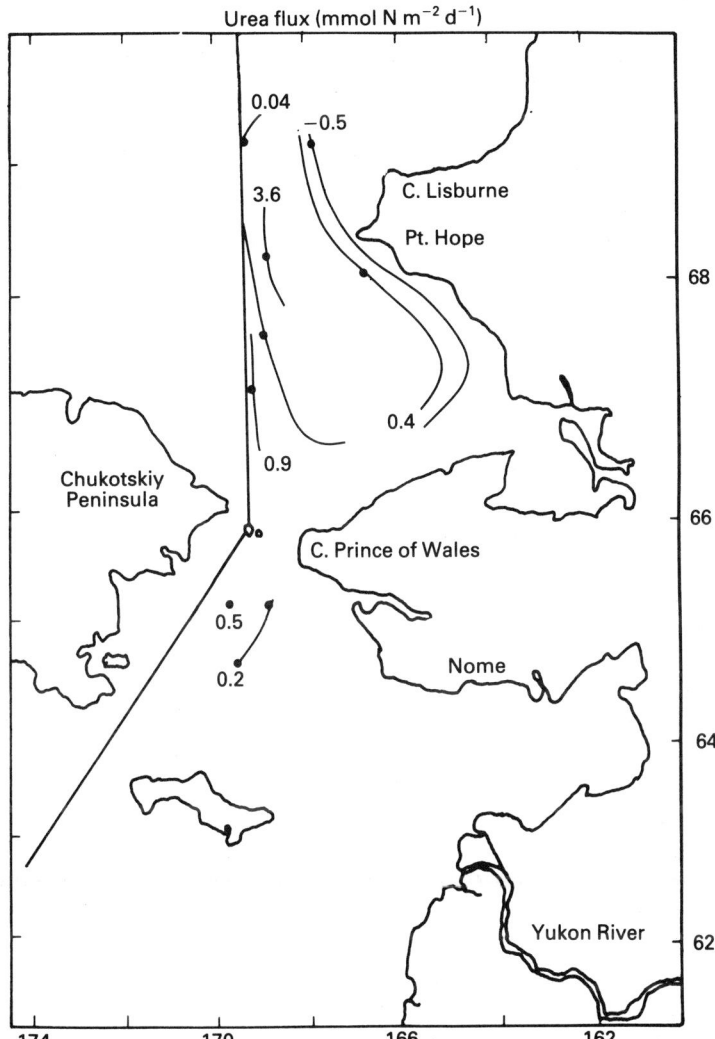

Fig. 6—Urea flux from sediments from the North Bering and Chukchi Seas (Lund *et al.*, unpublished data).

In the present context, the high flux of urea from the sediments at sites A, B, D and E is indicative of the dominant role of the macrofauna in these regions (Fig. 6). The turnover of urea in the sediments is a significant source of ammonium, and should be distinguished from the ammonium that is derived from the direct bacterial degradation of sediment detrital matter.

QUALITY OF SEDIMENT DETRITUS

All sediment detritus is not of the same quality. Detritus at some locations is degraded by bacteria at a more rapid rate than the detritus at other sites. In general, there is a relationship between the quantity of detritus deposited to the sediment and the organic content in that sediment. In the Bering–Chukchi sediments the situations seems to be more complicated, and the sand content (defined as the percentage of the dry weight composed of grain size 125 to 250 μm), whose distribution is shown in Fig. 3, has a negative correlation (Fig. 7B) with organic carbon content. It seems that a high sand content favours a high rate of decomposition. Not surprisingly, there is a good negative correlation ($r^2 = 0.70$) between sand content and porosity (Fig. 7A) and a positive correlation ($r^2 = 0.77$) between porosity and organic carbon content (Fig. 7C). The most important factor in determining the accumulation of organic detritus in the sediment may be the extent to which bacteria participate in the degradation of the freshly sedimented material. If a significant proportion of the bacterial cell is degraded very slowly (e.g. cell walls), and if bacterial growth has an average efficiency of 30%, then for every 100 g of sedimenting detritus degraded by bacteria, up to 30 g of that substrate would appear to be poorly degradable. In many models of sediment mineralization this would be attributed to poorly degradable material that was originally present. When the microbial component is replaced by direct macrofaunal feeding (Table 3), there may be less accumulation of organic carbon.

Although there is not a good correlation between the C:N ratio of detritus and other sediment parameters, there is a general tendency for lower C:N ratio detritus to be present in the zones dominated by the amphipods (North Bering) and bivalves (South Chukchi) than in the immediate surroundings, as seen in Fig. 8. This would again be consistent with the presence of fresh, low C:N ratio detritus in zones of faunal activity and old, high C:N ratio detritus in regions dominated by bacterial activity. There seems to be a general tendency for C:N ratios in sediments (and soils) to stabilize around 10, very seldom being greater than 12. This suggests the operation of a mechanism which determines the production of residue with a definite C:N ratio, irrespective of the C:N of the input material. It is likely that this residue is degraded bacterial cells, and more specifically bacterial cell wall components.

These sediments, particularly those in the North Bering sea, are subject to another type of disturbance, which is claimed to have a marked effect on their character (Nelson and Johnson, 1987). Whales and walruses create considerable disturbance of the sediment, and whales scoop up mouthfuls of sediment, leaving pits 2.5 m long, 1.5 m wide and 10 cm deep. Conservatively, 5.6% (12 000 km^2) of the whale feeding area is resuspended each year. The authors suggest that this is an important mechanism for removing fine silt particles from these sediments, and thus making them better suited to the amphipods.

MINERALIZATION OF DETRITUS IN MICROCOSMS

The relationship between C:N ratio and detrital degradability was further investigated in aquaria experiments (Kristensen and Blackburn, 1987). The microcosms consisted of sieved (1.5 mm mesh) surface sediment in 4.5 cm diameter Plexiglass cores. Three treatments were used: aerobic, anaerobic and aerobic containing *Nereis* polychaetes. A number of measurements were made during a 94 day incubation, but the main results that will be discussed here relate to actual

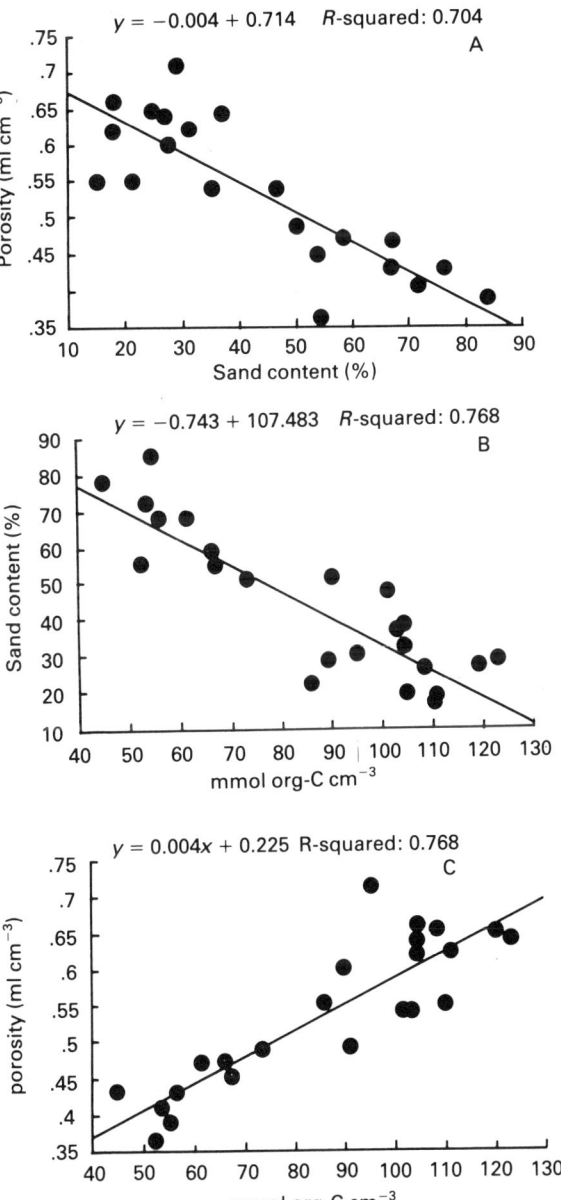

Fig. 7—Relationship between content of sand, organic carbon and porosity in sediments from the North Bering and Chukchi Seas (Blackburn *et al.*, unpublished data).

C:N ratios

+ leg II stations 85　　　ᵡ leg II stations 86
♦ leg III stations 85　　o leg III stations 86
　　　　　　　　ᴋ Kaj stations 86

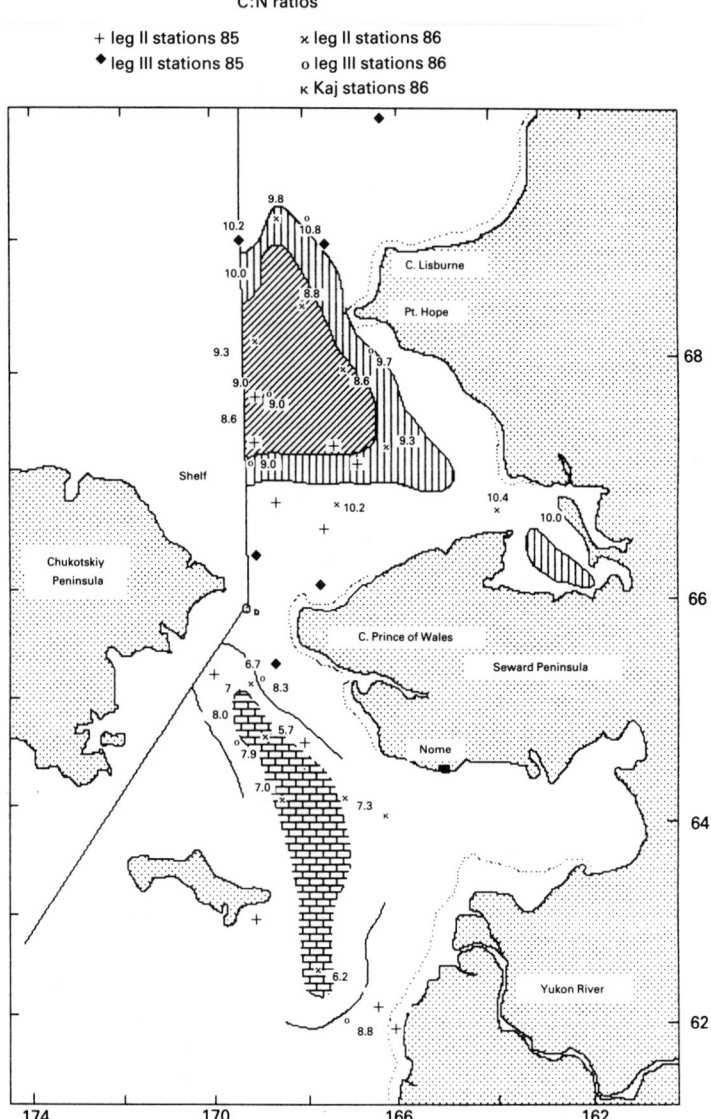

Fig. 8—C:N ratios in sediment organic matter from the North Bering and Chukchi Seas (Blackburn *et al.*, unpublished data).

changes which occurred in the organic detritus during that time. There was a loss of organic-C and organic-N due to mineralization; these losses are expressed as percentages of the starting material in Fig. 9.

There was a preferential mineralization of organic-N; in all treatments nitrogen losses exceeded those of carbon. Mineralization was greatest in the sediments containing *Nereis*, and surprisingly the anaerobic sediments had the next greatest losses. The C:N ratio of the mineralized material was always less than the original

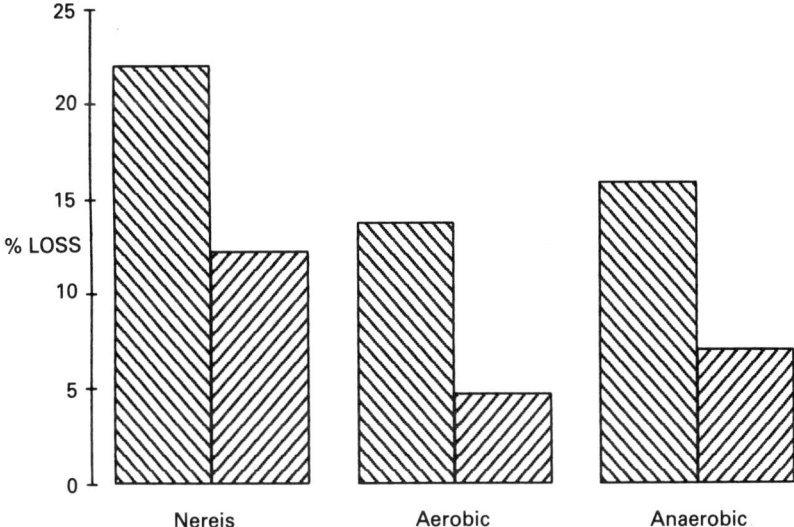

Fig. 9—Percentage of original organic-N (first column) and organic-C (second column) mineralized after 94 days in aquarium sediments designated *Nereis*, aerobic and anaerobic, as described in the text (data from Kristensen and Blackburn, 1987).

detritus (Fig. 10), and the aerobic sediments which had the smallest total losses also had the lowest C:N ratio in the substrate used (approximately 4 compared with 6 for the *Nereis* sediments). Rate measurements at the end of the 94 day incubation indicated that the material undergoing mineralization in the *Nereis*, anaerobic and aerobic cores had C:N ratios of approximately 16, 4 and 4, respectively. It seemed that the most degradable material in the original detritus had a very low C:N ratio and that this had not all been mineralized in the aerobic and anaerobic treatments. It had, however, disappeared in the *Nereis* sediments, and apparently the N-stripped residue was more degradable than the bulk of the remaining organic detritus.

The carbon losses in the aerobic and *Nereis* sediments were very close to the carbon losses calculated from O_2 consumption. The latter measurements were made at frequent intervals and showed an exponential decrease with time. This exponential decrease in the rate of mineralization has been used to construct a simulation model (Fig. 11) which is a more dynamic representation of the course of mineralization in the *Nereis* sediments than the one which was previously presented (Kristensen and Blackburn, 1987). The initial rate of C-mineralization and the rate at 94 days were derived from the rate of O_2 consumption. It is assumed that the C:N ratio of the substrate initially degraded was 4.0, and that the rate of N-mineralization was thus 1/4 that of the C-mineralization rate. The rate at 96 days was experimentally determined, and it was assumed that there was an exponential decrease to that rate. The decrease in both rates with time is illustrated in the top of Fig. 11. The accumulated production of C-mineralization and N-mineralization are plotted against incubation time in Fig. 11B and C:N ratios are plotted in Fig. 11C. The C:N ratio of the products begins at 4.0 and increases gradually to the predicted value of 6 at 94 days. The C:N ratio of the substrate also begins at 4, but increases

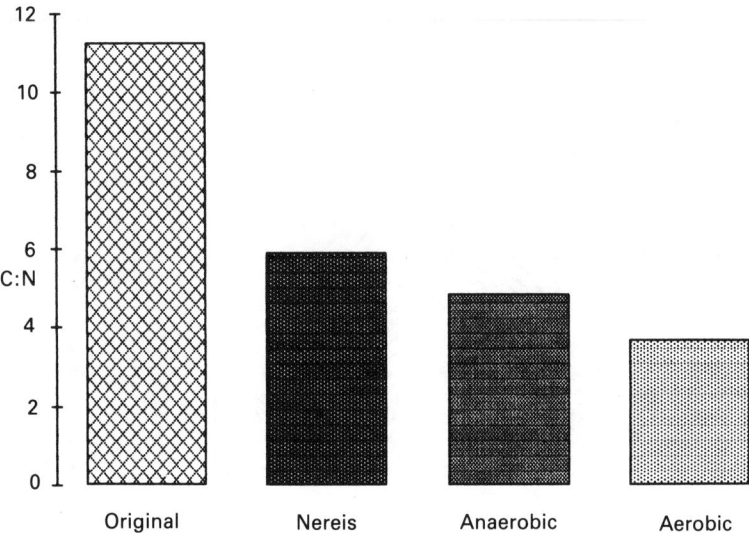

Fig. 10—C:N ratios in the aquarium experiment. The first column represents the original sediment. The columns designated *Nereis*, anaerobic and aerobic represent the C:N ratio of the organic matter that had been mineralized in each of these treatments (data from Kristensen and Blackburn, 1987).

dramatically to 16 at 94 days. Shortly after this time it is predicted that almost no further mineralization of organic-N would occur; material of a very high C:N ratio undergoes degradation. The model predicts that, at 300 days, the C:N ratios of the products would be approximately 11, corresponding to that of the bulk starting material. At this time 10000 mmol $C m^{-2}$ would have been oxidized or 29% of the carbon initially present.

Additional experiments (Blackburn, 1987a) indicated that at 94 days the bacteria in the *Nereis* sediment were incorporating carbon with a very poor efficiency of 0.3, considering that the sediment was well oxidized. It was suggested that the low efficiency was related to the high C:N ratio of the detritus. This low efficiency of incorporation by bacteria growing on high C:N ratio substrate has previously been noted (Linley and Newell, 1984).

Experiments some years ago (Blackburn, 1980) indicated that in natural sediments material of a low N:C ratio was utilized by bacteria below 4 cm depth (Fig. 12). It was postulated at that time that preferential N-mineralization of fresh substrate had occurred at the sediment surface. As the partially degraded, low N:C fresh detritus was mixed downward by bioturbation or physical mixing, it was more rapidly attacked and degraded than the old bulk detrital material of a higher N:C content. The result of this mineralization was a net uptake of ammonium in these strata, something that would not have been predicted from the high N:C ratios of the bulk detritus.

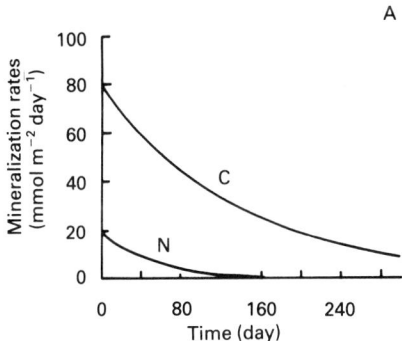

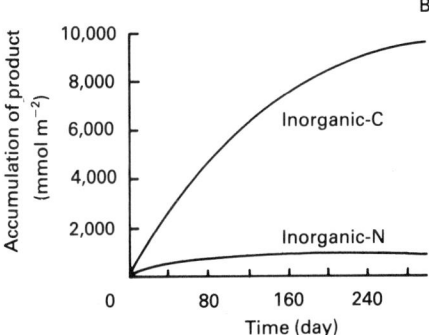

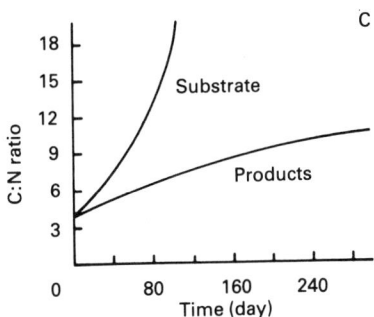

Fig. 11—Model of nitrogen and carbon mineralizations in *Nereis* aquarium sediment. See text for explanation.

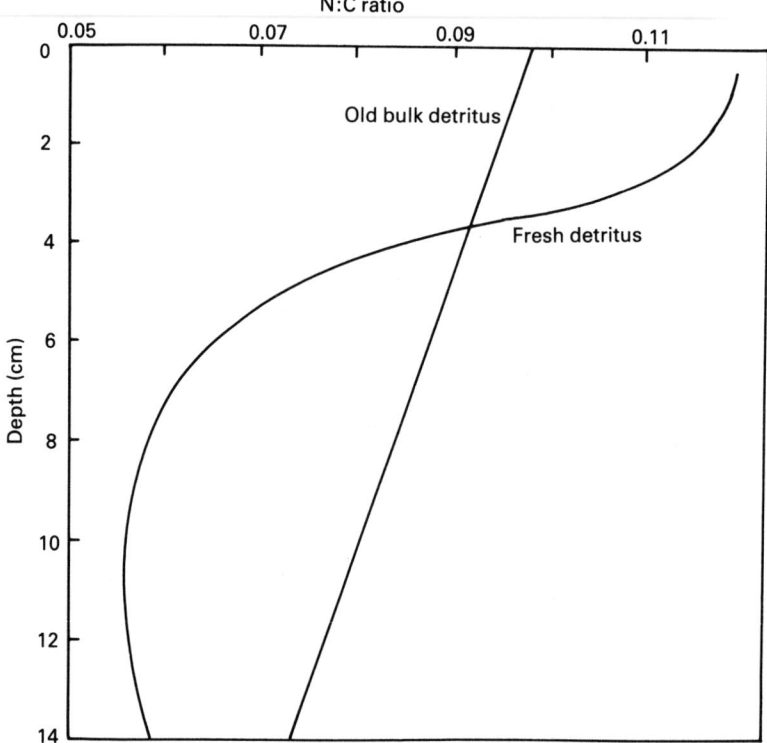

Fig. 12—N:C ratios in total detritus and in detritus undergoing mineralization (redrawn after Blackburn, 1980).

CONCLUSIONS

Quite a lot of information may be gained about the rates of mineralization of sediment detritus, and thus about the potential of a sediment system to produce biomass. The amount of biomass depends on the efficiency with which carbon is incorporated by the metabolizing organism. The methods that are available for measuring rates of mineralization are not specific for bacteria and the rates of C-oxidation are for the whole community, unless the macrofauna have been removed. Specific processes, e.g. SO_4^{2-} reduction, sulphide oxidation, nitrogen fixation and most denitrification, can be attributed to prokaryotes exclusively, but the oxidation of reduced carbon compounds is performed by a wide range of eukaryotes in the sediment. Thus, even though growth rates of most heterotrophic bacteria cannot be determined in sediment, much information may be obtained about bacterial processes, their relationship to detrital quality, reoxidation of metabolic products, and release of nutrients to the overlying water. These are important considerations and have resulted in a greater understanding of food webs in sediments, even if we do not know who eats whom, and in what quantity.

The quality of the detrital material reaching the sediment, and in the sediment, obviously has an influence on the rate and extent of bacterial biomass formation. It is likely that the organic material entering the sediment has a definite proportion that can be degraded, irrespective of who eats it. It is the rate of degradation, not amount of degradation, that is variable. The presence of macrofauna increases the rate of degradation by some mechanism other than by the direct introduction of O_2. In some situations, an increase in the rate of mineralization is due to macrofaunal digestion of the input material, while in others it may be due to the mechanical exposure of fresh surfaces to bacterial colonization. There is the implication that macrofaunal digestion gives a more complete digestion, but it is more probable that the bacterial cells, produced as a result of microbial digestion, are themselves responsible for the accumulation of slowly degrading material.

REFERENCES

Berner, R.A. 1980. *Early diagenesis: a theoretical approach*. Princeton University Press, Princeton.

Blackburn, T.H. 1979. Method for measuring rates of NH_4^+ turnover in anoxic marine sediments, using a $^{15}N-NH_4^+$ dilution technique. *Applied and Environmental Microbiology*, **37**: 760–765.

Blackburn, T.H. 1980. Seasonal variations in the rate of organic-N mineralization in anoxic marine sediments. In *Biogeochimie de la matière organique a l'interface eau–sediment marin*, pp. 173–183. Edition du CNRS, Paris.

Blackburn, T.H. 1987a. Microbial processes of N- and C-cycles in marine sediments. In *Proc. 6th international congress of microbial ecology, Ljubljana, August 1986* (ed. F. Megusar). In press.

Blackburn, T.H. 1987b. Benthic mineralization and bacterial production. In *Nitrogen cycling in coastal, marine environments*, (eds. T.H. Blackburn and J. Sørensen). Wiley, New York, in press.

Dando, P.R., Southward, A.J., Southward, E.C. and Barrett, R.L. 1986. Possible energy sources for chemoautotrophic prokaryotes symbiotic with invertebrates from a Norweigan fjord. *Ophelia*, **26**: 136–150.

Dobbs, F.C. and Scholly, T.A. 1986. Sediment processing and selective feeding by *Pectinaria koreni (Polychaeta: Pectinariidae)*. *Marine Ecology Progress Series*, **29**: 165–176.

Fenchel, T. and Blackburn, T.H. 1979. *Bacteria and mineral cycling*. Academic Press, New York.

Jørgensen, B.B. 1979. The sulfur cycle of a coastal marine sediment (Limfjorden, Denmark). *Limnology and Oceanography*, **22**: 814–832.

Jørgensen, B.B. and Sørensen, J. 1985. Seasonal cycles of O_2, NO_3^- and SO_4^{2-} reduction in estuarine sediments: the significance of an NO_3^- reduction maximum in spring. *Marine Ecology Progress Series*, **24**: 65–74.

Karl, D.M. and Winn, C.D. 1984. Adenine metabolism and nucleic acid synthesis: applications to microbiological oceanography. In *Heterotrophic activity in the sea* (eds. J.E. Hobbie and P.J.LeB. Williams), pp. 197–215. Plenum Press, New York.

Karl, D.M., Burns, D.J., Orrett, K. and Jannasch, H.W. 1984. Thermophilic microbial activity in samples from deep-sea hydrothermal vents. *Marine Biology Letters*, **5**: 227–231.

Kristensen, E. and Blackburn, T.H. 1987. The fate of organic carbon and nitrogen in experimental marine sediment systems: influence of bioturbation and anoxia. *Journal of Marine Research*, in press.

Linley, E.A.S. and Newell, R.C. 1984. Estimates of bacterial growth yields based on plant detritus. *Bulletin of Marine Science*, **35**: 409–425.

Moriarty, D.J.W. and Pollard, P.C. 1981. DNA synthesis as a measure of bacterial productivity in seagrass sediments. *Marine Ecology Progress Series*, **5**: 151–156.

Moriarty, D.J.W. and Pollard, P.C. 1982. Diel variation of bacterial productivity in seagrass (*Zostera capricorni*) beds measured by rate of thymidine incorporation into DNA. *Marine Biology*, **72**: 165–173.

Nelson, C.H. and Johnson, K.R. 1987. Whales and walruses as tillers of the sea floor. *Scientific American*, (February).

Tenore, K.R. 1987. Nitrogen in benthic food chains. In *Nitrogen cycling in coastal, marine environments* (eds. T.H. Blackburn and J. Sørensen). Wiley, New York, in press.

Williams, P.J.leB. 1984. Bacterial production in the marine food chain: the emperor's new suit of clothes? In *Flows of energy and materials in marine ecosystems* (ed. M.J.R. Fasham), pp. 271–299. NATO Conference Series, Series IV: Marine Sciences, Plenum Press, New York.

3

Ecology on the microscale

Julian W.T. Wimpenny, and **Adrian Peters**, Department of Microbiology, University College, Newport Road, Cardiff CF2 1TA, UK

The marine environment contains a number of spatially heterogeneous ecosystems including sediments and microbial film where microbes proliferate in physico-chemical gradients. Such systems may be investigated *in situ*, in microcosms or by the use of experimental and mathematical models. The latter route is discussed in more detail, in particular the use of gel-stabilized models and multistage continuous culture systems to investigate sediment ecosystems and laboratory film fermenters applied to natural biofilm formation.

INTRODUCTION

70% of the worlds' surface is covered with water and the latter provides a huge and varied range of ecosystems for the growth of living creatures including microorganisms. Seas and oceans are, except on the very largest scale, physically and chemically rather homogeneous: this cannot be said, however, of several ecosystems associated with the marine environment which are spatially quite heterogeneous. In such systems physical and chemical gradients dominate each habitat and lead to differentiation of the microbial species present. There are three main habitats which fall into this category:
(i)   marine sediments
(ii)   microbial film or biofilm or marine fouling
(iii) marine benthic and detrital communities
Of these the first two have been the subject of considerable research and will be discussed in more detail. Rather little is known about benthic and detrital communities at least as far as spatial heterogeneity is concerned.

It is impossible to do justice to a subject like 'Ecology on the microscale' in a brief presentation such as this. I have chosen to isolate one main component of the subject and that is spatial heterogeneity and the value of experimental models applied in particular to sediment and film communities. Attention will be paid to determinants of spatial position in heterogeneous ecosystems.

Investigations into the ecology of microbial communities follow a number of different strategies depending largely on the training, background, research philosophy and general inclinations of the investigator.

Ecologists at one extreme pursue a 'holist' route which demands studying a natural system where it lies. Simplification or fragmentation of the system is unacceptable since quintessential detail will be lost, rendering any conclusion meaningless. At the other end of the scale there are simplifiers. The latter cannot cope with the high degree of variability and complexity shown by natural systems. This group is captivated by the simplicity and power of steady state monocultures. Most microbial ecologists are somewhere between these extremes; however, peace to groups of whatever inclination for as the philosopher Feyerabend observed (1975) when discussing the scientific process:

> 'There is only one principle that can be defended under all circumstances and in all stages of human development. It is the principle that anything goes.'

This statement may sound a shade too cynical; however, it is clear that science envelops a multitude of different approaches to particular problems before arriving at a generally acceptable consensus. Each route has its advantages and disadvantages.

The simplification of natural ecosystems can be achieved in at least two ways — by the use of microcosms on the one hand and of models on the other. A microcosm is a small piece of a natural ecosystem brought into the laboratory and investigated under controlled conditions. It is normally enclosed and no longer in contact with the habitat from which it was taken. It may be modified in a predetermined fashion from the original ecosystem; however, it generally retains most of the properties of the latter including its complexity.

A model, in contrast, is an abstraction, often drastically simpler than the natural system. Where the microcosm is a homologue of the system from which it came models can only be analogues. Modelling divides into two complementary parts: laboratory-based experimental systems on the one hand and formal mathematical constructions on the other. Experimental models aim to simplify by selecting one or a few parameters to investigate, either ignoring or holding constant the remainder. Modelling is an evolutionary process starting simply but moving towards complexity and verisimilitude with the process it aims to elucidate. At its simplest one might model a marine sediment by isolating a single sulphur-oxidizing species and investigating its behaviour as a function of one factor using a chemostat. This would be useful academically but rather remote from the sediment itself. Evolution of the model might suggest using more than one species and more than one substrate in the chemostat. Alternatively, one might develop a gel-stabilized system containing a sulphur oxidizer and a sulphate-reducing bacterium growing in opposing gradients of oxygen and acetate. This model now incorporates solute diffusion gradients and the elements of a sulphur cycle and will constitute a good if restricted model of a marine sediment. This exercise can be made more complex and more real by incorporating further species, substrates and perhaps particulate material. At length the gel model will begin to approach the structural and functional complexity of the system it aims to simulate.

Mathematical modelling seeks to reproduce the behaviour of a natural system as a system of equations describing each of the key factors of the system. If a mathematical model *does not* replicate the real system adequately then assumptions of the model must be wrong. It may be argued that the function of mathematical models, indeed of all models, is to test the rigour of a hypothesis allowing the latter to be modified if it fails to fit comparisons with reality. It will be apparent that this view has its roots in Popperian falsification (Popper, 1968).

Billen (1982) has formulated mathematical models of sedimentary ecosystems based on experimental data from sandy sediments of the North Sea. He has stressed the importance of caution in the use of models:

'Modelling, however, cannot work miracles. The value of the results it provides always reflect the value of the data or hypothesis used. It is therefore important to use them in close combination with direct experimental methods for determining the mechanisms and the rate of microbial activity in sediments.'

My research group in Cardiff have developed a number of laboratory growth systems which fall midway between the simplicity of steady state homogeneous fermenters and the extreme complexity of the natural ecosystem. These models all incorporate spatial heterogeneity at one level or another and can be classified as follows:

(1) open steady state gradient systems
(2) closed gel-stabilized gradient systems
(3) microbial film fermentation systems
(4) multidimensional gradient systems
(5) the microbial colony.

These systems have been reviewed before (Wimpenny, 1981; Wimpenny, 1982; Wimpenny *et al.*, 1983; Wimpenny, 1984; Wimpenny and Waters, 1984; Wimpenny and Waters, 1987).

Although these models have not been applied extensively to the marine environment the suitability of some of them will be discussed in the following sections.

THE SEDIMENT ECOSYSTEM

There is probably no such thing as a 'standard marine sediment'. Sediments lie at the base of all marine environments. They can be divided into 'littoral', 'continental shelf' and 'abyssal' depending on their relation with land masses. Each sediment is derived from a number of sources which include land masses, biologically produced plant and animal remains, cosmic and volcanic materials and material generated by salts coming out of solution in the water itself.

Sediments consist of particulate material and the interstitial water that surrounds it. The transfer of solutes within a sediment ecosystem is complex but dominated by molecular diffusion modulated by tortuosity factors. Other processes include the mixing of the entire sediment structure by the action of water currents or wave action, the mixing of interstitial water even where the sediment particles are stationary and finally bioturbation due to the burrowing activities of animals, in particular invertebrates. The presence of clay particles provides surfaces on which solute molecules can adsorb. Such adsorption may be reversible or irreversible and provides a buffered store of material which may be mobilized and used by the sediment microflora.

There are a number of orders of heterogeneity in the sediment ecosystem. First is vertical stratification from the normally oxidized upper layers through successively more reducing regions to the most anaerobic zones. A second order of heterogeneity occurs in the region of bioturbation where the activities of burrowing animals lead oxygen penetration into deeper anaerobic zones. A third order of heterogeneity is due to the development of anaerobic microsites in aerobic regions (Jørgensen, 1977). Further orders of heterogeneity are associated with particulate matter, for example the microscopic pH gradients due to diffuse double layers adjacent to clay mineral lattice structures.

Gradients in redox potential from oxidized near the surface to more reduced at lower levels are common. Wood (1964) has indicated that Eh values can range from $+650\,mV$ at the surface of some sediments down to $-350\,mV$ in the lower regions. Electron acceptors tend to be present in an electrochemical series in sedimentary ecosystems (Fenchel, 1969; Jørgensen, 1981). Oxygen microelectrodes show that oxygen penetrates only a few millimetres into the upper layers of some such systems (Revsbech *et al.*, 1980). Oxygenic photosynthetic species in the upper few millimetres of sediments can respond to illumination by generating peaks of oxygen supersaturation at points below the surface.

Gel-stabilized model systems
What are the advantages and disadvantages of gel-stabilized gradient models for studying natural ecosystems like sediments? A gel model is a closed systems for cells and for most solutes except for gaseous substrates that can exchange across a gas–liquid interface. It is therefore more akin to a batch culture than to a continuous flow system. In addition, it is 'fine grain' with a very high resolving capacity as judged by the formation of periodic growth structures in gels where more than 20 discrete bands have been seen under some circumstances (see later). The gel model is more specifically a good model of a sediment ecosystem because molecular diffusion is in both cases the main mechanism for solute transfer. The model in its simplest form lacks particulate matter and there is no provision for non-diffusive solute transport, for example by mechanical motion (wave and tidal action) or by bioturbation. A number of applications of the gel models are discussed below.

Growth of Beggiatoa *in a gel-stabilized model system*
The *Beggiatoa* group of marine and freshwater sediment organisms are seen by Nelson and his colleagues (Nelson *et al.*, 1986a, b) as 'gradient' organisms that move by gliding motility to a point where opposing gradients of sulphide and oxygen meet. At this point they proliferate as thin opaque discs or veils. They have been investigated using gel-stabilized model systems as follows. Gel diffusion tubes contained a source layer made up with approximately 8.5 mM Na_2S in 10 ml which was about 17 mm deep. This layer was solidified with 0.8% w/v agar. The growth

zone was made up with 30 ml of basal medium plus 0.25% w/v agar. This was inoculated on the surface with a *Beggiatoa* culture in semi-solid agar. The authors employed microelectrodes for determining oxygen, sulphide and pH values. Results of one such experiment in the presence and absence of *Beggiatoa* show that the bacterial plate forms a few millimetres below the agar surface and the authors suggested that the organism is fundamentally microaerophilic. Gradients of sulphide and oxygen were determined using microelectrodes and concentrations of both solutes fell to zero within the bacterial plate itself. Shallower gradients are seen in the uninoculated controls as a result primarily of the slow chemical reaction between the two solutes.

The authors determined a number of kinetic parameters for the system. Doubling time at the sulphide:oxygen interface was 11 h; yield on sulphide was 8.4 g dry weight per mole and the diffusion coefficients for sulphide and oxygen in the gel were respectively 1.39×10^{-5} cm^{2}s^{-1} and 2.03×10^{-5} cm^{2}s^{-1}. Enough data were available from these results to model this system using a gel simulation computer program written for my group by S. Jaffe. The latter shows that the *Beggiatoa* plate forms quite deep in the agar at the intersection of the oxygen and sulphide gradients. Reducing the source sulphide concentration by a factor of ten in the simulation moves the plate even deeper into the gel, perhaps emphasizing the point. There is therefore no evidence based on *position* alone to suggest that the organisms are unduly oxygen sensitive. Position in the simulations and probably the model system as well is almost certainly due to a balance between relative solute concentrations, diffusion coefficients, dimensions of the system and yield coefficients.

The *Beggiatoa* system gives clear and beautiful results. In a gradient system, if two solutes, both needed for growth, are approaching one another from diametrically opposite directions, growth is possible only where these counter gradients meet. One can conclude from such experiments that spatially heterogeneous ecosystems contain two major types of space: *zones of reaction* where all the essential nutrients needed for growth are present and *transfer zones* where one or more essential nutrient is absent but through which other solutes are free to diffuse. Other regions can be discerned in many heterogeneous systems, for example the *impermeable zone* through which solutes cannot pass: this corresponds to the solid fraction of soils or sediments. Finally there are *pool zones* where solutes are retained by adsorption to components of the system. All these divisions are of course absent from and irrelevant to homogeneous growth systems used in the laboratory.

An estuarine sediment system model

R.A. Herbert and his colleagues (MacFarlane *et al.*, 1984) have employed a gel-stabilized model to simulate a Tay estuary sediment. The gel model was set up in a sterile 1 l Quickfit reaction vessel. 100 ml of sterile basal salts medium which contained in addition 10 mM $(NH_4)_2SO_4$, 0.05 g l^{-1} cycloheximide and 1% w/v agar was poured into the container. Once this had set, 100 ml of well-mixed material from the upper 5 cm of a Kingoodie Bay sediment was added to the system. Finally 600 ml of basal salts medium containing 0.3% agar was poured carefully over the sediment and the system allowed to set. The gel was covered with foil and incubated at 20 °C. The sole carbon source for these systems came from the sediment itself. Sampling took place at 0, 4, 8, 12 and 13 weeks after inoculation. pH, oxygen, *Eh*, NO_2^-, NO_3^- and NH_4^+ profiles were measured and population densities of nitrifying, nitrate-respiring and sulphate-reducing bacteria were estimated.

Results of this experiment were in good agreement with actual measurements in Kingoodie Bay sediments themselves. The system was judged to be a useful laboratory model which gave reproducible results and seemed suitable for further development. The authors point out that the only major difference between the natural habitat and the laboratory model was in the lack of bioturbation and in the absence of gas or solute exchange in the closed model when compared to the real sediment. It should be observed that although these comments are true, the system *is* quite different from the sediment itself in that the gel lacks substantial amounts of the particulate components that characterize sediments. It is even more interesting therefore, that the model and the natural sediment behave so similarly.

Investigations of pollutant action on sediments using gel models by Morgan and Watkinson (1987) have used a gel-stabilized model to investigate the effects of 3-nitrophenol and 1:1 hexadecane–naphthalene on the biochemistry and microbiology of freshwater and marine sediment communities. Numerous parameters of these systems were measured, including oxygen, pH and ion gradients and populations of sulphate-reducing, nitrogen-recycling, xenobiotic-metabolizing and heterotrophic bacteria.

Use of gel-stabilized model systems in Cardiff

Gel-stabilized model systems have been used by my group for a number of applications (Wimpenny *et al.*, 1981a), in particular to reproduce some of the properties of the ecosystem found in the water base in an oil storage tank and to investigate the phenomenon of periodic growth. Gel-diffusion experiments can be constructed in any suitable container, from large beakers and measuring cylinders to test-tubes. We have generally employed a two-layer system divided into a source zone containing the diffusible solute of interest and a growth region lacking the source component but containing a cell inoculum and all the other basal nutrients. The lower layer also contained 1% w/v agar which is sufficient to form a solid resilient gel. Source and growth layers are normally but not necessarily equal in volume. The upper layer has the same basal nutrient profile as the source zone but is semi-solid, containing 0.4% w/v agar. The upper layer is cooled to about 45 °C and mixed with the cell inoculum before it is poured into the container. The container is covered but generally not sealed, especially if the upper layers are to be aerobic. In long term experiments some evaporative losses occur which can be obviated, if necessary, by sealing the container and passing a water-saturated gas over the gel surface.

Sampling the gels is straightforward. The simplest approach is to remove gel cores with a sterile cork borer. In practice the agar, although 'sloppy', will hold together as an intact core if it is handled carefully. A gel slicer can be used to produce gel sections down to about 1 mm thick.

Growth can be estimated in the slices either as total counts by extinction measurements or as viable counts. In either case each slice should be homogenized to release as many cells from the gel matrix as possible. Experiments have shown that

10–12 strokes of a Teflon and glass homogenizer are generally sufficient. We have measured total growth by following absorbance values in phosphate buffer at pH 7.0. Viable counts can then be made following normal procedures. In addition the gel can easily be scanned using a commercially available gel scanner. If scanners are to be used there are advantages to growing the cultures in selected optically clear tubes. To follow growth each tube should be marked so that the beam passes through the same part of the tube each time. In many modern computer-controlled gel scanners base-line 'noise' from the tube itself can be subtracted from the final sample.

Certain physico-chemical parameters may be assessed if suitable electrodes are available. It is comparatively easy to buy or construct 'needle' electrodes for pH, pO_2, redox potential and sulphide activity. These are deployed using a micromanipulator.

If other components need to be measured this can be done in gel slices. Each slice is collected individually. The gel can be diluted and melted. The presence of low concentrations of agar does not generally interfere with most routine assays.

Growth in opposing gradients of essential solutes
Paracoccus denitrificans and *Pseudomonas aeruginosa* were grown in gel-stabilised gradients at the expense of oxygen entering the system from the surface and of a range of substrates diffusing upwards from a source layer at the base of the tube (H.J. Ewers and J.W.T. Wimpenny, unpublished observations). Substrate concentration was varied in the source layer. Some of the results are shown in Fig. 1. The position of the growth band depends on the inner substrate concentration. That is the greater the concentration of substrate the higher up the gel the band of growth appears. In each experiment the position of the band changes as might be predicted assuming a moving boundary solution to the reaction–diffusion equations governing growth.

In most of the experiments single growth bands were observed especially in the early phases of growth. In some cases, especially with *P. denitrificans*, multiple banding was seen. In almost all experiments prolonged incubation led to multiple banding. The reason for multiple banding is not known at present. In old cultures especially, one should not ignore the possibility that cell lysis and regrowth occur. In addition other physico-chemical changes were probably taking place in the medium, in particular the accumulation of inhibitors. Banding phenomena have been reported by us before and will be discussed in the next section.

Periodic growth phenomena in nature and in gel-stabilized model systems
There are many examples of periodic growth behaviour in nature. These can include fungal growth rings ranging from fairy rings in grassy fields to very clear rings of spore formation in Petri dish cultures in the laboratory. *Proteus vulgaris* often swarms across the surface of the agar in a series of sometimes beautiful rings. Perfil'ev and Gabe (1969) developed capillary methods for observing microbial growth *in situ*. They used the capillary 'peloscope' to examine growth in a sediment ecosystem and showed that *Gallionella* grew as a series of closely packed discrete bands.

The first clear demonstration of periodic growth in gel-stabilized culture media was by Williams (1938a, b, 1939a, b). Williams likened the growth bands to the spectrum of a coloured compound which appears as a series of dark absorption bands

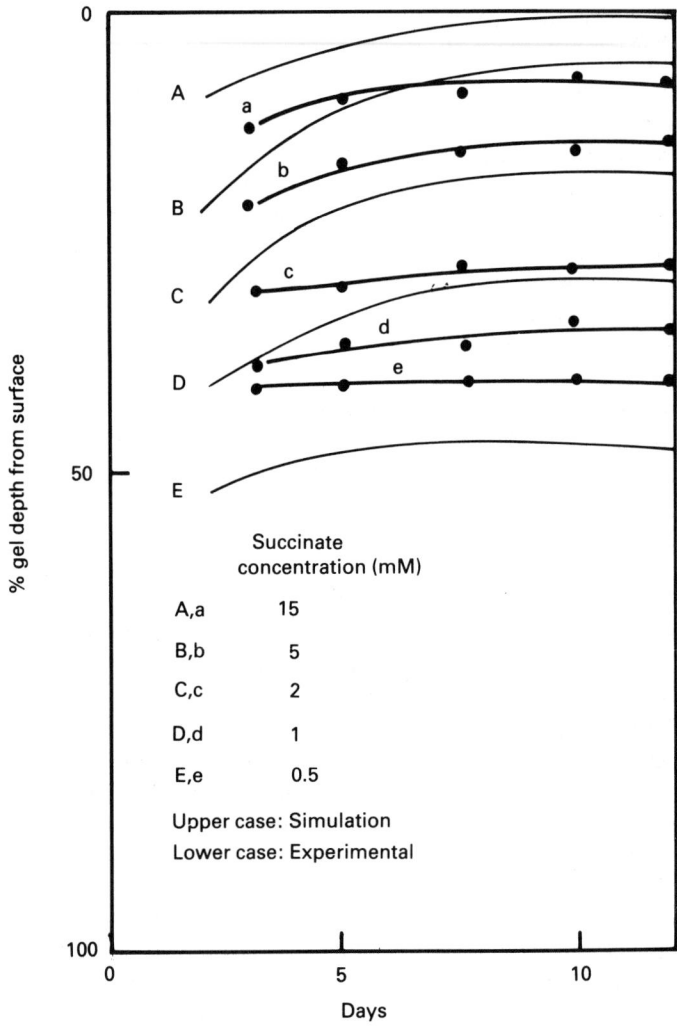

Fig. 1—Growth of *Pseudomonas aeruginosa* in oxygen:succinate gradients. Experimental and computer simulation results. Total gel length: 8 cm; source zone 4 cm; succinate concentration from 0.5 to 15 mM. The following values were used in the simulation: $D_{\text{succinate}} = 0.029\,\text{cm}^2\,\text{h}^{-1}$; $D_{\text{oxygen}} = 0.072\,\text{cm}^2\,\text{h}^{-1}$; K_s succinate $= 0.02\,\text{mM}$; $K_{sO_2} = 0.02\,\text{mM}$; $1/Y_{\text{succ}} = 0.0096\,\text{mM}\,\text{mg}^{-1}$; $1/Y_{O_2} = 0.03\,\text{mM}\,\text{mg}^{-1}$; $u_{\max} = 0.43$ h^{-1}; initial cell concentration $= 2\,\text{mg}\,\text{l}^{-1}$; oxygen concentration at interface $= 0.25\,\text{mM}$. Peak position was determined at different times up to 12 days.

when seen through a spectroscope and he therefore gave them the name 'bacterial spectrums'.

Periodic growth bands in agar-stabilized systems have also been noted for streptomycete cultures by Nitsch and Kutzner (1973). Multiple bands were seen in 'control' agar shake cultures with no added carbon source. The authors concluded that very low nutrient levels were needed for band formation.

A range of bacteria was examined for growth in such gel systems. It was concluded that facultatively anaerobic and certain obligately aerobic bacteria generated a pattern consisting of more than one growth band, but that most strict aerobes did not.

In practice, cultures of *Bacillus cereus* produced bands most prolifically and later work focused on this organism (Coombs and Wimpenny, 1982). When grown aerobically in a gel-gradient system with glucose as diffusing solute this organism generates a family of growth bands. The first of these forms in about the middle of the growth zone while later bands are produced at successively higher positions in the gel.

A number of experiments were performed to investigate the phenomenon in greater detail. The following observations were made. Increasing the concentration of glucose in the source zone pushed the bands higher up the gel. Similarly, reducing the concentration of the basal CYS medium in both layers did the same thing. Changing the strain of *B. cereus* and the type and concentration of agar had only minor effects on the position of the bands. Band formation was only apparent when oxygen was present and when there was an actual gradient of glucose in the gel. Thus preincubation or the incorporation of glucose throughout both layers abolished the phenomenon. It was deduced that motility was unimportant since band formation was still present at agar concentrations in the upper layer at which motility was almost certainly impossible. Early on in the incubation period oxygen was rapidly depleted. At the same time growth in the lower parts of the system led to major changes in the pH of the system which quickly fell to around 5.1. Unexpectedly, growth at the expense of oxygen near the surface of the gel led to pH changes in the opposite direction. Here the pH became alkaline at around pH 7.8–8.0. It was assumed that this was due to the aerobic oxidation of amino acids leading to the liberation of free ammonia.

pH was implicated as one trigger mechanism for band formation. Thus phosphate buffer was added to the gels at a range of different concentrations. It was clear from the results that the higher the buffering capacity, the shallower the pH gradient, the more growth took place and the deeper into the gel were the bands. A second experiment used three gel systems. The first was the standard system incubated aerobically, the second was the same system incubated in the absence of oxygen, where no bands were seen, whilst the third had an additional layer of gel incorporating a source of alkali poured on top. This vessel was incubated anaerobically. Band formation was now apparent in this last vessel, as was the pH gradient pattern always seen when bands were being formed.

Computer simulations of growth in diffusion systems

'Simple' growth in solute gradients
Microbial growth in opposing essential solute gradients has been investigated using a simple mathematical model assuming that growth rates are determined by the Monod growth model, one Monod relationship for each essential substrate. At the same time it is assumed that solutes are translocated by Fickian diffusion. The equations for the model are given in the Appendix, p. 79. In addition the values of each parameter used in the model are described in Table 1.

Parameters to be varied concern first the cell (maximum specific growth rate, affinity for substrate and cell yield) and second external variables including diffusivity and solute concentration. It has been assumed that a substrate such as glucose is

Boundary conditions:
   Left-hand boundary impermeable to solutes and cells
   Right-hand boundary impermeable to S_1 and cells
   Right-hand boundary concentration of S_2 fixed at 0.25 mM

System dimensions:
   Source zone   30 mm
   Growth zone 30 mm

Initial concentration:
   Source zone: $S_1 = 1$ mM; $S_2 = 0$ mM; Cells $= 0$ mg l^{-1}
   Growth zone: $S_1 = 0$ mM; $S_2 = 0$ mM; Cells $= 10$ mg l^{-1}

Diffusion coefficients:
   Cells $= 0$ cm^2 h^{-1}
   S_1 $= 0.024$ cm^2 h^{-1}
   S_2 $= 0.072$ cm^2 h^{-1}

Cell parameters:
   u_{max} $= 0.2$ h^{-1}
   $1/Y_{(S_1)}$ $= 0.02$ h^{-1}
   $1/Y_{(S_2)}$ $= 0.02$ h^{-1}
   $K_{s(S_1)}$ $= 0.02$ mM
   $K_{s(S_2)}$ $= 0.02$ mM

Table 1   Assumptions and parameters of the model.

diffusing upwards from a source region whilst a second substrate, assumed to be oxygen, is diffusing from a constant source at the surface of the gel system. The program used in these simulations was written for the GEC 4090 minicomputer by S. Jaffe and made use of NAG mathematical library routines to solve the differential equations involved.

A standard set of results (Fig. 2) illustrates growth at the intersection of the two gradients as a function of time. Each of the key parameters has been varied systematically and the results showing position of the growth peak plotted against values for the parameter in Fig. 3. It is at once obvious that what may seem to be critical cell parameters like growth rate and substrate affinity are actually unimportant space determinants. Absolutely critical to growth position are diffusion coefficient and substrate concentrations. On the surface it is surprising that yield coefficient is a key space determinant. It should be understood that there is a close relationship between *substrate concentration* and *cell yield*. If yield for one of the two limiting substrates is very low balanced growth can only occur nearer its source just as if under *normal* yield conditions the substrate was present in low concentrations. These simulations emphasize the importance of environmental factors in determining growth position and suggest that most microbes in spatially structured environments are growing under diffusion-limited conditions.

Another major space determinant is the presence of inhibitory agents. Fig. 4 shows the results of a simulation where growth is at the expense of a single substrate coming from one direction. At the same time an inhibitor is diffusing from the same direction. The actual results will depend on the diffusivity, concentration and toxicity of the inhibitor. Assuming that concentration only is varied it is clear that growth position is dependent on this factor.

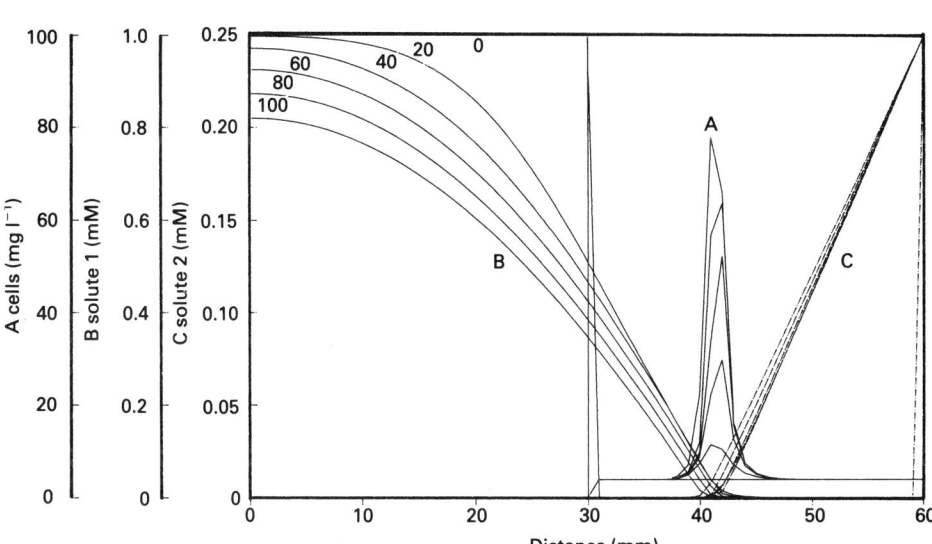

Fig. 2—Simulation of microbial growth in opposing solute gradients. Data are plotted every 20 h for 100 h. Conditions are the 'standard' set described in Table 1.

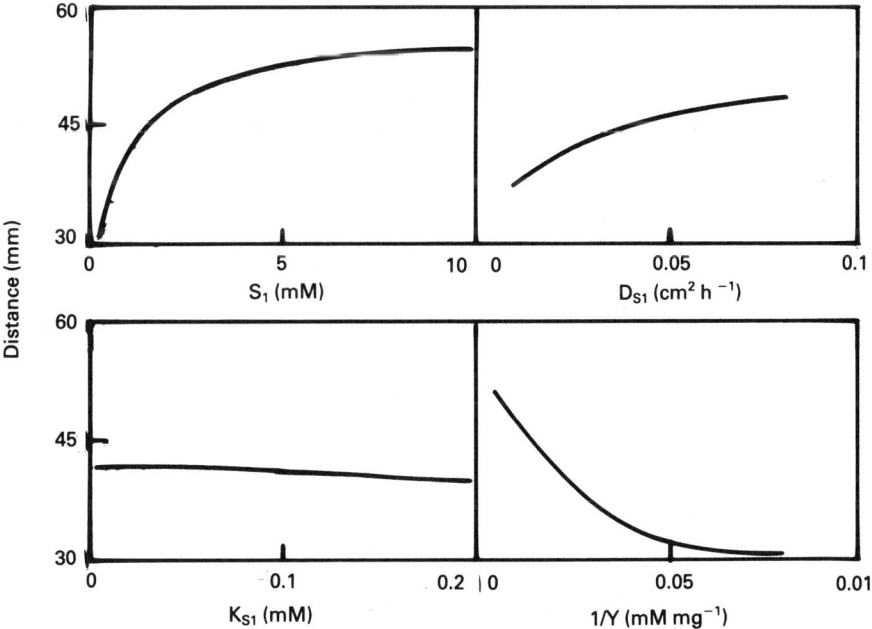

Fig. 3—Simulation of microbial growth in opposing solute gradients. The following parameters were varied: (a) the concentration of internal substrate S_1; (b) diffusion coefficient for S_1; (c) affinity constant of the cell for S_1; (d) reciprocal growth yield for the cell on S_1. The simulation was for 100 h growth and the line indicates the position of the growth peak. Other parameters as in Table 1.

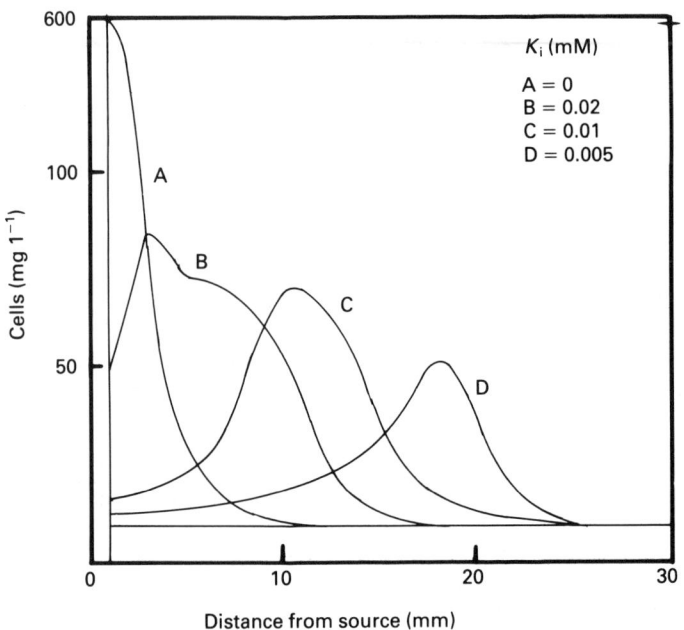

Fig. 4—Inhibition as a determinant of position. In this simulation cell growth is dependent on a single solute diffusing from the left-hand boundary. This solute is also a non-competitive growth inhibitor. In each simulation the affinity constant (K_s) is 0.02 mM. The inhibition constant (K_i) is varied. Other parameters as in Table 1.

So far these simulations consider only a single organism. A natural question concerns competition between bacteria. Do competition kinetics which apply in homogeneous environments also apply in spatially heterogeneous systems? Fig. 5 confirms that they do. Organisms with lower Ks values for both substrates are outcompeted at the growth position as would be predicted. If affinity constants are altered so that a second organism has a higher affinity for one substrate and a lower affinity for the second when compared with the first cell both cells will grow and there will be a *small* change in growth position for the two competing organisms. Once more actual position is highly dependent on the environmental factors but within a narrow zone there is spatial differentiation.

Periodic growth in solute gradients

The mechanism for the observed band formation in cultures of *B. cereus* growing on an amino-acid-containing medium in opposing oxygen and glucose gradients was not immediately obvious. S. Jaffe developed mathematical models of growth in gel-stabilized gradient systems. Each model incorporated growth of a single organism in a diffusion system containing opposing gradients of two nutrients. Fickian diffusion was assumed as was simple Monod growth kinetics. No combination of the latter with inhibition, lag, death or any other obvious growth function ever gave periodic patterns similar to those seen in the experiments. Periodic behaviour could be initiated in the simulation under the following

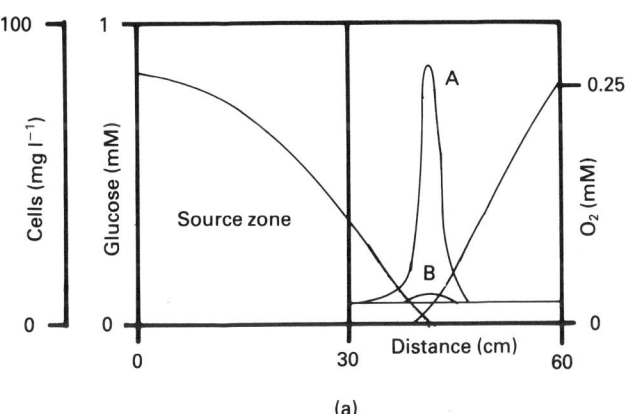

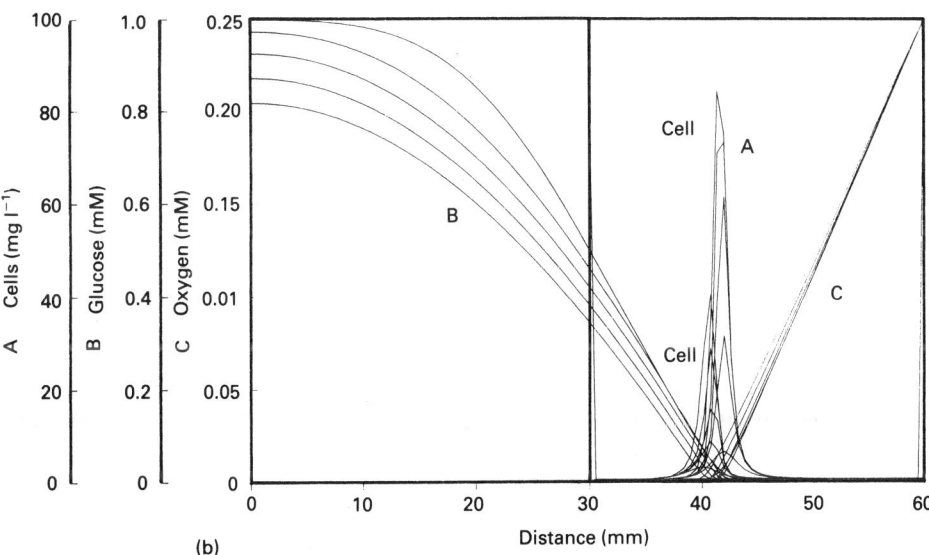

Fig. 5—Simulation of competition between two bacteria in opposing solute gradients. (a) Affinity constants for both substrates are 0.01 mM for A and 0.02 mM for B. Results after 100 h. (b) Set of results where both K_s values for X are 0.02 mM: K_{S_1} and K_{S_2} for Y are 0.2 mM and 0.002 mM respectively. Results are plotted every 20 h for 100 h. Other parameters as in Table 1.

conditions. Cells were assumed to exist in two states, active and inactive. Activation of inactive cells was assumed to be asymmetric with inactivation. Thus growth took place until conditions altered far enough to lead to its cessation. However, growth could only *start* again at a different level of the critical parameter. The latter could be *concentration* of a critical solute or it could be manifest as *time*. In this case there would be an asymmetry between the time it takes to stop growing (presumably short) and the time it takes to start growing once more (presumably a traditional and longish lag phase). Such a theory (Wimpenny *et al.*, 1984) is broadly in line with that suggested by Hoppensteadt and Jager (1979) and Hoppensteadt *et al.* (1984) who were interpreting circular periodicities seen in plate cultures of *Escherichia coli* histidine auxotrophs, grown on a glucose:buffered salts medium, when a small amount of concentrated histidine was placed at the centre of the plate.

The existence of such periodic structures as those discussed in this article adds another level of complexity to the manifestation of microbial growth in spatially heterogeneous systems. The fact that such structures have been seen in marine sediments means that we cannot afford to disregard periodic phenomena in discussing the ecology of microorganisms.

Other sediment models

Multistage continuous culture systems
Biochemical processes in marine sediments can be regarded as a sequence of reactions falling in order of the redox potential of the major electron acceptors present. Since nutrients enter the sediment ecosystem from its surface it is possible to regard a multistage continuous culture system as a resonable sediment model. Thompson et al. (1983) have used a five-stage continuous flow system to investigate carbon flow in anaerobic microbial communities. These workers indicated that spatial separation of functional groups of bacteria associated with sulphate reduction, methanogenesis and acidogenesis took place in such a system and that this separation was characteristic of that seen in a natural sediment.

Steady state gradient systems
While unidirectional continuous culture systems seem to be good models of sediment processes they naturally do not include any reverse flow of materials such as would occur in the natural ecosystem. Thus in the sediment reduced products of metabolism will diffuse from deeper strata towards the surface. The most useful model therefore ought to incorporate bidirectionality.

Lovitt and Wimpenny (1979, 1981a, b) have described such a device which they called the 'gradostat'. Gradostats are usually constructed with five vessels which are linked together so that their contents are transferred from right to left and from left to right simultaneously. Media enter the system from reservoirs located at both ends of the array. At the same time culture leaves the system from the two end vessels. Assuming that solutes A and B enter the system from reservoirs 1 and 2 respectively, that flow rates in each direction and that all vessel volumes are the same and finally that the two solutes are unchanged in the process, opposing linear stepped gradients for each solute will form under steady state conditions. Theoretical and practical characteristics of the gradostat have already been described (Wimpenny and Lovitt, 1984).

The gradostat as described above is not a perfect model for investigating sediment systems precisely because it has inputs at each end of the vessel array. However, if the input and output lines from one end are removed but bidirectional exchange is *retained* between all the vessels the system now becomes an analogue of diffusive systems with single-ended inputs and outputs (Fig. 6). In fact such systems include not only sediments but also microbial films attached to inanimate substrata from which they derive no biologically active solutes. R. Earnshaw and I have investigated some of the properties of a sediment community growing in a single ended gradostat. In this model the first vessel into which nutrients entered was also aerated to simulate

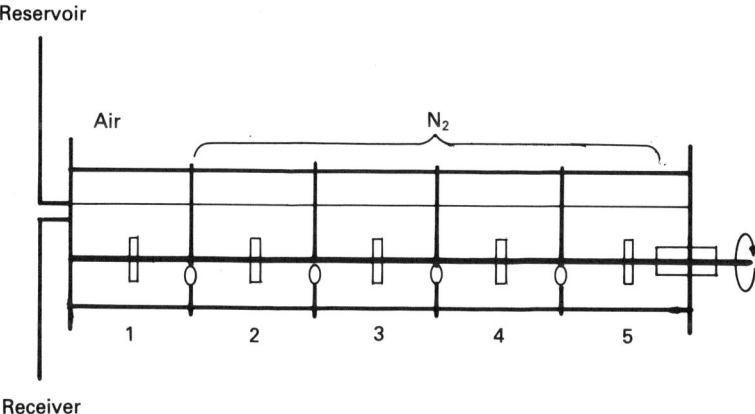

Fig. 6—Diagram of a single-ended diffusion-coupled gradostat.

the narrow region a few millimetres thick which is aerobic in many natural sediment communities. Gradients of redox potential and of sulphate and sulphide concentration were all measured (Fig. 7). The lowest redox potential was found in vessel 2 suggesting more metabolic activity in this position. The highest sulphide concentration appeared in the third vessel whilst the population of sulphate reducing bacteria was highest in vessel 4. Most aerobic and anaerobic heterotrophs were present in vessels 1 and 2. Although these were only preliminary experiments they point in general in the right direction and suggest that the single-ended gradostat is a suitable model for natural single-ended spatially-heterogeneous ecosystems.

A version of the gradostat having more than five vessels is now under development and this ought to make the spatial resolution of the system greater. Naturally, the more vessels are linked bidirectionally, the closer the gradostat becomes to a pure diffusive system like the gel-stabilized model. The difference is that the gradostat retains its characteristic openness, allowing the system to operate under steady state conditions.

MICROBIAL FILM

Most initially clean surfaces immersed in water for any period will be colonized by a community of microorganisms. Colonization may be unimportant economically where it is restricted to natural marine surfaces like rocks, animals or plants; however, it becomes a costly nuisance where microbial film forms on ship hulls and marine installations like oil drilling platforms. 'Fouling' by microbes can lead to the colonization of the surface by higher animals (ZoBell, 1943), the latter causing increased frictional resistance and the slowing of vessels through the water. Fouling may lead to phenomena that range from impairing vision through submarine periscopes on the one hand to establishing corrosion cells in steel plates on the other (Characklis, 1981; Hamilton, 1985, 1987).

An initially clean surface immersed in an aqueous solution is rapidly coated with a

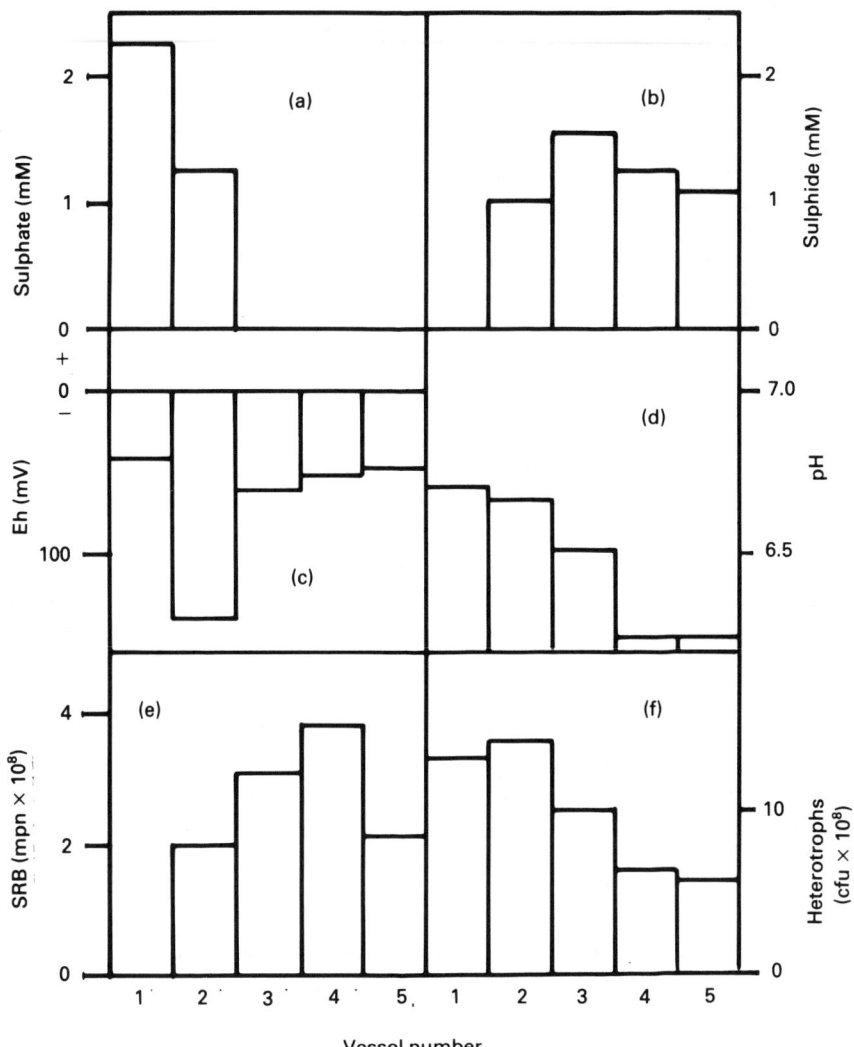

Fig. 7—The gradostat illustrated in Fig. 6 was inoculated with a freshwater sediment community and fed with a glucose-mineral salts medium into the left-hand vessel. This vessel was aerated whilst the remaining vessels were gassed with nitrogen. The system was incubated at 20 °C and allowed to approach the steady state. Parameters measured included (a) sulphate; (b) sulphide; (c) redox potential; (d) pH; (e) sulphate-reducing bacteria (SRB); (f) total heterotrophic bacteria.

conditioning film consisting of organic molecules. Pioneer organisms attach to the surface, first loosely then more firmly. These cells proliferate, forming microcolonies which gradually spread out over the surface. At the same time one or more colonizing species produces extracellular polymer which helps to stabilize the developing structure. The microcolonies eventually coalesce and the structure thickens into a *bona fide* film. Other organisms may become incorporated into the structure as secondary colonizers. Biochemical activities deplete the film of certain solutes, in particular oxygen. This means that parts of the film may become anaerobic and hence suitable loci for the growth of anaerobic microbes. Eventually, perhaps as a result of a combination of film thickness and the activity of anaerobic species, the film may detach and slough away from the original surface. The cycle of attachment–growth–sloughing off will then start again (Howell and Atkinson, 1976).

Under turbulent flow conditions the film seldom exceeds about 1 mm in thickness and may be considerably less than this. Film is largely water, dry matter ranging from 10 to 50 mg cm^{-3}; higher values are a function of shear conditions. The volatile film fraction ranges from 1.9–3.2% and the fixed non-volatile material from 0.8–11.8% of the total film weight. Considerable inorganic matter is associated with film. This includes iron, silicon, aluminium, calcium, magnesium and sometimes large amounts of manganese. Solute diffusivity in microbial film is reduced. Rather variable values appear in the literature (Characklis, 1981); however, values for oxygen from 50% to 90% of that in pure water may be appropriate.

The film community may be very complex and its actual structure is dependent on prevailing nutritional and hydrodynamic conditions. In a model film reactor system Pederson (1982) found that the greatest species diversity appeared after 25 days incubation in an unsupplemented basic water medium. The community showed bacteria in aggregates; solitary, stalked and filamentous bacteria as well as a range of protozoal species. Costerton (1979) and Jones *et al.* (1969) have investigated microbial successions in films in more detail and include photomicrographic analysis of specific systems.

Hamilton (1987) has stressed the importance of considering microbial film as a community of organisms with a certain *functional* homogeneity in spite of its structural heterogeneity:

> 'The biofilm community is capable of activities other than those only of the individual organisms, and these activities are dependant on the structural integrity of the film *in toto*.'

Whilst physico-chemical gradients have been assumed to be present in all film systems and their extent has been predicted by many workers there have been few experimental estimates of solute profiles in these structures. The most careful work was published by Bungay and his colleagues who used oxygen microelectrodes to determine oxygen gradients in a number of films from effluent treatment systems or from other natural fresh water environments (Bungay *et al.*, 1969; Bungay and Harold, 1971; Chen and Bungay, 1981 and Bungay and Chen, 1981).

The modelling philosophy has been applied to microbial film as well as to other natural microbial ecosystems. Film models have appeared in a number of discrete areas in microbiology. Often the *problems* have been very similar but the disciplines so disparate as to make communications between research groups minimal. The first great area is in oral medicine where dental plaque is a film of major importance. Plaque models (somethimes called 'artificial mouths') have been reviewed by Tatevossian (in press). A second area is in the water industry where the importance of film in removing organic pollutants on the one hand, and as a nuisance in blocking pipework on the other, cannot be overstated. A third area involves marine systems where bifouling of ship hulls is a traditional problem, whilst microbially induced corrosion is a more contemporary ailment of the oil and indeed of other process industries.

Wimpenny *et al.* (1983), Hamilton (1987) and Peters and Wimpenny (in press, submitted) have briefly discussed the types of film fermenters that have been described in the literature. Particularly relevant to the marine situation is the Robbins device (McCoy *et al.*, 1981) discussed in some detail by Hamilton (1987). This experimental system consists of a spool tube section which can be incorporated into a recirculating loop which is fed with nutrients and a bacterial inoculum. Film forms on replaceable studs or on removable sections of the tube. Costerton and Lashen (1984) have used such a system to investigate biocide effects primarily in oil-field situations. Hamilton and his colleagues (Tanner *et al.*, 1985) have examined the use of biocides *in situ* and in laboratory simulations as well as engaging in fundamental studies of biofilm itself using modified forms of the Robbins device.

Characklis and his co-workers have used different experimental systems. The annular reactor which incorporates removable glass sections subjects microbial film to controlled high shear fields. This device has been used to investigate the relationship between bacterial adsorption, polysaccharide production and growth rate (Robinson *et al.*, 1984; Nelson *et al.*, 1985). More recently Characklis (personal communication) has used rectangular capillary reaction vessels with microscopy and image analysis to monitor all stages of adsorption, growth, detachment, cell separation and desorption during early film formation.

The Cardiff constant depth film fermenter

Our design philosophy in Cardiff was to construct a totally enclosed system capable of aseptic operation under any gassing or nutrient regime, which generated film which could be maintained in a quasi-steady state. The latter was to be achieved by allowing the film to form on a surface which had been recessed a chosen amount within the film pan. The film pans were themselves all located in a PTFE disc which slowly rotated below a scraper bar. Any growth which exceeded the calibrated depth was removed by the bar. It was felt that the advantages of such a system were two-fold.

(1) That the system would operate under approximately steady state conditions. This meant that film could be maintained indefinitely and that perturbing the system would lead to meaningful responses.
(2) That microelectrode experiments could be carried out with confidence since the film depth was known.

The first versions of this fermenter were applied particularly to dental plaque model systems (Coombe *et al.*, 1982). After a further period of development the final version was conceived (Peters and Wimpenny, in press). It incorporates a Teflon disc in which 15 separate film pans are fitted. Each film pan is drilled with five holes in which can be located plugs which constitute the substratum on which the film is grown. 75 separate film plugs are therefore present in the fermenter at the start of each run. Substrate plugs are recessed into their holes using a steel template. This recess forms the space within which the film grows and of course its depth can be altered using different templates so that a range of film dimensions can be investigated in a single run. The upper surface of the rotating disc passes beneath a spring-loaded Teflon scraper bar attached to the steel head plate of the fermenter.

The main disc passes through an autoclavable bearing system and is driven by a small geared synchronous motor. The film system is totally enclosed within a standard QVF glass section and is autoclavable. The film pan is irrigated with sterile medium and there are inlets and outlets for medium and gas mixtures. The system can be sampled aseptically during a run by removing individual pans through the sampling port using a stainless steel tool designed for the job.

The film fermenter was used initially to investigate film produced by a river water consortium isolated by A. Peters. The organisms were grown in very dilute media containing only traces of glucose as carbon and energy source. Film growth was monitored by measuring the increase in protein in each film pan as a function of time. Results showed that growth was slow and almost certainly substrate limited. The film doubled in mass every 60 h until a steady state protein level was reached after about 20 days. The films could be removed from the fermenter and oxygen gradients measured within them using microelectrodes with an appropriate positioning device. Films were also fixed and examined using scanning and transmission electron microscopy.

The constant depth film fermenter described here has yet to be applied to problems in the marine environment; however, it ought to have considerable potential in tackling some of the economically more pressing microbial film problems in this area. For instance, the use of 75 separate film pan surfaces simultaneously should allow the effects of different surfaces to be compared in a well-controlled manner. If the surfaces are steel, corrosion processes as they relate to film structure and function seem a clear line to investigate.

Other possible film models

It seems clear on the surface that the best model of a microbial film must surely be a microbial film. Whilst this is likely to be true it ought to be possible to use other structured models to investigate properties of biofilms. A simple definition of a biofilm is that it consists of a community of microbes entrapped in a polymer matrix. This is also a fair description of the gel-stabilized model system already discussed. Gel-stabilized models are generally much thicker than biofilms and this poses some interesting problems and possibilities. If all the organisms present in a microbial film are transferred to an agar gel layer, fed at the surface with nutrients characteristic of the film environment, in what way will the community resemble the original film? Is it possible to stretch a film community and for that stretched system to give useful information as to the operation of the film? It seems likely that the species order across the gel-stabilized film system will reflect that found in the film. Certainly anaerobic species will be located beneath aerobes. Interactions between groups will proceed, but more slowly. Certain types of close physical associations, for example the syntrophic association characteristic of the two species that together make up '*Methanobacillus omelianskii*', are not susceptible to stretching in this way. It is by no means clear how close cooperating partners in the *general* process of interspecific hydrogen transfer need to be for the process to operate adequately. Provided that a good sink for hydrogen exists, diffusion could presumably lower the hydrogen concentration sufficiently to allow acetogenic metabolism to proceed satisfactorily, even where the two groups of bacteria were located in different spaces.

IN DEFENCE OF EXPERIMENTAL MODELS

The model builder is often exposed to criticism by 'real' ecologists since the latter feel, with some justification, that models assume a life of their own quite separate from, and irrelevant to, reality. It seems clear that the mathematical modeller may sometimes be justly criticized on this score. This is because families of equations are many steps removed from the actual behaviour of microorganisms. Thus the construction of a numerical model assumes a great deal about the behaviour of a microbial pure culture *before* it builds on these assumptions with further conjecture about interactions between different species and with the physico-chemical environment. It is also a highly seductive art because it is easy to formulate a family of differential equations, to solve them using a computer and then to regard the results as having some primal significance whether or not they match up with reality. It is much easier than actually performing the necessary experiments. Often mathematicians or theoretical biologists are quite unequipped for such experimentation anyway and are really dependent on good working relationships with practicing microbial ecologists. Finally the 'objectivity' and formalism of mathematical models seem to be high status attributes to the community of biologists, some of whom may be unable to understand parts or all of the arguments. It must be stressed finally that mathematical models can be of first-rate importance to subjects like microbial ecology. It is simply that the onus is on the modeller to prove that this is so.

With one major proviso the same type of criticism can be applied to the experimental modeller. The latter does perform genuine laboratory experiments although these can be regarded in two quite different lights. On the one hand they may be designed to throw light on a genuine natural ecosystem. The modeller may attempt to sell these assertions to the ecologist but it is surely a case of *caveat emptor* and buyer and seller should both beware! On the other hand the experimental system has an identity of its own as an ecosystem in just the same way that a chemostat has its own internal validity. In each case microbial growth occurs in response to the environment in which organisms find themselves. The type of spatially heterogeneous models described here may be attacked for not being homologues of natural systems; however, they surely ought not to be criticized if they throw light on microbial growth and interactions in the physico-chemical gradients of heterogeneous environments. My own feeling, expressed in other communications, is that spatially heterogeneous models lie somewhere between homogeneous laboratory growth systems and the natural environment. From this position they should cast light in both directions!

ACKNOWLEDGEMENTS

The work reported here owes a great deal to the experimental work of, and discussions with, the following: Bob Lovitt, Steve Jaffe, Philip Coombs, Richard Earnshaw, Rachael Coombe and Adolf Tatevossian. To all of them we express our gratitude.

Appendix. Equations for the model described on p. 67.

For two substrate simulations:

$$\frac{\partial S_1}{\partial t} = D_{S_1} \frac{\partial^2 S_1}{\partial x^2} - \frac{1}{Y_{S_1}} \cdot C \cdot \mu max \cdot F(S_1, S_2)$$

$$\frac{\partial S_2}{\partial t} = D_{S_2} \frac{\partial^2 S_2}{\partial x^2} - \frac{1}{Y_{S_2}} \cdot C \cdot \mu max \cdot F(S_1, S_2)$$

$$\frac{\partial C}{\partial t} = D_C \frac{\partial^2 C}{\partial x^2} - C \cdot \mu max \cdot F(S_1, S_2)$$

The growth function $F(S_1, S_2)$ is the product of two Monod equations:

$$F(S_1, S_2) = \frac{S_1}{(K_{S(S_1)} + S_1)} \cdot \frac{S_2}{(K_{S(S_1)_2} + S_1)}$$

Where a single substrate is also an inhibitor:

$$\frac{\partial S}{\partial t} = D_S \frac{\partial^2 S}{\partial x^2} - \frac{1}{Y_S} \cdot C \cdot \mu max \cdot F(S_1)$$

The growth function is:

$$F(S) = \frac{S}{(K_S + S)} \frac{K_i}{(K_i + S)}$$

In these models S_1, S_2 and S are substrate concentrations; C is the cell concentration; Y_{S_1}, Y_{S_2} and Y_S are yield coefficients; $K_{S(S1)}$, $K_{S(S2)}$ and K_S are substrate affinity coefficients; K_i is an inhibition constant; μmax is specific growth rate and x and t represent distance and time respectively. D_S, D_{S_1}, D_{S_2} and D_C are diffusion coefficients for substrates and cells.

REFERENCES

Billen, G. 1982. Modelling the processes of organic matter degradation and nutrient recycling in sedimentary systems. In *Sediment Microbiology, Special Publications of the Society for General Microbiology*, pp. 15–22.

Bungay, H.R. and Chen, Y.S. 1981. Dissolved oxygen profiles in photosynthetic microbial slimes. *Biotechnology and Bioengineering*, **23**: 1893–1895.

Bungay, H.R., Whalen, W.J. and Sanders, W.M. 1969. Microprobe technique for determining diffusivities and respiration rates in microbial slime systems. *Biotechnology and Bioengineering*, **11**: 765–772.

Bungay, R.R., III. and Harold, D.M. 1971. Simulation of oxygen transfer in microbial slimes. *Biotechnology and Bioengineering*, **13**: 569–579.

Characklis, W.G. 1981. Fouling biofilm development: a process analysis. *Biotechnology and Bioengineering*, **23**: 1923–1960.

Chen, Y.S. and Bungay, H.R. 1981. Microelectrode studies of oxygen transfer in trickling filter slimes. *Biotechnology and Bioengineering*, **23**: 781–792.

Coombe, R.A., Tatevossian, A. and Wimpenny, J.W.T. 1982. Bacterial thin films as *in vitro* models for dental plaque. In *Surface and Colloidal Phenomena in the Oral Cavity: Methodological Aspects* (ed. R.M. Frank and S.A. Teach), pp. 239–249. IRL Press Ltd, London.

Coombs, J.P. and Wimpenny, J.W.T. 1982. Growth of *Bacillus cereus* in a gel-stabilised nutrient gradient system. *Journal of General Microbiology*, **128**: 3093–3101.

Costerton, J.W. 1979. The mechanism of primary fouling of submerged surfaces by bacteria. Presented at Condenser Biofouling Control Symposium Atlanta, Georgia, USA, March 25–29th, 1979.

Costerton, J.W. and Lashen, E.S. 1984. The inherent biocide resistance of corrosion-causing biofilm bacteria. *Materials Performance*, **23**(2): 13–16.

Fenchel, T. 1969. The ecology of marine microbenthos IV. Structure and function of the benthos ecosystem, its chemical and physical factors and the microfauna communities with special reference to the ciliated protozoa. *Ophelia*, **6**: 1–182.

Feyerabend, P.K. 1975. *Against Method: Outline of an Anarchistic Theory of Knowledge*. New Left Books, London.

Hamilton, W.A. 1985. Sulphate-reducing bacteria and anaerobic corrosion. *Annual Reviews of Microbiology*, **39**: 195–217.

Hamilton, W.A. 1987. Biofilms: microbial interactions and metabolic activities. *Symposia of the Society for General Microbiology*, **41**: 361–385.

Hoppensteadt, F.C. and Jager, W. 1979. Pattern formation by bacteria. In *Biological Growth and Spread* (ed. W. Jager, H. Rost and P. Taylor), pp. 38, 68–81. *Lecture Notes in Biomathematics*, Vol. 55. Springer-Verlag, Berlin.

Hoppensteadt, F.C., Jager, W. and Poppe, C. 1984. A hysteresis model for bacterial growth patterns. In *Modelling of Patterns in Time and Space* (ed. W. Jager and J. D. Murray). *Lecture Notes in Biomathematics*, Vol. 55. Springer-Verlag, Heidelberg.

Howell, J.A. and Atkinson, B. 1976. Sloughing of microbial film on trickling filters. *Water Research*, **18**: 307–315.

Jones, H.C., Roth, I.L. and Sanders, W.M. (1969). Electron microscopic studies of a slime layer. *Journal of Bacteriology*, **99**: 316–325.

Jørgensen, B.B. 1977. The sufhur cycle of a coastal marine sediment. *Limnology and Oceanography*, **22**: 814–832.

Jørgensen, B.B. 1981. Mineralisation and the cycling of carbon, nitrogen and sulphur in marine sediments. In *Contemporary Microbial Ecology* (ed. D.C. Ellwood, J.N. Hedger, M.J. Latham, J.M. Lynch and J.H. Slater), pp. 239–257. Academic Press, London.

Lovitt, R.W. and Wimpenny, J.W.T. 1979. The gradostat: a tool for investigating microbial growth and interactions in solute gradients. *Society for General Microbiology Quarterly*, **6**: 80.

Lovitt, R.W. and Wimpenny, J.W.T. 1981a. The gradostat, a bidirectional compound chemostat, and its application in microbiological research. *Journal of General Microbiology*, **127**: 261–268.

Lovitt, R.W. and Wimpenny, J.W.T. 1981b. Physiological behaviour of *Escherichia coli* grown in opposing gradients of glucose and oxygen plus nitrate in the gradostat. *Journal of General Microbiology*, **127**: 269–276.

Macfarlane, G.T., Russ, M.A., Keith, S.M. and Herbert, R.A. 1984. Simulation of microbial processes in estuarine sediments using stabilised systems. *Journal of General Microbiology*, **130**: 2927–2933.

McCoy, W.F., Bryers, J.D., Robbins, J. and Costerton, J.W. 1981. Observations on fouling biofilm formation. *Canadian Journal of Microbiology*, **27**: 910–917.

Morgan, P. and Watkinson, R.J. 1987. The use of gel-stabilised systems to model the effects of pollutants on sediment populations. Oral paper given at the 108th Meeting of the Society for General Microbiology, St. Andrews, Scotland.

Nelson, C.H., Robinson, J.A. and Characklis, W.G. 1985. Bacterial adsorption to smooth surfaces: rate, extent, and spatial pattern. *Biotechnology and Bioengineering*, **27**: 1662–1667.

Nelson, D.C., Jørgensen, B.B. and Revsbech, N.P. 1986a. Growth pattern and yield of a chemoautotrophic *Beggiatoa* sp, in oxygen–sulphide microgradients. *Applied and Environmental Microbiology*, **52**: 225–233.

Nelson, D.C., Revsbech, N.P. and Jørgensen, B.B. 1986b. Microxic–anoxic niche of *Beggiatoa* spp.: Microelectrode survey of marine and freshwater strains. *Applied and Environmental Microbiology*, **52**: 161–168.

Nitsch, B. and Kutzner, H.J. 1973. Wachstum von streptomycetin in schuttelagar-kultur: eine neue methode zur festellung des c-quellen-spektrums. *Symposium on Technische Mikrobiologie, Berlin*, Vol. 3, 481–486.

Pederson, K. 1982. Method for studying microbial biofilms in flowing water systems. *Applied and Environmental Microbiology*, **43**: 6–13.

Perfil'ev, B.V. and Gabe, D.R. 1969. *Capillary Methods of Investigating Microorganisms* (English translation). Oliver and Boyd, Edinburgh.

Peters, A. and Wimpenny, J.W.T. (in press). A constant-depth laboratory model film fermenter. In *A Handbook of Laboratory Model Systems for Microbial Ecosystem Research* (ed. J.W.T. Wimpenny). CRC Press, Boca Raton, USA.

Peters, A.C. and Wimpenny, J.W.T. (submitted for publication). A constant depth laboratory model film fermenter. *Biotechnology and Bioengineering*.

Popper, K.R. 1968. *The Logic of Scientific Discovery*. Hutchinson, London.

Revsbech, N.P., Jørgensen, B.B. and Blackburn, T.H. 1980. Oxygen in the sea bottom measured with a microelectrode. *Science*, **207**: 1355–1356.

Robinson, J.A., Trulear, M.G. and Characklis, W.G. 1984. Cellular reproduction and extracellular polymer formation by *Pseudomonas aeruginosa* in continuous culture. *Biotechnology and Bioengineering*, **26**: 1409–1417.

Tanner, R.S., Haack, T.K., Semet, R.F. and Greenley, D.E. 1985. A mild steel tubular flow system for biofilm monitoring. *U.K. Corrosion '85* pp. 259–269.

Tatevossian, A. (in press). Film fermenters in dental research. In *Handbook of Laboratory Model Systems in Microbial Ecosystem Research* (ed. J.W.T. Wimpenny). CRC Press, Boca Raton, USA.

Thompson, L.A., Nedwell, D.B., Balba, M.T., Banat, I.M. and Senior, E. 1983. The use of multiple-vessel, open flow systems to investigate carbon flow in anaerobic microbial communities. *Microbial Ecology*, **9**: 189–199.

Williams, J.W. 1938a. Bacterial growth 'spectrum' analysis. I. Methods and application. *The American Journal of Medical Technology*, **4**: 58–61.

Williams, J.W. 1938b. Bacterial growth 'spectrums'. II. Their significance in pathology and bacteriology. *American Journal of Medical Technology*, **14**: 642–645.

Williams, J.W. 1939a. Growth of microorganisms in shake cultures under increased oxygen and carbon dioxide tensions. *Growth*, **3**: 21–33.

Williams, J.W. 1939b. The nature of gel mediums as determined by various gas tensions and its importance in growth of microorganisms and cellular metabolism. *Growth*, **3**: 181–196.

Wimpenny, J.W.T. 1981. Spatial order in microbial ecosystems. *Biological Reviews*, **56**: 295–342.

Wimpenny, J.W.T. 1982. Responses of microbes to physical and chemical gradients. *Philosophical Transactions of the Royal Society*, **297**: 497–515.

Wimpenny, J.W.T. 1985. Novel growth systems. *Microbiological Sciences*, **2**: 53–60.

Wimpenny, J.W.T. and Lovitt, R.W. 1984. The investigation and analysis of heterogeneous environments using the gradostat. In *Microbiological Methods for Environmental Biotechnology* (ed. J.M. Grainger and J.M. Lynch). SAB Technical Series, Vol. 19, pp. 295–312. Academic Press, London.

Wimpenny, J.W.T. and Waters, P. 1984. Growth of micro-organisms in gel-stabilized two-dimensional diffusion gradient systems. *Journal of General Microbiology*, **130**: 2921–2926.

Wimpenny, J.W.T. and Waters, P. 1987. The use of gel-stabilized gradient plates to map the responses of microorganisms to three or four environmental factors varied simultaneously. *FEMS Microbiology Letters*, **40**: 263–267.

Wimpenny, J.W.T., Coombs, J.P., Lovitt, R.W. and Whittaker, A. 1981. A gel-stabilised model ecosystem for the investigation of microbial growth in spatially ordered solute gradients. *Journal of General Microbiology*, **127**: 277–287.

Wimpenny, J.W.T., Lovitt, R.W. and Coombs, J.P. 1983. Laboratory model systems for the investigation of spatially and temporally organised microbial ecosystems. *Symposium of the Society for General Microbiology*, **34**: 67–117.

Wimpenny, J.W.T., Jaffe, S. and Coombs, J.P. 1984. Periodic growth phenomena in spatially organized microbial systems. In *Modelling of Patterns in Time and Space, Lecture Notes in Biomathematics*, Vol. 55 (ed. W. Jager and J.D. Murray), pp. 388–405. Springer-Verlag, Heidelberg.

Wood, E.J.F. 1964. *Marine Microbial Ecology*. Reinhold, New York.

ZoBell, C.E. 1943. The effect of solid surfaces upon bacterial activity. *Journal of Bacteriology*, **46**: 39–56.

4

Contribution of symbiotic chemoautotrophs to the nutrition of benthic invertebrates

Eve C. Southward, Marine Biological Association, Citadel Hill, Plymouth PL1 2PB, UK

Symbiotic chemoautotrophic bacteria occur in tissues of certain worms and bivalve molluscs that live in marine environments where sulphide is present, i.e. reducing sediments and deep-sea hydrothermal springs. Cells containing bacteria occur internally in Pogonophora, epidermally in bivalve molluscs and oligochaetous annelids. The bacterial cell walls are of Gram-negative type, and there are periplasmic sulphur vesicles in several symbiont forms. Successful isolation and culture of symbiotic bacteria outside the host animal has not yet been reported; hence there has been little progress with classification and description of species. There is evidence that the bacteria are chemoautotrophic. Carbon-dioxide-fixing and sulphide-oxidizing enzymes have been identified in many symbiont-containing tissues and CO_2-fixation rates measured. Some of the larger animals have transport systems for CO_2, sulphide and oxygen, the three main requirements of sulphide-oxidizing bacteria. Methanotrophy has been detected in three animals living at sites where methane has been measured or is suspected to occur. The symbionts in these animals resemble type I methanotrophs in fine structure.

Many of the host animals have either no gut or a very reduced gut. Others have comparatively inefficient means of gathering and digesting food. The bacteria are thought to contribute to the nutrition of the host by production of extracellular organic compounds; in some cases complete bacteria are digested by host cells. Measurement of $\delta[^{13}C]$values in host tissues with and without symbionts indicates that chemoautotrophy makes a contribution to tissue carbon and that carbon is transferred from symbiont to host. It has been estimated that the symbionts may contribute from 50% to 100% of the host carbon, depending on species and circumstances.

INTRODUCTION

The discovery of symbiosis between chemoautotrophic bacteria and marine invertebrates is very recent (Cavanaugh *et al.*, 1981) compared with the long history of investigation of symbioses between photosynthetic organisms and marine invertebrates, or heterotrophic bacteria and insects (Henry, 1967). Just as photosynthetic symbionts need light for growth, chemosynthetic symbionts need reduced inorganic compounds such as sulphide or methane; they also need oxygen. Usually the reduced compounds occur together with oxygen only in narrow boundary zones between oxic and anoxic environments, because of the rapidity of chemical oxidation (Jørgensen, 1982). The animals which harbour chemosynthetic bacteria are worms and molluscs which live in places where they can bridge the gap between oxic and anoxic conditions to obtain reduced compounds and oxygen and supply them to their bacteria. They extend the range available to such bacteria and provide them with shelter from predation by other organisms. The animals obtain nourishment (either from bacterial exudations or by digesting some of their symbionts) from an energy source which is not utilized directly by animals without symbionts. They also gain the ability to colonize environments high in sulphide, which are toxic to most other animals.

The environments exploited are widespread on the ocean floor, ranging from black or grey reducing sediments on the shore or in shallow water, especially seagrass beds, to continental shelf, slope or even deep-sea trench sediments; any sediments with a redox discontinuity close enough to the sediment surface for an animal to contact both the oxic and anoxic zones. Waste disposal sites, such as sewer outfalls and pulp mill effluents, where organic matter accumulates and black sulphidic sediments develop, are places where some symbiont-containing animals can flourish, providing that they can obtain enough oxygen. Hydrothermal vents at rifts in the ocean floor are typically colonized by large and spectacular worms and molluscs which contain chemoautotrophic bacteria. These animals live on hard rock surfaces in places where geothermally modified seawater, rich in H_2S, mixes with cold ocean bottom water containing oxygen (e.g. Grassle, 1986). It was the observation of these flourishing hydrothermal vent communities in ocean depths (2500 m) where animals are usually scarce and small for lack of food that led to the discovery of sulphide-exploiting symbioses.

The word *symbiosis* means living together and it is usually applied to associations where both partners benefit. Though both partners are symbionts, a custom has grown up of terming the animal partner the *host* and the microbial partner the *symbiont*, and it is convenient to follow the custom in this review.

CHEMOSYNTHESIS

Chemosynthesis is the biosynthesis of organic compounds from CO_2, using energy from chemical oxidations, in contrast to photosynthesis where light is the energy source. Chemolithotrophy is the general ability to use reduced inorganic compounds as energy sources; chemoautotrophy refers to the assimilation of CO_2 (Jannasch and Mottl, 1985). Autotrophy in the sense of generation of most cellular carbon from

Electron donor	Electron acceptor	Type of organism	Isolated from vent fluids[a]
$S^{2-}, S^0, S_2O_3^{2-}$	O_2	Sulphur-oxidizing bacteria	+
$S^{2-}, S^0, S_2O_3^{2-}$	NO_3^-	Denitrifying and sulphur-oxidizing bacteria	−
H_2	O_2	Hydrogen-oxidizing bacteria	+
H_2	NO_3^-	Denitrifying hydrogen bacteria	−
H_2	S^0, SO_4^{2-}	Sulphur- and sulphate-reducing bacteria	+
H_2	CO_2	Methanogenic and sulphate-reducing bacteria	+
NH_4^+, NO_2^-	O_2	Nitrifying bacteria	−
Fe^{2+}	O_2	Iron-oxidizing bacteria	−
(Mn^{2+})	O_2	Manganese-oxidizing bacteria	+
CH_4	O_2	Methylotrophic bacteria	+
CO	O_2	Carbon-monoxide-oxidizing bacteria	+

[a]Jannasch, 1985.

Table 1   Types of chemoautotrophic bacteria isolated from sediments

one-carbon compounds covers a wide spectrum of organisms (e.g. Strohl and Tuovinen, 1984), and much interest has been aroused by the recent discovery that hydrothermal vent effluents contain a variety of free-living chemoautotrophic bacteria (Table 1). In addition to free-living bacteria the vent waters sustain animals with intracellular bacteria that oxidize sulphur compounds (review by Jannasch, 1985). Other sites where methane and sulphide-rich fluids seep out of the sea floor are inhabited by closely related animals, some of which contain methane-oxidizing bacteria (Cavanaugh *et al.*, 1987; Childress *et al.*, 1986). It is possible that some of the other electron donors listed in Table 1 may also be used by symbionts, where sulphide and methane are scarce.

HOST ANIMALS

Pogonophora
It is probable that all members of this phylum of gutless worms have internal symbionts. More than 100 sediment-living species are known, from depths of about 20 to 9950 m (Ivanov, 1963; Southward, 1971). A few species live in rotting wood (family Sclerolinidae), and the larger vestimentiferans attach their tubes to hard substrates around warm hydrothermal vents or cool seeps (Jones, 1985). It has been suggested that Vestimentifera should be separated from Pogonophora at the phylum level (Jones, 1985), but they have so many similarities that they will be considered together here. *Riftia pachyptila* is the most spectacular vestimentiferan; at a diameter of over 30 mm and a length of over 1 m, it is likely to be the largest animal

sustained by chemoautotrophic symbionts. Sediment-living pogonophores have slender bodies, 0.1 to 2 mm in diameter and from 50 to more than 500 mm long. Their tubes are largely buried in muddy sediments, with the anterior end in oxygenated seawater and the posterior end well below the redox discontinuity. There is little circulation of water inside the tube but haemoglobin dissolved in the blood carries oxygen from the anterior tentacles down to the symbiont-containing cells in the trophosome, which forms a core to the posterior part of the body (Southward, 1982; Terwilliger et al., 1987). Reduced inorganic compounds can diffuse through the thin tube and epidermis of this elongated region to reach the symbionts, which are housed in vacuoles in the cells of the trophosome, where their environment must be closely controlled by the host. Ventimentiferan tubes are very thick-walled and probably form a barrier to diffusion, but vestimentiferans have a plume of fine tentacles which is held outside the tube, and the blood can carry oxygen, sulphide and carbon dioxide to the large and elaborate trophosome which fills most of the trunk (Arp et al., 1985).

Bivalve molluscs
Five families of bivalves include species with symbiotic bacteria in the gills. Only two of these families, the Lucinidae and Thyasiridae, are closely related to one another, and it appears that comparable symbioses have evolved at least four times among the Bivalvia (Table 2). All members of the families Solemyidae, Vesicomyidae and Lucinidae will probably prove to have gill symbionts, but in the Thyasiridae some species have them and others do not (Southward, 1986), while the family Mytilidae consists mainly of efficient filter feeders, with few symbiont-containing species. Although these bivalves have gills of different types (protobranch, filibranch and eulamellibranch) the site occupied by symbionts is the same in each case, in epidermal cells, termed bacteriocytes, covering the inner non-ciliated region of each gill filament and any tissue junctions between (Dando et al., 1985, 1986a; Fiala-Médioni and Métivier, 1986; Fiala-Médioni et al., 1986b; Le Pennec and Hily, 1984; Reid and Brand, 1986; Southward, 1986). In thyasirids the symbionts are in external blisters, protected by a cuticle made up of the adherent tips of microvilli, outside the cytoplasmic membrane of the bacteriocyte. In the other families the bacteria are housed singly or in groups in vacuoles inside the bacteriocyte (see Fig. 4). Bivalve gills are primarily food-collecting structures and gas exchange can take place over the body and mantle surface. Cilia on the outer margins of gill filaments force water between the filaments while filtering and transporting particles on the outer surface (Jørgensen et al., 1986). The bacteriocytes are swept by this flow of filtered water which continues posteriorly and out of the bivalve through an exhalent siphon or aperture. The basal surfaces of the bacteriocytes are in close contact with the central blood sinus of the gill filament, so the epithelium has fluid on both sides and ready access to both internal and external media. Most symbiont-containing bivalves are burrowers. They draw oxygenated water through their burrows, where it mixes with reduced inorganic compounds from the sediment (Dando et al., 1985, 1986a, b; Doeller, 1984; Reid, 1980). Possibly the probing foot may also take up sulphide etc. from the deeper layers of the sediment. On rocky surfaces near hydrothermal vents the vent clam *Calyptogena magnifica* (member of a burrowing family) lives in crevices which it probes deeply with its foot for sulphide in the

upwelling water (Arp *et al.*, 1984; Hessler *et al.*, 1985). The vent mussel, *Bathymodiolus thermophilus*, attaches itself with byssis threads like other mussels and takes in ready-mixed bottom water and vent water through an inhalent siphon (Kenk and Wilson, 1985; Smith, 1985).

Gastropod molluscs

Among the numerous unnamed species of limpet-shaped gastropods found on the rocks around hydrothermal vents (McLean, 1985), one has its gill covered by a dense aggregation of filamentous bacteria, some of which are endocytosed and digested by the cells, but it is not known whether the bacteria make any contribution to the nutrition of the limpet (de Burgh and Singla, 1984).

Oligochaeta

Gutless worms of the subfamily Phallodrilinae (family Tubificidae) have substantial numbers of bacteria living under the cuticle of the epidermal cells, increasing the apparent thickness of the epidermis although not inside individual cells (Giere, 1981, 1985a; Giere and Langheld, 1987; Richards *et al.*, 1982). *Phallodrilus leukodermatus* and *P. planus* live is subtidal carbonate sands in Bermuda, where they concentrate in the redox discontinuity layer, but can move down into sand smelling of H_2S and migrate freely between oxic and anoxic zones (Giere, 1985a; Giere *et al.*, 1982). Such gutless tubificids are important members of the fauna of coral reef sands in other parts of the world (Erséus, 1984) and chemoautotrophic symbionts may well be important to them, but the majority of oligochaetes have guts, feed normally and are not known to harbour symbiotic bacteria. One normal tubificid has been reported to be colonized by external filamentous bacteria, possibly linked to presence of sulphide in the sediment (Dubilier, 1986), but it is not known whether they contribute to the nutrition of the host.

Polychaeta

Many polychaetes inhabit burrows in anoxic sediments but they usually have functional digestive systems. One gutless polychaete, *Astomus taeniodes*, has been described (Jouin, 1979) and would be worth further investigation.

Polychaetes of hydrothermal vent regions include the Pompeii worms (family Alvinellidae), with normal guts and feeding tentacles, which build masses of tubes on the chimneys through which hot water flows, termed black and white smokers (Desbruyères *et al.*, 1985; Desbruyères and Laubier, 1986). *Alvinella pompejana* and *A. caudata* have numerous bacteria attached to their body surface and parapodia and there is evidence that these bacteria are autotrophic. It is not clear whether there is any direct transfer of organic compounds from the bacteria to the host worms, which have been shown to eat bacteria from the water and from their own tube walls (Desbruyères *et al.*, 1983; Gaill *et al.*, 1984; Alayse-Danet *et al.*, 1985, 1986; Baross and Deming, 1985). A serpulid polychaete (*Laminatubus alvini* Ten Hove and Zibrowius), also from the hydrothermal vents, has filamentous bacteria connected to the wall of the mid-gut (Desbruyères *et al.*, 1985) which require investigation.

	Bacteria observed	Evidence[a] of chemoautotrophy
Subclass Protobranchia		
Order Solemyoidea		
Family Solemyidae		
Solemya velum Say	2,3	2,3
S. reidi Bernard	9,18,19	9
Subclass Lamellibranchia		
Superorder Pteroidea		
Family Mytilidae		
Bathymodiolus thermophilus Kenk & Wilson	2,12,16,17	10
Mytilids	4,5	4,5
Superorder Veneroidea		
Superfamily Lucinacea		
Family Lucinidae		
Ctena orbiculata (Montagu)	—	20
Codakia orbicularis (L.)	1	1
Lucina floridana (Conrad)	11	11
L. pectinata (Gmelin)	24	—
L. radians (Conrad)	14	20
L. costata (d'Orbigny)	14	20
L. nassula (Conrad)	—	1
L. multilineata Tuomey & Holmes	14	20
L. tenuisculpta Carpenter	19	10
Anodontia philippiana Reeve	14	20
Lucinoma annulata (Reeve)	23	10
L. borealis (L.)	7,21	7
Myrtea spinifera (Montagu)	6,12	6
Family Thyasiridae		
Thyasira flexuosa (Montagu)	19,21	8
T. sarsi (Philippi)	21	8
T. gouldi (Philippi)	21	—
T. equalis (Verrill & Bush)	21	8
Thyasira spp.	21	—

Superfamily Glossacea		
Family Vesicomyidae		
Calyptogena magnifica Boss & Turner	2,3,12,13	22
C. pacifica Dall	2	10
C. elongata Dall	23	—
C. ponderosa Boss	—	15

References: 1, Berg and Alatalo, 1984; 2, Cavanaugh, 1983; 3, Cavanaugh, 1985; 4, Cavanaugh et al., 1987; 5, Childress et al., 1986; 6, Dando et al., 1985; 7, Dando et al., 1986a; 8, Dando and Southward, 1986; 9, Felbeck, 1983; 10, Felbeck et al., 1981; 11, Fisher and Hand, 1984; 12, Fiala-Médioni, 1984; 13, Fiala-Médioni and Métivier, 1986; 14, Giere, 1985b; 15, Kennicutt et al., 1985; 16, Le Pennec and Hily, 1984; 17, Le Pennec et al., 1985; 18, Powell and Somero, 1985; 19, Reid and Brand, 1986; 20, Schweimanns and Felbeck, 1985; 21, Southward, 1986; 22, Tuttle, 1985; 23, Vetter, 1985; 24, Wittenberg, 1985.

[a] Evidence of chemoautotrophy — ribulosebisphosphate activity and/or unusual $^{13}C:^{12}C$ ratio.

Table 2 Classification of bivalve molluscs with symbiotic bacteria in the gills

Other invertebrates

Among the small animals living in sulphidic sediments (the sulphide biome of Fenchel and Reidl, 1970), some are able to detoxify sulphide, apparently without the help of bacteria (Powell *et al.*, 1980). Another fixes CO_2 under anaerobic conditions (Maguire and Boaden, 1975), but it is not known whether bacteria are involved. Symbiotic bacteria have been observed in or on other small infaunal animals, notably inside gutless Nematoda and Turbellaria, and on the cuticle of normally feeding nematodes (Ott *et al.*, 1982). Some of these bacteria are suspected of being sulphur oxidizers but there is no evidence yet of autotrophy.

BACTERIA

The identification of prokaryotic cells in the vestimentiferan *Riftia* was first accomplished by Cavanaugh *et al.* (1981) using SEM to show the coccoid shape, TEM to demonstrate the fine structure of the cell envelope and the absence of nuclear membrane, and the DNA-specific fluorescent stain DAPI to show the presence of DNA in the cells. Biochemical analysis of trophosome tissue showed the presence of liposaccharide cell wall material characteristic of Gram-negative bacteria. It was estimated that the trophosome tissue of *Riftia* contained 3.7×10^9 of these cells per gram of wet weight. Similar tests on the gills of the coastal bivalve *Solemya velum* gave similar results (Cavanaugh, 1983). Thus, there is good evidence that Gram-negative bacteria are abundant in the trophosome of *Riftia* and the gills of *Solemya*, and also that they are absent from the rest of the body of these animals. The discovery of the CO_2-fixing enzyme ribulosebisphosphate carboxylase in the symbiont-containing tissues indicates that the bacteria are chemoautotrophic (Felbeck, 1981; Cavanaugh, 1983), since this enzyme occurs only in autotrophic plants and bacteria and is absent from animal tissues. Investigation of the distribution of stable carbon isotopes in the tissues of *Riftia* indicated the transfer of bacterially fixed carbon to the animal cells (Rau, 1981b).

Subsequent investigations of other host species have relied mainly on TEM to show the presence of bacteria, and on assay of ribulosebisphosphate carboxylase to demonstrate autotrophy (see references in Cavanaugh, 1985; Giere and Langheld, 1987; Southward, 1982, 1986; Table 2).

The fine structure of the symbionts has been studied in at least three vestimentiferan pogonophores, eight sediment-living pogonophores, about 25 bivalve molluscs and three oligochaetes. There is much variation in shape and size, from spherical cells of 3–5 μm diameter in *Riftia* and smaller spheres in some lucinid bivalves, to thick rods (1–2 μm × 2–10 μm) in lucinids, solemyids and the pogonophore *Sclerolinum brattstromi*, while thin rods (< 0.5 μm × 1–2 μm) are found in thyasirid and mytilid bivalves. Some pogonophores contain long thin rods (0.5 μm × 2–5 μm) (Fig. 1). Symbionts can vary in size and shape within one host, even within one cell. Where both rods and spheres occur, as in *Riftia*, *Solemya* and *Phallodrilus* for example, they may represent different growth stages of one species, as seems more clearly the case in some lucinids (Southward, 1986). Two deep-water thyasirids contain mixed populations of two very distinct rod-shaped bacteria (Fig. 2), differing in size, shape and cell wall detail, which are more likely to be separate species (Southward, 1986). Living bacteria removed from the host cells do not

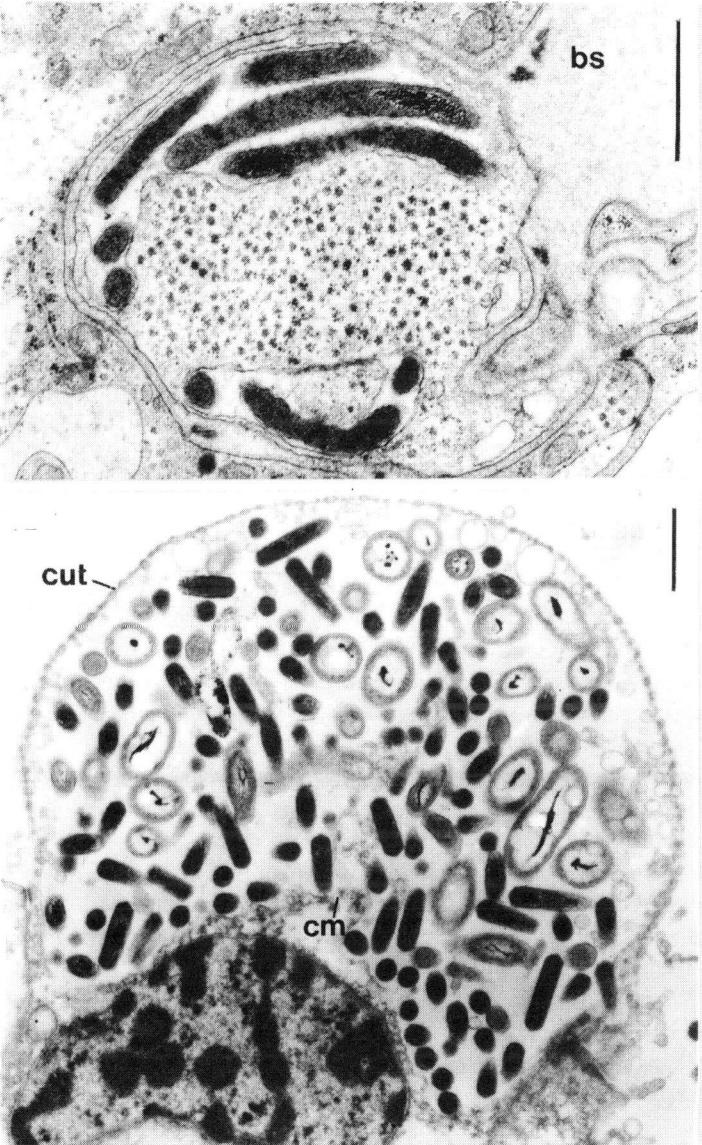

Fig. 1—Bacteria in vacuoles in trophosome cell of *Siboglinum fiordicum*. bs, blood sinus. Scale bar 1 μm.

Fig. 2—Two types of bacteria in space between cuticle and cytoplasmic membrane of gill cell of *Thyasira* sp. cm, cytoplasmic membrane; cut, cuticle. Scale bar 1 μm.

appear motile, nor have flagella been described, but in one of the gutless oligochaetes, *Phallodrilus albidus*, the rod-shaped bacteria have unusual filamentous appendages (Richards *et al.*, 1982).

The cell envelope of symbionts consists of a thin outer cell wall and an inner cytoplasmic membrane, as in Gram-negative bacteria (for details see Cavanaugh (1985), Fig. 1, Felbeck *et al.* (1983a), Fig. 4, Southward (1982), Fig. 4, and Southward (1986), Fig. 10). In some symbionts vesicles are formed by invaginations of the cytoplasmic membrane, which can be as much as 1 μm in diameter and may be visible with the light microscope as bright specks in living bacteria. Viewed with polarizing filters the bright specks show a maltese-cross pattern characteristics of liquid crystals (Vetter, 1985). Freeze-fracture and freeze etching of the gills of *Lucinoma annulata* has revealed the globular contents of such vesicles (Vetter, 1985), and further isolation of these globules allowed energy-dispersive X-ray analysis to be carried out, showing them to consist almost entirely of sulphur (Vetter, 1985). During embedding for TEM sectioning the solvents remove elemental sulphur so that the periplasmic vesicles look empty (Figs. 3 and 4) and energy-dispersive X-ray analysis of sections does not show a concentration of sulphur in the vesicles (Southward, 1986). The fine structure is similar to that of sulphur vesicles identified in free-living sulphur bacteria (Hageage *et al.*, 1970; Strohl *et al.*, 1981). Judging from fine structure of the symbionts and analysis of whole gills for sulphur (p. 100), sulphur vesicles occur in the symbionts of most of the lucinid bivalves so far examined, some of the thyasirids and *Calyptogena elongata* (Southward, 1986; Vetter, 1985). *Solemya velum* and *Solemya reidi* symbionts generally lack such vesicles (Cavanaugh, 1983, 1985; Felbeck, 1983; Reid and Brand, 1986) but they can occur in *S. reidi* (Powell and Somero, 1985). Vestimentiferan symbionts contain small membrane-bound vesicles, about 0.1 μm in diameter, which look empty in TEM sections, but are not obviously connected to the cytoplasmic membrane (Cavanaugh, 1985). It is not known whether they contain sulphur. The presence of sulphur crystals in the trophosome of (formalin-fixed) *Riftia* was the first clue to the presence of sulphur-oxidizing bacteria (Cavanaugh *et al.*, 1981), but it may have been precipitated after oxidation of the sulphide which is present in the blood of this animal (Arp *et al.*, 1985). A study of the development of *Riftia* symbionts shows that the 'empty' vesicles appear at a stage when the bacteria have ceased to divide and begin to increase in volume (Bosch and Grassé, 1984b) but the membranes may be formed in the earlier stage, which contains flat, double, intracytoplasmic membranes (Bosch and Grassé, 1984a). Pogonophore symbionts usually lack intracytoplasmic membranes, but *Sclerolinum brattstromi* has flat, double membranes extending across some of the cells (Southward, 1982).

A different type of intracytoplasmic membrane arrangement is found in the symbionts of one pogonophore and two mytilid bivalves. In all these there are stacks of double, parallel membranes like those of type I methanotrophic bacteria (see Anthony, 1982). The symbionts of *Siboglinum poseidoni* (Flügel and Langhof, 1983; Schmaljohann and Flügel, 1987; Southward *et al.*, 1981) and those of two unnamed mytilids (Cavanaugh *et al.*, 1987; Childress *et al.*, 1986), are implicated in methane oxidation (p. 101).

In addition to intracytoplasmic membranes, symbiotic bacteria contain a variety of inclusions. The smallest are ribosomes, seen in sections of most symbionts as electron-dense bodies 10–50 nm in diameter. More irregular bodies up to 100 nm in

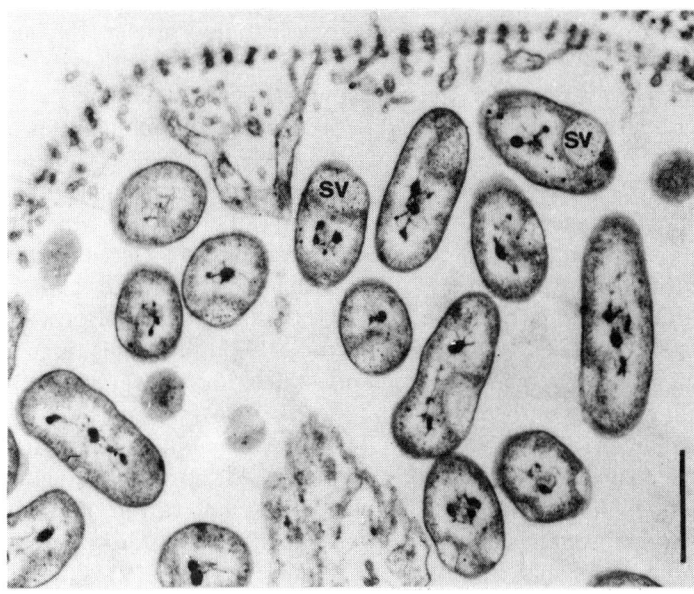

Fig. 3—Bacteria beneath cuticle of gill cell of *Thyasira sarsi*. sv, sulphur vesicle. Scale bar 1 µm.

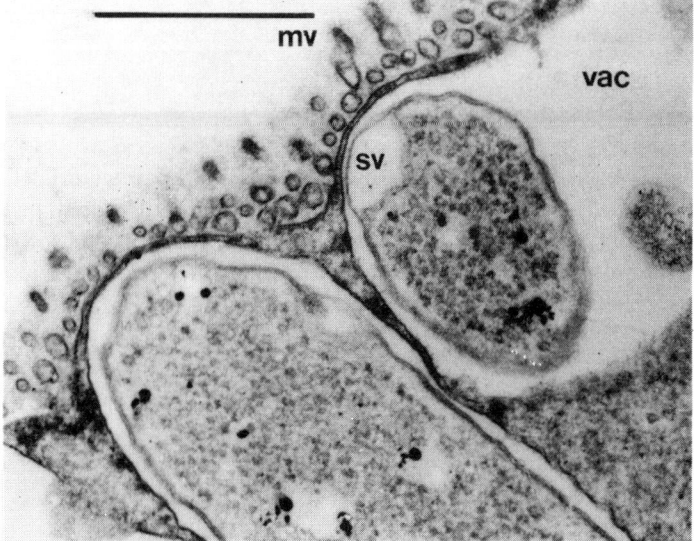

Fig. 4—Two bacteria close to surface of gill cell of *Myrtea spinifera*. Host cytoplasm surrounds the bacteria, enclosing them in vacoules. mv, microvillus; pg, polyphosphate granule; sv, sulphur vesicle; vac, vacuole. Scale bar 1 µm.

diameter may be glycogen (polyglucose). Glycogen has been identified in *Phallodrilus albidus* symbionts by the thiosemicarbazide–silver proteinate method (Richards *et al.*, 1982). Similar granules are common in some lucinid symbionts (Southward, 1986). Poly-β-hydroxybutyrate has been identified in *Phallodrilus leukodermatus*, presumably in the symbionts, which contain many electron-lucent vacuoles without very obvious boundary membrane, resembling sites of storage of this compound in other bacteria (Giere and Langheld, 1987; cf. Cohen-Bazaire and Kunisawa, 1963). Such electron-lucent vacuoles occur occasionally in the symbionts of *Siboglinum fiordicum* (Southward, 1982) and rather rarely in other pogonophore and mollusc symbionts. Much commoner inclusions are electron-dense spherical granules 30–200 nm in diameter. Their dense appearance and tendency to sublimate in the electron beam, leaving holes in the section, suggests that they are polyphosphate granules (cf. Jensen *et al.*, 1977) and this is confirmed by energy-dispersive X-ray analysis of sections, which shows the granules are composed mainly of phosphorus and calcium (Southward, 1986). 'Crystals' of electron-dense material have been seen in the symbionts of *Myrtea spinifera*, but their composition has not yet been investigated (Dando *et al.*, 1985). Carboxysomes might be expected in organisms which contain much ribulosebisphosphate carboxylase, but in general they seem to be absent in symbionts. Those labelled in Vetter's figure (1985) of the *Lucinoma annulata* symbiont look more like the polyphosphate granules identified in the *L. borealis* symbiont (Southward, 1986), while the inclusions in the symbionts of *Phallodrilus* spp. (Giere and Langheld, 1987) are rather small and indistinct, compared with typical carboxysomes, which are membrane-bound polyhedral bodies 100–700 nm in diameter (Allen, 1984; Shively, 1974).

RESERVE COMPOUNDS

Symbionts

The inclusions observed are normal reserve compounds in bacteria, though their occurrence depends on the type of metabolism and nutritional history of the organism. Elemental sulphur is a non-toxic intermediate product of oxidation of sulphide in chemolithitrophic and photolithotrophic bacteria, which may be stored safely and oxidized to sulphite or sulphate later, with the production of additional energy (Kelly, 1982; Vetter, 1985). Polyphosphate is an insoluble form of inorganic phosphate, which can be deposited at a time of phosphate surplus and release phosphate for ATP production when required (Shively, 1974). Glycogen is a convenient and easily mobilized carbohydrate store, and poly-β-hydroxybutyrate is a lipidic substance stored in abundance by some bacteria, though uncommon apparently in most of the symbionts.

Hosts

Among the bacteriocytes in the gills of some lucinids there are specialized cells, without bacteria, full of globules of storage material, probably protein, and glycogen granules (Dando *et al.*, 1985; Southward, 1986). Such storage cells do not occur in the gills of bivalves without symbionts. The trophosome of small pogonophores is made up of an inner layer of bacteriocytes and an outer layer of storage cells, containing lipid, protein and glycogen deposits (Southward, 1982). In both cases it appears that extracellular products of the symbionts may be laid down in animal cells close to the site of production.

CULTURE OF SYMBIONTS

It is perhaps not surprising that the chemoautotrophic symbionts have proved exceedingly difficult to isolate in culture so far, since heterotrophic symbionts of arthropods, which are common and well known, are very difficult (Dasch *et al.*, 1984). The chief problem for both lies in predicting the environment inside the animal cell. Where the symbionts are in intracellular vacuoles the host will control pH, oxygen tension, the form in which sulphur reaches the symbiont, other inorganic ion concentrations, and may supply some essential organic contribution as well. The symbionts are likely to be adapted to a narrow range of growth conditions under such circumstances. In the gutless oligochaetes and the thyasirid bivalves the symbionts live under the cuticle but outside the animal cell. It seems likely that these symbionts will be exposed to more variation in substrate and oxygen concentration, and will be less completely controlled by the host, so these extracellular symbionts may be the most promising material for isolation attempts especially since extracellular, luminous bacteria, symbiotic in epidermal pockets in fish and cephalopods, have been isolated in pure cultre (e.g. Kuwae *et al.*, 1982).

Attempts to grow bacteria from vestimentiferan trophosome or mollusc gills in media enriched with sulphide or thiosulphate seem to have produced cultures of a variety of bacteria, but none is readily identifiable with the symbiont in question (Jannasch, 1985; Jannasch and Nelson, 1984). Cultures growing on methane have been produced from the small pogonophore *Siboglinum poseidoni* and from the mud it lives in. Both isolates have the same fine structure as the symbiont described from the trophosome of this species (Schmaljohann and Flügel, 1987), so the authors suggest that a free-living methanotroph has been acquired by the pogonophore. More work is needed to confirm the identity of the isolates with one another and with the internal symbiont.

CLASSIFICATION OF SYMBIONTS

In the absence of data on the requirements of microorganisms in culture it is difficult to classify them and unwise to attempt to describe and name the species (Kreig and Holt, 1984). Those being discussed here are all Gram-negative chemolithotrophic or methanotrophic bacteria. Most are sulphur oxidizers. Comparison with known free-living species awaits the isolation and cultivation of sufficient material for immunological and biochemical comparisons.

A start has been made in studies of DNA and RNA. Nelson *et al.* (1984) extracted DNA from the trophosome and vestimentum of *Riftia pachyptila*. They found that 90% of the trophosome DNA was prokaryotic (i.e. from the symbionts) and had a base ratio of 58 mol% G + C. The genetic complexity is typical of free-living bacteria ($M_r = 2.14 \times 10^9$). Similar work on *Bathymodiolus* gill symbionts found a base ratio of 28–33 mol% $G + C$ and genome size of $1.9–2.0 \times 10^9$ (Belkin *et al.*, 1986). Analysis of 5S ribosomal RNA nucleotide sequences in symbiont-containing tissues of *Riftia pachyptila*, *Calyptogena magnifica* and *Solemya velum* compared the rRNA of all six organisms, hosts and symbionts, with one another (Lane *et al.*, 1985; Olsen *et al.*, 1986; Stahl *et al.*, 1984). It was concluded that the three symbionts are related to one another, but the relationship is not as close as the relationship between the hosts (which belong to two different animal phyla), and the symbioses seem to have been established independently. The symbionts are affiliated to a group of eubacteria known as 'group III purple photosynthetic bacteria', but which also includes several colourless sulphur-oxidizing bacteria and some heterotrophs (*Escherichia coli* and *Pseudomonas aeruginosa*) according to Lane *et al.* (1985). Both rRNA and DNA investigators agree that each host species contains a single species of bacterium.

EVIDENCE FOR CHEMOAUTOTROPHY

Enzymes

Two diagnostic enzymes of the Calvin–Benson cycle of CO_2 fixation are detectable in symbiont-containing tissues, but not in other parts of the same animals. Ribulosebisphosphate carboxylase (EC 4.1.1.39, also referred to as ribulose-1,5-bis-phosphate carboxylase, ribulose-1,5-diphosphate carboxylase, RuBPCase or RubisCo) has been found in most symbiont-containing species investigated, except methane oxidizers (references in Table 2; Cavanaugh, 1985; Felbeck *et al.*, 1981, 1983b; Southward *et al.*, 1981, 1986). Phosphoribulokinase (EC 2.7.1.19, also referred to as ribulose-5-P kinase or Ru-5-P kinase) has been assayed in fewer species, but has been detected in vestimentiferans, small pogonophores and bivalve molluscs. These enzymes can be released from symbiont-containing tissues by grinding, sonication or osmotic shock and since they cannot be found in tissues of the animals it is assumed they are located in the symbiotic bacteria. Enzyme activity has been reported relative to fresh weight of tissue assayed, which may be the whole animal or part of the animal. Since the density of bacteria varies with species of host and with condition, comparison of rates of activity in different species is not necessarily very useful. However, it has been estimated that for *Solemya velum* the ribulosebisphosphate carboxylase activity per milligram of bacterial protein is equivalent to that measured in *Thiobacillus neapolitanus*, a free-living obligate chemoautotrophic sulphur-oxidizing bacterium (Cavanaugh, 1983).

Enzymes involved in sulphur oxidation in free-living bacteria (reviews: Kelly, 1982; Trüper and Fischer, 1982) have been found in symbiont-containing tissues (e.g. Felbeck *et al.*, 1981; Dando *et al.*, 1985). Two of these are restricted to bacteria and seem to be only in the symbionts: adenylylsulphate reductase (EC 1.8.99.2, also referred to as APS reductase) and sulphate adenylyltransferase (ADP) (EC 2.7.7.5). Three others are known to occur in animal tissues as well as in bacteria: sulphate adenylyltransferase (EC 2.7.7.4, also known as ATP sulphurylase), thiosulphate sulphur transferase (EC 2.8.1.1, also known as rhodanese) and sulphide oxidase. Sulphur-oxidizing enzymes have been found in vestimentiferans, small pogonophores, various bivalves and gutless oligochaetes (Table 3), with the

Species	Adenylylsulphate reductase	Sulphate adenylyltransferase	Sulphate adenylyltransferase (ADP)	Rhodanese	Sulphide oxidase
Bivalvia					
Solemya reidi	5	5	—	5	9
Bathymodiolus thermophilus	—	—	—	—	10
Lucina floridana	8	8	—	8	—
L. tenuisculpta	—	5	—	5	—
Lucinoma annulata	—	5	—	—	—
L. borealis	3	2	2	—	—
Myrtea spinifera	2	4	4	—	—
Thyasira flexuosa	4	4	4	—	—
T. sarsi	4	—	—	—	—
Pogonophora					
Riftia pachyptila	5	5,7	—	—	10
Lamellibrachia barhami	—	5	11	—	—
Siboglinum fiordicum	11	11	—	—	—
S. ekmani	11	—	—	—	—
S. atlanticum	11	—	—	—	—
Oligobrachia gracilis	11	—	—	—	—
Sclerolinum brattstromi	11	—	—	—	—
Annelida					
Phallodrilus leukodermatus	—	6	—	—	6
P. planus	—	6	—	—	—

References: 1, Arp *et al.*, 1984; 2, Dando *et al.*, 1985; 3, Dando *et al.*, 1986a; 4, Dando and Southward, 1986; 5, Felbeck *et al.*, 1981; 6, Felbeck *et al.*, 1983b; 7, Fisher and Childress, 1984; 8, Fisher and Hand, 1984; 9, Powell and Somero, 1985; 10, Powell and Somero, 1986a; 11, Southward *et al.*, 1986.

Table 3   Records of the occurrence of sulphur-oxidation enzymes in symbiont-containing species

highest activity in symbiont-containing tissues, but some activity in other parts of the body. For example, the foot of *Myrtea spinifera* shows some sulphate adenylyltransferase activity (Dando *et al.*, 1985), and Powell and Somero (1985, 1986b) have evidence of sulphide-oxidizing activity in the mitochondria of gill and foot tissue in *Solemya reidi*. Body wall muscles of *Riftia* and foot and mantle tissues of *Calyptogena magnifica* and *Bathymodiolus thermophilus* also have sulphide-oxidizing activity (Powell and Somero, 1986a).

The enzymes listed above are effective in oxidizing sulphide and generating ATP. Other evidence for ATP production during sulphide oxidation is provided by the presence in *Lucinoma borealis* gill of a form of cytochrome *c* similar to the cytochrome *c*-551 which is involved in electron transport in some other sulphur-oxidizing bacteria (Dando *et al.*, 1986a; Lu and Kelly, 1984). ATP generation, stimulated by sulphide and sulphite, has been observed in gill extracts of *L. borealis*, while thiosulphate had no effect (Dando *et al.*, 1986a). Examination of three vent species for the same reaction (Powell and Somero, 1986a) showed that intact bacteria from *Bathymodiolus* gill are stimulated by thiosulphate only; lysed bacteria from *Calyptogena* gill are stimulated by sulphite only, while lysed bacteria from *Riftia* trophosome are stimulated by sulphide and sulphite. This suggests that the species deliver reduced sulphur to their symbionts in different forms.

Carbon dioxide fixation

The accumulation and metabolism of ^{14}C from labelled bicarbonate have been studied in several symbiont-containing animals (Table 4). In the gill of *Soleyma velum* CO_2 fixation is stimulated by sulphide and thiosulphate (Cavanaugh, 1983). Sulphide is known to enhance CO_2 fixation in *Siboglinum fiordicum* and *Myrtea spinifera* (Dando *et al.*, 1985; Southward *et al.*, 1986). In the bivalve *Lucinoma borealis* the greatest effect of sulphide is on the gills of animals which have been starved of sulphide, and whose gill symbionts have lost most of their stored sulphur (Dando *et al.*, 1986a). Laboratory experiments on vent animals involve keeping them at high pressure, because they do not survive long at 1 atm. Felbeck (1985a) kept some *Riftia* at 110 atm for a week and measured $^{14}CO_2$ fixation by small animals and pieces of trophosome. All assimilated ^{14}C and extracts were used to trace the metabolism of CO_2 (see p. 99). The effect of sulphide was not examined. Experiments with the vent mussel *Bathymodiolus in situ* show the incorporation of $^{14}CO_2$ into the gills (Fiala-Médioni *et al.*, 1986a). Freshly collected *Bathymodiolus* gill and *Riftia* trophosome were used by Belkin *et al.* (1986) to investigate the effects of sulphide and thiosulphate on CO_2 fixation in homogenates. Sulphide enhances fixation in *Riftia* and thiosulphate enhances it in *Bathymodiolus*, as the same compounds stimulate ATP production in these species (p. 98). A 'purified' bacterial fraction was found to have three times the CO_2-fixing activity of the crude homogenate of *Bathymodiolus* gill, indicating that much of the activity was localized in the bacteria.

Invertebrates in general fix CO_2 into acids of the Krebs citric acid cycle (Hammen and Osborne, 1959; Rau *et al.*, 1986) so it is useful to have comparative studies of animals with and without symbionts from the same locality or habitat (Cavanaugh, 1983; Felbeck, 1983; Dando *et al.*, 1985, 1986a). Gills from symbiont-free species fix CO_2 in the absence of sulphide but most fix less in the presence of sulphide. It seems

Species and tissue	Temperature (°C)	Fixation rate (μmol CO_2 g^{-1} wet weight h^{-1})	Sulphur compound[a] and concentration	Reference
Bivalvia				
Solemya velum				
gill	22	4.5 –4.6	S, 0.2 mM	1
		1.6 –9.1	T, 1.0 mM	1
		0.35–0.7	0	1
Solemya reidi				
gill	—	5	0	4
Solemya reidi				
whole	7.5	0.8	S, 40 μM	7
Lucinoma borealis				
normal gill	17	0.56	S, 100 μM	3
		0.46	0	3
starved gill		0.32	S, 100 μM	3
		0.19	0	3
Myrtea spinifera				
gill	12	0.18–0.25	S, 30 μM	2
		0.03–0.06	0	2
Calyptogena magnifica				
gill	27	0.05–0.07	T, 18 mM	9
		0.06–0.08	0	9
Pogonophora				
Riftia pachyptila				
plume (no symbionts)	8	16	0	5
Siboglinum fiordicum				
whole	12	0.36	S, 10 μM	8
		0.25	0	8
Siboglinum atlanticum				
whole	8	0.01–0.05	0	8
Annelida				
Phallodrilus leukodermatus				
whole	—	7.1	0	6

References: 1, Cavanaugh, 1983; 2, Dando *et al.*, 1985; 3, Dando *et al.*, 1986a; 4, Felbeck, 1983; 5, Felbeck, 1985a; 6, Felbeck *et al.*, 1983b; 7, Fisher and Childress, 1986; 8, Southward *et al.*, 1986; 9, Tuttle, 1985.

[a]Sulphur compounds: S = Na_2S; T = $Na_2S_2O_3$.

Table 4   Rate of fixation of CO_2 by symbiont-containing animals or tissues

probable that animal cells in symbiont-containing species also fix some CO_2. Metabolic pathways in the gills of *Solemya reidi* have been studied by Felbeck (1983). Short incubations of the gills with ^{14}C bicarbonate showed that the label first appeared in malate, and within 1 min much of it moved to aspartate. Pulse chase experiments with the whole animal showed the subsequent movement to glutamate, alanine and succinate. Initial fixation of ^{14}C into malate also occurs in *Riftia*, in the tentacular plume (Felbeck, 1985a). The labelled malate is carried in the blood to the trophosome, where the symbionts are, and ^{14}C appears in succinate, aspartate, alanine and glutamate, all within 30 min. Both experiments were done without sulphide and the symbiont contribution to initial fixation may be greater in life. It is clear from the *Riftia* experiment that the initial fixation to malate is done by the animal cells. When trophosome tissue is incubated separately with ^{14}C-malate, labelled succinate, aspartate, glutamate and alanine appear in the tissue, while $^{14}CO_2$ appears in the water (Felbeck, 1985a). The CO_2 released by decarboxylation of malate becomes available to the ribulosebisphosphate fixation pathway in the bacteria. It seems probable that *Solemya* also transports malate to its symbionts. The oligochaete *Phallodrilus leukodermatus* has its symbionts on the outer surface, and they seem to incorporate CO_2 directly, since ^{14}C from bicarbonate appears within 5 min in 3-phosphoglyceric acid, sugars and sugar phosphates as well as in malate (Felbeck *et al.*, 1983b). Initial fixation of CO_2 has not been examined in pogonophores, but ^{14}C from labelled bicarbonate appears after 1 h, or longer, in aspartate and glutamate. The posterior end of the body, which contains the symbionts and storage cells, accumulates more ^{14}C in insoluble form and contains an unidentified soluble compound which is scarcely represented in the anterior end (Southward *et al.*, 1986). Felbeck (1985b) has found equal concentrations of free D-aspartate and L-aspartate in the body gill of *Solemya*. He suggests that D-aspartate may be the carrier for CO_2 or a shuttle compound, and that it can be racemized to the L form by aspartate racemase, which is present in the tissues.

The transfer of carbon from symbiont to host in *Solemya reidi* is probably fairly substantial, to judge from experiments by Fisher and Childress (1986). Using autoradiography and liquid scintillation counting they showed that ^{14}C from labelled bicarbonate is first fixed in the gills, mainly in the bacteriocytes, and later transferred to other parts of the body. Stable carbon isotope ratios in tissues with and without symbionts in various species confirm this (p. 104).

Sulphide oxidation and binding

Increased oxidation of sulphide in seawater containing living *Solemya reidi* has been demonstrated by Felbeck (1983). Isolated gills of *Myrtea spinifera* take up and metabolize ^{35}S-labelled sulphide, depositing some in bound forms in the tissues (Dando *et al.*, 1985). A small proportion of this is elemental sulphur. The symbionts in *M. spinifera* have sulphur storage vesicles (p. 92) and analysis of whole gills has demonstrated up to 0.08% elemental sulphur in the total wet weight of the gill (Dando *et al.*, 1985). The amount of elemental sulphur in the gills of this and other bivalve species varies with the sulphide content of the habitat and on the treatment of the animals after collection. It can be consumed within a few days in well-oxygenated conditions (Dando *et al.*, 1985, 1986a, b; Vetter, 1985).

Riftia blood has a remarkable sulphide binding capacity (Arp *et al.*, 1985). *R. pachyptila* lives in rock crevices where diluted vent fluid emerges with a sulphide content of about 0.2 mM, yet the sulphide content of the blood can be as high as 4 mM. Dialysis experiments have shown that *Riftia* blood can accumulate sulphide from a 0.1 mM solution to reach a level of 3 mM within 12 h. Sulphide is bound by haemoglobin (Arp *et al.*, 1987) and is carried to the trophosome in this form, protecting the animal from sulphide poisoning and preventing the chemical oxidation of sulphide in transit (Arp *et al.*, 1985). Alberic (1986) has evidence of the involvement of taurine in the transport of sulphide in *Riftia*. The symbiotic bacteria on *Riftia* trophosome are able to remove sulphide from the blood (Fisher and Childress, 1984). *Riftia* haemoglobin is dissolved in the blood coelomic fluid and has a high affinity for oxygen (Terwilliger and Terwilliger, 1985; Wittenberg, 1985). The vent clam *Calyptogena magnifica* also has haemoglobin in its blood, but inside erythrocytes, and sulphide binding in the blood of this species takes place in the serum. The sulphide-binding component has a molecular weight greater than 10 000 but is apparently not a protein (Arp *et al.*, 1984). Blood may transport sulphide from the foot to the gill in this species. Sulphide oxidation has been noted in animal cells as well as symbionts and may be located in mitochondria in *Solemya* (Powell and Somero, 1985, 1986a, b).

Sulphur isotopic compositions of vent animals show that *Riftia* and *Calyptogena* must obtain their tissue sulphur from hydrothermal sulphide rather than seawater sulphate (Fry *et al.*, 1983), which fits well with the biochemical and metabolic observations.

EVIDENCE FOR METHANOTROPHY

The pogonophore *Siboglinum poseidoni* and two unnamed mussels contain symbionts whose fine structure is comparable with that of methane-oxidizing bacteria (p. 92). In *S. poseidoni*, which is found in the Skagerak in places where methane seeps are suspected, Schmaljohann and Flügel (1987) report that ^{14}C from labelled methane is incorporated in the tissues; they have isolated methanotrophic bacteria from the animal and its environment.

Mussels, collected alive from a hydrocarbon seep region in the Gulf of Mexico, consumed methane and oxygen rapidly under experimental conditions, and produced CO_2 (Childress *et al.*, 1986). Isolated gills of these mussels consumed methane in the presence of oxygen and this activity was inhibited by acetylene. The foot and mantle did not consume methane so the gill symbionts must be responsible. Animals known to possess sulphur-oxidizing symbionts did not consume methane under the same conditions.

Mussels from a much deeper site in the Gulf of Mexico, in a hypersaline seep region, were assayed for enzymes involved in methylotrophic metabolism (Cavanaugh *et al.*, 1987). Homogenates of frozen material were used. Methanol dehydrogenase was detected in gills but not in foot or mantle samples; NAD-linked formate dehydrogenase activity and NAD- and glutathione-linked formaldehyde dehydrogenase activities were present but variable. Ribulosebisphosphate carboxylase, the key enzyme of autotrophic carbon assimilation (p. 96) was absent, while a high activity of hexulosephosphate synthase in the gills, but not the foot or mantle, indicates that the symbionts employ the ribulose monophosphate pathway of carbon assimilation, which is typical of type I methanotrophic bacteria.

There is evidence that the trophosome bacteria in *Riftia* may be able to use methane or methyl compounds, though they normally use sulphide (Fisher and Childress, 1984).

NITROGEN ASSIMILATION

In gutless host animals and those with reduced guts nitrogen will have to be obtained from the surrounding seawater or sediment pore water. Pogonophores can take up amino acids from low environmental concentrations (Southward and Southward, 1980; Southward *et al.*, 1979). *Solemya reidi* can take up some amino acids via the gills (Felbeck, 1983) as can *Bathymodiolus* (Fiala-Médioni *et al.*, 1986a), and non-symbiotic bivalves normally have an efficient uptake mechanism for amino acids via the gills (p. 103). Bacteria usually assimilate nitrogen as ammonia, so the symbionts should be able to recycle ammonia produced as waste by the host animals. Sediment-living animals have access to abundant ammonia in pore water (e.g. Dando *et al.*, 1986b) but transport systems have not been investigated. Assimilation of nitrate is possible, as shown by the activity of nitrate reductase in *Riftia*, bivalves and *Phallodrilus* (Berg and Alatalo, 1984; Felbeck *et al.*, 1981, 1983) and of nitrate reductase (EC 1.7.7.1) in *Lucina floridana* (Fisher and Hand, 1984) but these enzymes might alternatively be used in obtaining oxygen for the symbionts. Fixation of N_2 has not been demonstrated; the acetylene reduction method gave a negative result for the trophosome of *Riftia* (Fisher and Childress, 1984).

Nitrogen isotope ratios in vent animals suggest the use of a local source of inorganic nitrogen rather than sedimentary organic nitrogen (Rau, 1985).

QUANTIFYING THE CONTRIBUTION OF THE SYMBIONTS

The large numbers of bacteria in the animals under discussion do not seem to harm the tissues; on the contrary, the tissues holding the bacteria are enlarged compared with those of non-symbiotic animals, they have a well-developed blood circulation and often contain specialized storage cells. In spite of the burden of bacteria the animals have a reduced ability to feed normally, yet many of them grow large. This is particularly noticeable at hydrothermal vents, where the clam *Calyptogena magnifica* can reach 26 cm in length (Boss and Turner, 1980) and the mussel *Bathymodiolus thermophilus* can reach 18 cm (Kenk and Wilson, 1985). The vestimentiferan *Riftia* can grow to over 1 m long and more than 3 cm wide (Jones, 1985). The coastal symbiont-containing bivalve *Codakia orbicularis* grows as fast as *Mercenaria mercenaria* and can reach a length of 7 cm (Berg and Alatalo, 1984; Berg *et al.*, 1985).

There is evidence from fine structure that symbiotic bacteria are digested by host bacteriocytes in phagocytic vacuoles, particularly in thyasirid bivalves and gutless oligochaetes (Giere and Langheld, 1987; Southward, 1986) and to a lesser extent in pogonophores, including *Riftia* (Bosch and Grassé, 1984b; Southward, 1982). Bivalves with intracellular symbionts seem to rely more on 'milking' their symbionts (Southward, 1986).

Sulphide tolerance and requirement for different reduced sulphur compounds varies most obviously among the hydrothermal vent animals (p. 100) and these differences match their habitat preferences (Fustec *et al.*, 1987; Hessler *et al.*, 1985). *Riftia* lives in vent fluids which contain sulphide, *Calyptogena* favours crevices which have sulphide in their depths and *Bathymodiolus* lives in the coolest conditions, often without measureable sulphide. Smith (1985) compared the nutritional status of populations of *Bathymodiolus* at different distances from an effluent source. Those close to the source had higher tissue weight and were breeding when those farther away were lighter and not breeding. Twice as much particulate organic matter was available close to the source and sulphide was not detectable. Smith concluded that particulate food was more important than production by symbionts to this mussel. which has a functional gut (Kenk and Wilson, 1985; Le Pennec *et al.* 1985). However, since CO_2 fixation in *Bathymodiolus* gill is stimulated by thiosulphate, not by sulphide (Belkin *et al.*, 1986), chemosynthesis may be more important than he suspected (see also Johnson *et al.*, 1986).

Estimation of the value of symbionts' contribution to the nourishment of *Bathymodiolus* would require the measurement of the intake of particulate food, the rate of uptake of dissolved organic compounds (DOC) and the rate of fixation of CO_2. Since the gutless animals have abandoned particulate food they are easier examples with which to begin. Pogonophores can accumulate DOC such as amino acids, sugars and fatty acids through their tube wall into the epidermis, and transport and metabolize these compounds (e.g. Little and Gupta, 1968, 1969; Southward and Southward, 1968, 1981). Only one species, *Siboglinum ekmani*, is likely to be able to obtain enough energy in this way to sustain its oxidative requirement. *S. ekmani* is extremely thin, so has a high surface to volume ratio. Other small pogonophores may obtain, at the most, 50% of their metabolic needs from DOC, while large species are likely to obtain even less. The giant vestimentiferans, notably *Riftia*, are rather stout animals and their numerous tentacles are unlikely to contribute very much nourishment in the form of DOC. It would be useful to have measurements of uptake rates and environmental concentrations of dissolved amino acids etc. for these large forms.

Gutless oligochaetes are long and slender, like the smallest pogonophores, but the thick layer of external symbionts may block or intercept DOC diffusion to the epidermal cell membrane beneath. The uptake of glucose and other sugars has been demonstrated in *Phallodrilus leukodermatus* (Liebezeit *et al.*, 1983) and glucose can be metabolized (Felbeck *et al.*, 1983b), but information on uptake rates is lacking. Dissolved amino acids are found in high concentrations in the habitat of *P. leukodermatus* (Giere *et al.*, 1982); if this species has an uptake rate comparable with that of the non-symbiotic oligochaete *Enchytraeus albidus*, which can obtain nearly 40% of its oxidative requirement from $50\,\mu M$ amino acids (Siebers and Bulnheim, 1977), it may be able to live mainly by epidermal uptake of DOC, while its symbionts enable it to tolerate sulphide in the sediment.

Solemyid bivalves, which have either no gut or a very reduced one, have been suspected in the past of either digesting particles in the mantle cavity or absorbing DOC from the water passing through their gills (Reid and Bernard, 1980). There does not seem to be any support for the theory of external digestion. Felbeck (1983) has shown that amino acids can be taken up by *S. reidi*, but gives no data on uptake rates. Many bivalves without symbionts are known to be able to accumulate amino acids from low external concentrations, particularly via the gills. Wright (1982) has demonstrated that dissolved free amino acids can provide a potentially useful nutritional supplement for the mussels *Mytilus californianus* and *Modiolus demissus*, which can obtain 40% and 58% respectively of their calculated oxidative requirements from an external concentration of 2 µM glucine. More information is needed on rates of uptake by *Solemya* spp. and on concentrations of DOC within the burrows. Symbiont-containing animals, such as solemyids and lucinids, which live in organic-rich sediments, may have access to DOC concentrations which could supply a useful source of nutrition (e.g. Jørgensen *et al.*, 1981). However, inorganic carbon fixed in the gills is definitely transferred to other parts of the body (Fisher and Childress, 1986).

In addition to trying to quantify the input rates from the various potential energy sources it is useful to measure stable carbon isotope ratios in the animals and the carbon sources. There are two naturally occurring stable isotopes of carbon: about 1.1% of the heavier isotope ^{13}C is mixed with the predominant ^{12}C in atmospheric carbon dioxide. During the enzymatic fixation of CO_2 by plants or bacteria there is discrimination against ^{13}C and the organic matter formed is depleted in ^{13}C, the amount of depletion relative to the original CO_2 depending on the type of organism. Animal tissues reflect the average ^{13}C content of their diet, with very slight enrichment in ^{13}C, so measurement of $^{13}C:^{12}C$ ratios is widely used in food source studies (e.g. Fry and Sherr, 1984; Fry *et al.*, 1984; Jackson *et al.*, 1986; Rau *et al.*, 1983). Values are expressed as $\delta[^{13}C]$, calculated relative to an international limestone standard:

$$\delta[^{13}C] = \left(\frac{[^{13}C]/[^{12}C] \text{ sample}}{[^{13}C]/[^{12}C] \text{ standard}} - 1 \right) \times 1000$$

Values of $\delta[^{13}C]$ for biological materials are usually negative, but become less so during progress through a food chain of several trophic levels (Fry *et al.*, 1984). Normal deep-sea animals have $\delta[^{13}C]$ values similar to those of sedimenting organic carbon (Table 5) but hydrothermal vent animals have markedly different $\delta[^{13}C]$ values (see Table 8) and it is clear that locally produced food sources are involved.

Interpretation of data on symbiont-containing species requires information on $\delta[^{13}C]$ values for potential carbon sources, only some of which is available at present (see Tables 5–9). Chemoautotrophic bacteria may fractionate carbon isotopes to a greater degree than plants (Estep *et al.*, 1978; Fry and Sherr, 1984; Rau, 1981b, 1985; Rau and Hedges, 1979; Spies and DesMarais, 1983). If they use respired CO_2 which has a $\delta[^{13}C]$ value of about −18‰ their organic carbon could have a value as low as −40‰. Methanotrophic bacteria using biogenic methane, which is already greatly depleted (Table 5), can achieve extremely negative $\delta[^{13}C]$ values. Organic matter which is unusually depleted in ^{13}C could in turn be a source of similarly depleted CO_2 and dissolved organic carbon, which would complicate interpretation of animals' values still further, for example at the Florida Escarpment deep seep area (see Table 9).

In the simplest situation, coastal bivalves which feed only on organic particles (e.g. phytoplankton) have $\delta[^{13}C]$ values of about −17 to −19‰, while symbiont-containing gills have values of −24 to −31‰ and the bodies of the same symbiont-containing species have values 1 to 4‰ less depleted than their gills, indicating a mixture of chemosynthetic and photosynthetic food sources (Table 6). *Codakia orbicularis* is unusual in having very similar values in gill and body. It has

Source	δ[^{13}C] (‰)	Reference
Atmospheric CO_2	−7	2
Dissolved inorganic carbon in seawater:		
Surface	+2	5
Deep ocean	0	5
Sediment carbonates	+5 to −5	2
Dissolved inorganic carbon in pore water of sediments	+18 to −20	2
Marine plankton	−17 to −22	4
Autotrophic bacteria	−20 to −37	3
Benthic animals:		
Filter feeders	−17 to −21	3
Carnivores and scavengers	−15 to −18	3
Methane:		
Natural gas	−21 to −76	2
Biogenic in shallow water environments	−51 to −75	1
Methanotrophic bacteria	10–20‰ more negative than the methane consumed	3

References: 1, Burke and Sackett, 1986; 2, Deines, 1980; 3, Fry and Sherr, 1984; 4, Galimov, 1985; 5, Kroopnick, 1985.

Table 5 ^{13}C:^{12}C ratios in the marine environment

been suggested that, where the body δ[^{13}C] ranges from −23.4 to −28.2‰, half or more of the body carbon could be derived from bacterial autotrophy (Spiro *et al.*, 1986).

The shallow-water pogonophore *Siboglinum fiordicum* has a δ[^{13}C] value of −35‰ (Table 7), about 10‰ more negative than the bodies of the bivalves *Myrtea spinifera* and *Lucinoma borealis*, which share its habitat. The pogonophore has no gut and must depend more on autotrophically fixed carbon than the bivalves, in spite of its capacity for uptake of DOC. Pogonophores from greater depths have δ[^{13}C] values of about −45‰, which suggests almost total dependence on production by the symbionts, again in spite of ability to take up DOC (Southward *et al.*, 1986; Southward and Southward, 1987).

Close to hydrothermal vents there are various possible carbon sources with different δ[^{13}C] values (Table 8), but no values are available for suspended bacteria in the effluent. If the shell carbonate δ[^{13}C] values of +2 and +3‰ for *Calyptogena* and *Bathymodiolus* reflect approximately the value for the CO_2 used by the symbionts (the presence of symbionts may have a small effect on the δ[^{13}C] of shell carbonate (Jones *et al.*, 1986)), their tissue values of −32 to −34‰ are relatively a little more depleted than those of coastal bivalves. Without knowledge of the δ[^{13}C] values for the autotrophic bacteria growing in the vent fluid and used by the bivalves as their particulate food there can be no calculation of the relative importance of the various carbon sources.

The δ[^{13}C] for *Riftia* of about −11‰, differing as it does from the values for the bivalves living nearby, and showing much less depletion than would be expected

Location and species	Symbionts present	Depth (m)	$\delta[^{13}C]$ (‰)		Reference
			Gill	Rest of body	
England					
Lucinoma borealis	+	0	−28.1	−25.9	3
			−29.0	−25.3	3
Dosinia lupinus	−	0	−17.6	−16.8	3
			−17.2	−16.9	3
Norway					
Lucinoma borealis	+	33	−28.8	−24.1	3
Myrtea spinifera	+	33	−24.2	−23.4	3
Arctica islandica	−	33	−18.8	−18.9	3
Thyasira flexuosa	+	55	−29.3	n.d.	3
Thyasira sarsi	+	90	−31.0	−28.2	3
Bahamas					
Codakia orbicularis	+	0	−23.9	−23.2 to −23.8	1
			−28.3	−28.1	1
			whole animal		1
Lucina nassula	+	0	−23.0		1
California					
Solemya reidi	+	100	−30.0		2

References: 1, Berg and Alatalo, 1984; 2, Felbeck, 1983; 3, Spiro *et al.*, 1986.

Table 6 $^{13}C:^{12}C$ ratios in coastal bivalves

for an animal apparently entirely dependent on internal autotrophic bacteria, has been explained by Rau (1981b, 1985) as probably due to CO_2 limitation. The bacteria are crowded into an internal tissue mass where the supply of CO_2 in the blood may be only just sufficient and at the same time it will be difficult for carbon dioxide enriched in ^{13}C to diffuse away from the bacteria; therefore, there will be a smaller degree of fractionation than there would be if there was free diffusion of CO_2 (Deines, 1980). The single measurement of $\delta[^{13}C]$ for the vent polychaete *Alvinella* is also about −11‰. This worm builds masses of tubes on chimneys carrying effluent and apparently feeds on bacteria growing on and around its tubes (Baross and Deming, 1985). Whether CO_2 is limited in this environment is uncertain. Obviously more data are needed (see Rau, 1985). It is possible that the low pH of fluids leaving the hydrothermal system (e.g. Edmond and Von Damm, 1985) affects the CO_2–HCO_3^- balance in a way that influences the isotope fractionation by the bacteria growing in it (cf. Degens *et al.*, 1968), so that they may be less depleted in ^{13}C than expected.

At 'cool seeps' along continental margins some extremely negative $\delta[^{13}C]$ values have been found in the vent-type faunas. The values for the mytilids from two sites (Table 9) are apparently the result of methanotrophy by the symbionts in these mussels (p. 101). Both methane and sulphide are probably present in these environments and it is likely that the animals with the less negative $\delta[^{13}C]$ values will be found to contain sulphur-oxidizing symbionts.

Location and species	Depth (m)		δ[^{13}C] (‰)	Reference
Norwigian fjords				
Siboglinum fiordicum	33	Whole animal	−35.5	2
Siboglinum ekmani	700	Whole animal	−45.3	2
European continental shelf				
North Biscay	1850			
Siboglinum atlanticum		Whole animal	−45.3	1
		Tube	−43.3, −42.1	1
Organic carbon in sediment			−20.4, −20.5	1
Sediment-feeding polychaetes			−18.3, −18.5	1
Carnivorous polychaete			−16.1	1
South Biscay	1700			
Siboglinum atlanticum		Postannular region	−43.8	2
		Anterior region	−44.7	2
		Tubes	−43.3, −42.1	2
Sediment carbonate			+0.9 ± 0.1	2
Pore-water:				
Bicarbonate			−17.8 ± 3.3	2
Free amino acids			−18.1 ± 2.3	2

References: 1, Southward *et al.*, 1981; Spiro *et al.*, 1986.

Table 7 ^{13}C:^{12}C ratios in small Pogonophora and their habitat

AVAILABILITY OF ENERGY SOURCES

Sulphide concentrations as high as 250 μM H$_2$S occur in hydrothermal vent effluents at temperatures suitable for *Riftia* and the two vent bivalves (Grassle, 1986; Johnson *et al.*, 1986; Lilley *et al.*, 1983). Each species seems to prefer a different concentration and all die after the effluent ceases to flow (Hessler *et al.*, 1985). Intertidal and shallow-water habitats with reported 'high-sulphide' levels contain more than 25 μM (references in Dando *et al.*, 1986b), but there are other interesting habitats which have little or no dissolved free sulphide (< 1 μM) but are able to support flourishing sulphur-oxidizing symbiont associations (Dando *et al.*, 1985, 1986a, b; Southward *et al.*, 1986; Spiro *et al.*, 1986).

Low levels of methane are present in hydrothermal vent fluids but do not appear to be used by symbionts, possibly because of the lack of systems for concentration and transport of methane (Arp *et al.*, 1985). Methanotrophic symbionts have only been found at sites where methane concentrations are suspected of being much higher, such as hydrocarbon seep area in the Gulf of Mexico (p. 101; Table 9). Analysis of blood gas in vestimentiferans from this site showed up to 142 μM methane which

Species	Tissue	δ[^{13}C] (‰)	Reference
With chemoautotrophic symbionts			
Riftia pachyptila	Trophosome	−10.9, −11.1	4
	Vestimentum	−10.9, −11	4
	'Tissue'	−10.9	10
Bathymodiolus thermophilus	Foot and mantle	−32.7 to −33.6	6
	'Tissue'	−32.7 to −33.9	10
Calyptogena magnifica	Mantle	−32.0 to −32.7	4
External symbionts only			
Alvinella pompejana	Trunk	−11.2	2
Predator or scavenger			
Bythograea thermydron	Muscle	−13.7 to −17.6	5
Hydrothermal fluid:			
Carbon dioxide		−5.1, −7	1, 9
Methane		−15 to −17.6	9
Carbonate in skeletons of vent animals:			
Bathymodiolus shell		+2.8	10
Calyptogena shell		+3	7
Calyptogena shell		+5.3 to −1.5	3
Barnacle shell		−1.0	3
Serpulid tube		−1.64	8
Alvinocaris exoskeleton		−4 to −11.5	8
Munidopsis exoskeleton		−3 to −4	8

References: 1, Craig *et al.*, 1980; 2, Desbruyères *et al.*, 1983; 3, Killingley *et al.*, 1980; 4, Rau, 1981b; 5, Rau, 1985; 6, Rau and Hedges, 1979; 7, Turekian *et al.*, 1979; 8, Van Dover, 1986; 9, Welhan and Craig, 1983; 10, Williams *et al.*, 1981.

Table 8 ^{13}C:^{12}C ratios in hydrothermal vent animals and their habitat (Galapagos Spreading Centre and East Pacific Rise; depth about 2500 m)

may reflect the environmental concentration. The vestimentiferans probably have sulphur-oxidizing symbionts, but mussels from the same site have methanotrophic symbionts and have been shown to be able to consume methane from concentrations of 20–200 μM.

Reduced inorganic compounds other than sulphur may be used by some deep-water pogonophores and thyasirid bivalves, which live in sediments poor in sulphide and methane but richer in ammonia, which could support nitrifying symbionts (Southward *et al.*, 1986).

CONCLUSIONS

The evidence now available clearly supports the hypothesis that several groups of animals can gain a substantial part of their nutrition from internal chemoautotrophic bacteria, between 50% and 100% depending on species and circumstances.

Interesting lines for future research include identification of compounds exchanged between symbionts and hosts and, allied to this, study of mechanisms of

Species	Tissue	$\delta[^{13}C]$ ‰	Reference
Continental slope of Gulf of Mexico:			
Hydrocarbon seep at 600 m depth			
Mytilid bivalve	Gill	−51.6 to −52.1	1
	Mantle	−51.3 to −52.3	1
Calyptogena ponderosa		−35.4 to −35.3	2
Lucinoma atlantis		−31.2 to −33.0	2
Gastropod		−31.5	2
Vestimentifera	'Flesh'	−27	2
	Tube	−28.1	2
Fish (various)		−17.2 to −17.9	2
Oil		−26	2
Methane		−45	2
Florida Escarpment:			
Saline seep at 3266 m depth			
Mytilid bivalve	'Tissue'	−74	4
Gastropod	'Tissue'	−59	4
Vestimentifera	'Tissue'	−42.7	4
Organic carbon in sediment		−80	4
Carbonate in escarpment lime-stone		+3.2	4
Mytilid shell carbonate		−4.8	4
Gastropod shell carbonate		−3.1	4
Oregon subduction zone:			
Cool vents at 2036 m depth			
Calyptogena sp.	Gill	−51.6	3
	Periostracum	−35.7	3
Solemya sp.	Periostracum	−31.0	3
Vestimentiferan	'Tissue'	−31.9	3
	Tube	−26.7	3
Calyptogena shell carbonate		−0.1	3
Solemya shell carbonate		+1.8	3
Methane-derived carbonate crusts etc.		−34.9 to −66.7	3

References: 1, Childress *et al.*, 1986; 2, Kennicutt *et al.*, 1985; 3, Kulm *et al.*, 1986; 4, Paull *et al.*, 1985.

Table 9 $^{13}C:^{12}C$ ratios at cool seeps and cool vents

control of symbiont multiplication by host cells. More $\delta[^{13}C]$ values are needed for carbon sources in the environment, especially at hydrothermal vents and cool seeps, which are so much more difficult to sample than coastal habitats.

REFERENCES

Alayse-Danet, A.M., Gaill, F. and Desbruyères, D. 1985. Preliminary studies on the relationship between the Pompeii worm, *Alvinella pompejana* (Polychaeta: Ampharetidae), and its epibiotic bacteria. In *Proceedings of the 19th European marine biological symposium* (ed. P.E. Gibbs), pp. 167–172. Cambridge University Press, Cambridge.

Alayse-Denet, A.M., Gaill, F. and Desbruyères, D. 1986. *In situ* bicarbonate uptake by bacteria–*Alvinella* associations. *Marine Ecology, Naples*, **7**: 233–240.

Alberic, P. 1986. Présence de quantités importantes de thiotaurine et d'hypotaurine dans les tissus de *Riftia pachyptila* (Pogonophore, Vestimentifère). *Compte rendu hebdomadaire des séances de l'Académie des Sciences (ser. III)*, **302**: 503–508.

Allen, M.M. 1984. Cyanobacterial cell inclusions. *Annual Review of Microbiology*, **38**: 1–25.

Anthony, C. 1982. *The biochemistry of methylotrophs*. Academic Press, London.

Arp, A.J., Childress, J.J. and Fisher, C.R. 1984. Metabolic and blood gas transport characteristics of the hydrothermal vent bivalve *Calyptogena magnifica*. *Physiological Zoology*, **57**: 648–662.

Arp, A.J., Childress, J.J. and Fisher, C.R. 1985. Blood gas transport in *Riftia pachyptila*. *Bulletin of the Biological Society of Washington*, **6**: 289–300.

Arp. A.J., Childress, J.J. and Vetter, R.D. 1987. The sulphide-binding protein in the blood of the vestimentiferan tube-worm, *Riftia pachyptila*, is the extracellular haemoglobin. *Journal of Experimental Biology*, **128**: 139–158.

Baross, J.A. and Deming, J.W. 1985. The role of bacteria in the ecology of black-smoker environments. *Bulletin of the Biological Society of Washington*, **6**: 335–371.

Belkin, S., Nelson, D.C. and Jannasch, H.W. 1986. Symbiotic assimilation of CO_2 in two hydrothermal vent animals, the mussel *Bathymodiolus thermophilus* and the tube worm *Riftia pachyptila*. *Biological Bulletin, Marine Biological Laboratory*, Woods Hole, MA., **170**: 110–121.

Berg, C.J. and Alatalo, P. 1984. Potential of chemosynthesis in molluscan mariculture. *Aquaculture*, **39**: 165–179.

Berg, C.J., Krzynowek, J., Alatalo, P. and Wiggin, K. 1985. Sterol and fatty acid composition of the clam *Codakia orbicularis*, with chemoautotrophic symbionts. *Lipids*, **20**: 116–120.

Bosch, G. and Grassé, P.-P. 1984a. Cycle partiel des bactéries chimioautotrophiques symbiotiques et leurs rapports avec les bactériocytes chez *Riftia pachyptila* Jones (Pogonophore Vestimentifère). I. Le trophosome et les bactériocytes. *Compte rendu hebdomadaire des seances de l'Académie des Sciences* (ser. III), **299**: 371–376.

Bosch, C. and Grassé, P.-P. 1984b. Cycle partiel des bactéries chimioautotrophiques symbiotiques et leurs rapports avec les bactériocytes chez *Riftia pachyptila* Jones (Pogonophore Vestimentifère). II. L'évolution des bactéries symbiotiques et des bactériocytes. *Compte rendu hebdomadaire des seances de l'Academie des Sciences (ser. III)*, **299**: 413–419.

Boss, K.J. and Turner, R.D. 1980. The giant white clam from the Galapagos Rift, *Calyptogena magnifica* species novum. *Malacologia*, **20**: 185–190.

de Burgh, M.E. and Singla, C.L. 1985. Bacterial colonization and endocytosis on the gill of a new limpet species from a hydrothermal vent. *Marine Biology*, **84**: 1–6.

Burke, R.A. and Sackett, W.M. 1986. Stable hydrogen and carbon isotopic compositions of biogenic methanes from several shallow aquatic environments. In *Organic marine geochemistry* (ed. M.L. Sohn), pp. 297–313. ACS Symposium Series 305.

Cavanaugh, C.M. 1983. Symbiotic chemoautotrophic bacteria in marine invertebrates from sulphide-rich habitats. *Nature, London*, **302**: 58–61.

Cavanaugh, C.M. 1985. Symbioses of chemoautotrophic bacteria and marine invertebrates from hydrothermal vents and reducing sediments. *Bulletin of the Biological Society of Washington*, **6**: 373–388.

Cavanaugh, C.M., Gardiner, S.L., Jones, M.L., Jannasch, H.W. and Waterbury, J.B. 1981. Prokaryotic cells in the hydrothermal vent tube worm *Riftia pachyptila* Jones: possible chemoautotrophic symbionts. *Science*, **213**: 340–342.

Cavanaugh, C.M., Levering, P.R., Maki, J.S., Mitchell, R. and Lidstrom, M.E. 1987. Symbiosis of methylotrophic bacteria and deep-sea mussels. *Nature, London*, **325**: 346–348.

Childress, J.J., Fisher, C.R., Brooks, J.M., Kennicutt, M.C., Bidigare, R. and Anderson, A.E. 1986. A methylotrophic marine molluscan (Bivalvia, Mytilidae) symbiosis: mussels fuelled by gas. *Science*, **233**: 1306–1308.

Cohen-Bazaire, G. and Kunisawa, R. 1963. The fine structure of *Rhodospirillum rubrum*. *Journal of Cell Biology*, **16**: 401–419.

Craig, H., Welhan, J.A., Kim, K., Poreda, R. and Lupton, J.E. 1980. Geochemical studies of the 21°N EPR hydrothermal fields. *Transactions of the American Geophysical Union (EOS)*, **61**: 992 (abstract).

Dando, P.R. and Southward, A.J. 1986. Chemoautotrophy in bivalve molluscs of the genus *Thyasira*. *Journal of the Marine Biological Association of the United Kingdom*, **66**: 915–929.

Dando, P.R., Southward, A.J., Southward, E.C., Terwilliger, N.B. and Terwilliger, R.C. 1985. Sulphur oxidising bacteria and haemoglobin in gills of the bivalve mollusc *Myrtea spinifera*. *Marine Ecology — Progress Series*, **23**: 85–98.

Dando, P.R., Southward, A.J. and Southward, E.C. 1986a. Chemoautotrophic symbionts in the gills of the bivalve mollusc *Lucinoma borealis* and the sediment chemistry of its habitat. *Proceedings of the Royal Society B*, **227**: 227–247.

Dando, P.R., Southward, A.J., Southward, E.C. and Barrett, R.L. 1986b. Possible energy sources for chemoautotrophic prokaryotes symbiotic with invertebrates from a Norwegian fjord. *Ophelia*, **26**: 135–150.

Dasch, G.A., Weiss, E. and Chang, W.-P. 1984. Endosymbionts of insects. In *Bergey's manual of systematic bacteriology*, Vol. 1 (eds. N.R. Krieg and J.G. Holt), pp. 811–833. Williams and Wilkins, Baltimore.

Degens, E.T., Guillard, R.R.L., Sackett, W.M. and Hellebust, J.A. 1968. Metabolic fractionation of carbon isotopes in marine plankton — 1. Temperature and respiration experiments. *Deep-Sea Research*, **15**: 1–9.

Deines, P. 1980. The isotopic composition of reduced organic carbon. In *Handbook of environmental isotope chemistry*, Vol. 1. *The terrestrial environment* (eds. P. Fritz and J.C. Fontes), pp. 328–406. Elsevier, Amsterdam.

Desbruyères, D. and Laubier, L. 1986. Les Alvinellidae, une famille nouvelle d'annélide polychètes infeodée au sources hydrothermales sous-marines: systematique, biologie et ecologie. *Canadian Journal of Zoology*, **64**: 2227–2245.

Desbruyères, D., Gaill, F., Laubier, L., Prieur, D. and Rau, G.H. 1983. Unusual nutrition of the 'Pompeii worm' *Alvinella pompejana* (polychaetous annelid) from a hydrothermal vent environment: SEM, TEM, ^{13}C and ^{15}N evidence. *Marine Biology*, **75**: 201–205.

Desbruyères, D., Gaill, F., Laubier, L. and Fouquet, Y. 1985. Polychaetous annelids form hydrothermal vent ecosystems: an ecological overview. *Bulletin of the Biological Society of Washington*, **6**: 103–116.

Doeller, J.E. 1984. A hypothesis for the metabolic behaviour of *Solemya velum*, a gutless bivalve. *American Zoologist*, **24**: 57A (abstract).

Dubilier, N. 1986. Association of filamentous epibacteria with *Tubificoides benedii* (Oligochaeta: Annelida). *Marine Biology*, **92**: 285–288.

Edmond, J.M. and Von Damm, K.L. 1985. Chemistry of ridge crest hot springs. *Bulletin of the Biological Society of Washington*, **6**: 43–47.

Erséus, C. 1984. Taxonomy and phylogeny of the gutless Phallodrilinae (Oligochaeta, Tubificidae), with descriptions of one new genus and twenty-two new species. *Zoologica Scripta*, **13**: 239–272.

Estep, M.F., Tabita, F.R., Parker, P.L. and Van Baalen, C. 1978. Carbon isotope fractionation by ribulose-1,5-bisphosphate carboxylase from various organisms. *Plant Physiology*, **61**: 680–687.

Felbeck, H. 1981. Chemoautotrophic potential of the hydrothermal vent tube worm *Riftia pachyptila* Jones (Vestimentifera). *Science*, **213**: 336–338.

Felbeck, H. 1983. Sulfide oxidation and carbon fixation by the gutless clam *Solemya reidi*: an animal-bacteria symbiosis. *Journal of Comparative Physiology*, **152**: 3–11.

Felbeck, H. 1985a. CO_2 fixation in the hydrothermal vent tube worm *Riftia pachyptila* Jones. *Physiological Zoology*, **58**: 272–281.

Felbeck, H. 1985b. Occurrence and metabolism of D-aspartate in the gutless bivalve *Solemya reidi*. *Journal of Experimental Zoology*, **234**: 145–149.

Felbeck, H., Childress, J.J. and Somero, G.N. 1981. Calvin-Benson cycle and sulphide oxidation enzymes in animals from sulphide-rich habitats. *Nature, London*, **298**: 291–293.

Felbeck, H., Childress, J.J. and Somero, G.N. 1983a. Biochemical interactions between molluscs and their algal and bacterial symbionts. In *The Mollusca*, Vol. 2, *Environmental biochemistry and physiology* (ed. P.W. Hochachka), pp. 331–358. Academic Press, New York.

Felbeck, H., Liebzeit, G., Dawson, R. and Giere, O. 1983b. CO_2 fixation in tissues of marine oligochaetes (*Phallodrilus leukodermatus* and *P. planus*) containing symbiotic, chemoautotrophic bacteria. *Marine Biology*, **75**: 187–191.

Fenchel, T.M. and Riedl, R.J. 1970. The sulfide system: a new biotic community underneath the oxidized layer of marine sand bottoms. *Marine Biology*, **7**: 255–268.

Fiala-Médioni, A. 1984. Mise en évidence par microscopie électronique a transmission de l'abondance de bactéries symbiotiques dans la branchie de mollusques bivalves de sources hydrothermales profondes. *Compte rendu hebdomadaire des séances de l'Academie des Sciences (ser. III)*, **298**: 487–492.

Fiala-Médioni, A. and Métivier, C. 1986. Ultrastructure of the gill of the hydrothermal vent bivalve *Calyptogena magnifica*, with a discussion of its nutrition. *Marine Biology*, **90**: 215–222.

Fiala-Médioni, A., Alayse, A.M. and Cahet, G. 1986a. Evidence of *in situ* uptake and incorporation of bicarbonate and amino acids by a hydrothermal vent mussel. *Journal of Experimental Marine Biology and Ecology*, **96**: 191–198.

Fiala-Médioni, A., Métivier, C., Herry, A. and Le Pennec, M. 1986b. Ultrastructure of the gill of the hydrothermal-vent mytilid *Bathymodiolus* sp. *Marine Biology*, **92**: 65–72.

Fisher, C.R. and Childress, J.J. 1984. Substrate oxidation by trophosome tissue from *Riftia pachyptila* Jones (phylum Pogonophora). *Marine Biology Letters*, **5**: 171–183.

Fisher, C.R. and Childress, J.J. 1986. Translocation of fixed carbon from symbiotic bacteria to host tissues in the gutless bivalve *Solemya reidi*. *Marine Biology*, **93**: 59–68.

Fisher, M. and Hand, C. 1984. Chemoautotrophic symbionts in the bivalve *Lucina floridana* from seagrass beds. *Biological Bulletin, Woods Hole Marine Biological Laboratory, Woods Hole, MA*, **167**: 445–459.

Flügel, H.J. and Langhof, I. 1983. A new hermaphroditic pogonophore from the Skagerrak. *Sarsia*, **68**: 131–138.

Fry, B. and Sherr, E.B. 1984. δ^{13}C measurements as indicators of carbon flow in marine and freshwater ecosystems. *Contributions in Marine Science*, **27**: 13–47.

Fry, B., Gest, H. and Hayes, J.M. 1983. Sulphur isotopic compositions of deep-sea hydrothermal vent animals. *Nature, London*, **306**: 51–52.

Fry, B., Anderson, R.K., Entzeroth, L., Bird, J.L. and Parker, P.L. 1984. ^{13}C enrichment and oceanic food web structure in the north-western Gulf of Mexico. *Contributions in Marine Science*, **27**: 49–63.

Fustec, A., Desbruyères, D. and Juniper, K. 1987. Deep-sea hydrothermal vent communities at 13° N on the East Pacific Rise: microdistribution and temporal variations. *Biological Oceanography*, **4**: 121–164.

Gaill, F., Desbruyères, D., Prieur, D. and Gourret, J.-P. 1984. Mise en évidence de communautes bactériennes epibiontes du 'ver de Pompei' (*Alvinella pompejana*). *Compte rendu hedomadaire des séances de l'Académie des Science (ser. III)*, **298**: 553–558.

Galimov, E.M. 1985. *The biological fractionation of isotopes*. Academic Press, New York.

Giere, 1981. The cutless marine oligochaete *Phallodrilus leukodermatus*. Structural studies on an aberrant tubificid associated with bacteria. *Marine Ecology — Progress Series*, **51**: 353–357.

Giere, O. 1985a. The gutless marine tubificid *Phallodrilus planus*, a flattened oligochaete with symbiotic bacteria. Results from morphological and ecological studies. *Zoologica Scripta*, **14**: 279–286.

Giere, O. 1985b. Structure and position of bacteria and symbionts in the gill filaments of Lucinidae from Bermuda. *Zoomorphology*, **105**: 296–301.

Giere, O. and Langheld, C. 1987. Structural organisation, transfer and biological fate of endosymbiotic bacteria in gutless oligochaetes. *Marine Biology*, **93**: 641–650.

Giere, O., Liebezeit, G. and Dawson, R. 1982. Habitat conditions and distribution pattern of the gutless oligochaete *Phallodrilus leukodermatus*. *Marine Ecology — Progress Series*, **8**: 291–299.

Grassle, J.F. 1986. The ecology of deep-sea hydrothermal vent communities. *Advances in Marine Biology*, **23**: 301–362.

Hageage, G.D., Eames, E.D. and Gherna, R.L. 1970. X-ray diffraction studies of the sulfur globules accumulated by *Chromatium* species. *Journal of Bacteriology*, **101**: 464–469.

Hammen, C.S. and Osborne, P.J. 1959. Carbon dioxide fixation in marine invertebrates: a survey of major phyla. *Science*, **130**: 1409–1410.

Henry, A. (ed.) 1967. *Symbiosis*, 2 Vols. Academic Press, New York.

Hessler, R.R., Smithey, W.M. and Keller, C.H. 1985. Spatial and temporal variation of giant clams, tube worms and mussels at deep-sea hydrothermal vents. *Bulletin of the Biological Society of Washington*, **6**: 411–428.

Ivanov, A.V. 1963. *Pogonophora*. Academic Press, London.

Jackson, D., Harkness, D.D., Mason, C.F. and Long, S.P. 1986. *Spartina anglica* as a carbon source for salt marsh invertebrates: a study using $\delta^{13}C$ values. *Oikos*, **46**: 163–170.

Jannasch, H.W. 1985. The chemosynthetic support of life and the microbial diversity at deep-sea hydrothermal vents. *Proceedings of the Royal Society B*, **225**: 277–297.

Jannasch, H.W. and Mottl, M.J. 1985. Geomicrobiology of deep-sea hydrothermal vents. *Science*, **229**: 717–725.

Jannasch, H.W. and Nelson, D.C. 1984. Recent progress in the microbiology of hydrothermal vents. In *Current perspectives in microbial ecology* (eds. M.J. King and C.A. Reddy), pp. 170–176. American Society for Microbiology Publications, Washington, DC.

Jensen, T.E., Sicko-Goad, L. and Ayala, R.P. 1977. Phosphate metabolism in blue–green Algae. III The effect of fixation and post-staining on the morphology of polyphosphate bodies in *Plectonema boryanum*. *Cytologia*, **42**: 357–369.

Johnson, K.S., Beehler, C.M., Sakamoto-Arnold, C.M. and Childress, J.J. 1986. *In situ* measurements of chemical distributions in a deep-sea hydrothermal vent field. *Science*, **231**: 1139–1141.

Jones, D.S., Williams, D.F. and Romanek, C.S. 1986. Life history of symbiont-bearing giant clams from stable isotope profiles. *Science*, **231**: 46–48.

Jones, M.L. 1985. On the Vestimentifera, new phylum: six new species, and other taxa, from hydrothermal vents and elsewhere. *Bulletin of the Biological Society of Washington*, **6**: 117–158.

Jørgensen, B.B. 1982. Ecology of the bacteria of the sulphur cycle with special reference to anoxic–oxic interface environments. *Philosophical Transactions of the Royal Society B*, **298**: 543–561.

Jørgensen, C.B., Famme, P., Kristensen, H.S., Larsen, P.S., Möhlenberg, F. and Risgard, H.O. 1986. The bivalve pump. *Marine Ecology — Progress Series*, **34**: 69–77.

Jørgensen, N.O.G., Blackburn, T.H., Hendriksen, K. and Bay, D. 1981. The importance of *Posidonia oceanica* and *Cymodoce nodosa* as contributors of free amino acids in water and sediment of seagrass beds. *Marine Ecology, Naples*, **2**: 97–112.

Jouin, C. 1979. Description of a freeliving polychaete without gut: *Astomus taenioides* n.gen., n.sp. (Protodrilidae, Archiannelida). *Canadian Journal of Zoology*, **57**: 2448–2456.

Kelly, D.P. 1982. Biochemistry of the chemolithotrophic oxidation of inorganic sulphur. *Philosophical Transactions of the Royal Society B*, **298**: 429–602.

Kenk, V.C. and Wilson, B.R. 1985. A new mussel (Bivalvia, Mytilidae) from hydrothermal vents in the Galapagos Rift Zone. *Malacologia*, **26**: 253–271.

Kennicutt, M.C., Brooks, J.M., Bidigare, R.R., Fay, R.R., Wade, T.L. and McDonald, T.J. 1985. Vent-type taxa in a hydrocarbon seep region on the Louisiana Slope. *Nature, London*, **317**: 351–353.

Killingley, J.S., Berger, W.H., MacDonald, K.C. and Newman, W.A. 1980. $^{18}O/^{16}O$ variations in deep-sea carbonate shells from the Rise hydrothermal field. *Nature, London*, **287**: 218–221.

Kreig, N.R. and Holt, J.G. (eds) 1984. *Bergey's manual of systematic bacteriology*, Vol. 1. Williams and Wilkins, Baltimore.

Kroopnick, P.M. 1985. The distribution of ^{13}C of CO_2 in the world oceans. *Deep-Sea Research*, **32**: 57–84.

Kulm, L.D., Suess, E., Moore, J.C., *et al.* 1986. Oregon subduction zone: venting, fauna and carbonates. *Science*, **231**: 561–566.

Kuwae, T., Fukasawa, S., Sasaki, T. and Kurata, M. 1982. Immunological comparisons among lipopolysaccharides from symbiotic luminous bacteria isolated from several marine animals. *Microbiology and Immunology*, **26**: 1181–1186.

Lane, D.J., Stahl, D.A., Olsen, G.J. and Pace, N.R. 1985. Analysis of hydrothermal vent-associated symbionts by ribosomal RNA sequences. *Bulletin of the Biological Society of Washington*, **6**: 389–400.

Le Pennec, M. and Hily, A. 1984. Anatomie, structure et ultrastructure de la branchie d'un Mytilidae des sites hydrothermaux du Pacifique oriental. *Oceanologica Acta*, **7**: 517–523.

Le Pennec, M., Prieur, D. and Lucas, A. 1985. Studies on the feeding of a hydrothermal-vent mytilid from the East Pacific Rise. In *Proceedings of the 19th European marine biology symposium* (ed. P.E. Gibbs), pp. 159–166. Cambridge University Press, Cambridge.

Liebezeit, G., Dawson, R., Felbeck, H. and Giere, O. 1983. Transepidermal uptake of dissolved carbohydrates by the gutless marine oligochaete *Phallodrillus leukodermatus* (Annelida). *Oceanis*, **9**: 205–211.

Lilley, M.D., Baross, J.A. and Gordon, L.I. 1983. Reduced gases and bacteria in hydrothermal fluids: the Galapagos Spreading Center and 21° N East Pacific Rise. In *Hydrothermal processes at seafloor spreading centers* (eds. P.A. Rona, K. Bostrom, L. Laubier and K.L. Smith), pp. 411–449. Plenum Press, New York.

Little, C. and Gupta, B.L. 1968. Pogonophora: uptake of dissolved nutrients. *Nature, London*, **218**: 873–874.

Little, C. and Gupta, B.L. 1969. Studies on Pogonophora. III. Uptake of nutrients. *Journal of Experimental Biology*, **51**: 759–773.

Lu, W.-P. and Kelly, D.P. 1984. Purification and characterization of two essential cytochromes of the thiosulphate-oxidizing multi-enzyme system from *Thiobacillus* A2 (*Thiobacillus versutus*). *Biochimica Biophysica Acta*, **765**: 106–117.

Maguire, C. and Boaden. P.J.S. 1975. Energy and evolution in the thiobios: an extrapolation from the marine gastrotrich *Thiodasys sterreri*. *Cahiers de Biologie Marine*, **16**: 635–646.

McLean, J.H. 1985. Preliminary report on the limpets at hydrothermal vents. *Bulletin of the Biological Society of Washington*, **6**: 159–166.

Nelson, D.C., Waterbury, J.B. and Jannasch, 1984. DNA base composition and genome size of the prokaryotic symbiont in *Riftia pachyptila* (Pogonophora). *Federation of European Microbiological Societies, Microbiology Letters*, **24**: 267–270.

Olsen, G.J., Lane, D.J., Giovannoni, S.J., Pace, N.R. and Stahl, D.A. 1986. Microbial ecology and evolution: a ribosomal RNA approach. *Annual Review of Microbiology*, **40**: 337–365.

Ott, J., Rieger, R. and Enderes, F. 1982. New mouthless interstitial worms from the sulfide system: symbiosis with prokaryotes. *Marine Ecology, Naples*, **3**: 313–333.

Paull, C.K., Jull, A.J.T., Toolin, L.J. and Linick, T. 1985. Stable isotope evidence for chemosynthesis in an abyssal seep community. *Nature, London*, **317**: 709–711.

Powell, M.A. and Somero, G.N. 1985. Sulfide oxidation occurs in the animal tissue of the gutless clam, *Solemya reidi*. *Biological Bulletin, Marine Biological Laboratory, Woods Hole, MA*, **169**: 164–181.

Powell, M.A. and Somero, G.N. 1986a. Adaptations to sulfide by hydrothermal vent animals: sites and mechanisms of detoxification and metabolism. *Biological Bulletin, Marine Biological Laboratory, Woods Hole, MA*, **171**: 274–290.

Powell, M.A. and Somero, G.N. 1986b. Hydrogen sulfide oxidation is coupled to oxidative phosphorylation in mitochondria of *Solemya reidi*. *Science*, **233**: 563–566.

Powell, E.N., Crenshaw, M.A. and Rieger, R.M. 1980. Adaptations to sulfide in sulfide system meiofauna. Endproducts of sulfide detoxification in three turbellarians and a gastrotrich. *Marine Ecology — Progress Series*, **2**: 169–177.

Rau, G.H. 1981a. Low $^{15}N/^{14}N$ in hydrothermal vent animals: ecological implications. *Nature, London*, **289**: 484–485.

Rau, G.H. 1981b. Hydrothermal vent clam and tube worm $^{13}C/^{12}C$. Further evidence of nonphotosynthetic food sources. *Science*, **213**: 338–340.

Rau, G.H. 1985. $^{13}C/^{12}C$ and $^{15}N/^{14}N$ in hydrothermal vent organisms: ecological and biogeochemical implications. *Bulletin of the Biological Society of Washington*, **6**: 243–247.

Rau, G.H. and Hedges, J.I. 1979. Carbon-13 depletion in a hydrothermal vent mussel: suggestion of a chemosynthetic food source. *Science*, **203**: 648–649.

Rau, G.H., Mearns, A.J., Young, D.R., Olsen, R.J., Schafer, H. and Kaplan, I.R. 1983. Animal $^{13}C/^{12}C$ correlates with trophic level in pelagic food webs. *Ecology*, **64**: 1314–1318.

Rau, G.H., Karl, D.M. and Carney, R.S. 1986. Does inorganic carbon assimilation cause ^{14}C depletion in deep-sea organisms? *Deep-Sea Research*, **33**: 349–357.

Reid, R.G.B. 1980. Aspects of the biology of a gutless species of *Solemya* (Bivalvia: Protobranchia). *Canadian Journal of Zoology*, **58**: 386–393.

Reid, R.G.B. and Bernard, F.R. 1980. Gutless bivalves. *Science*, **208**: 609–610.

Reid, R.G.B. and Brand, D.G. 1986. Sulfide-oxidizing symbiosis in lucinaceans: implications for bivalve evolution. *Veliger*, **29**: 3–24.

Richards, K.S., Fleming, T.P. and Jamieson, B.G.M. 1982. An ultrastructural study of the distal epidermis and the occurrence of subcuticular bacteria in the gutless tubificid *Phallodrilus albidus* (Oligochaeta: Annelida). *Australian Journal of Zoology*, **30**: 327–336.

Schmaljohann, R. and Flügel, H.J. 1987. Methane oxidizing bacteria in Pogonophora. *Sarsia*, in press.

Schweimanns, M. and Felbeck, H. 1985. Significance of the occurrence of chemoautotrophic bacterial endosymbionts in lucinid clams for Bermuda. *Marine Ecology — Progress Series*, **24**: 113–120.

Shively, J.M. 1984. Inclusion bodies of prokaryotes. *Annual Review of Microbiology*, **28**: 167–186.

Siebers, D. and Bulnheim, H.-P. 1977. Salinity dependence, uptake kinetics, and specificity of amino-acid absorbtion across the body surface of the oligochaete annelid *Enchytraeus albidus*. *Hegoländer wissenschaftlich Meeresuntersuchungen*, **29**: 473–492.

Smith, K.L. 1985. Deep-sea hydrothermal vent mussel; nutritional state and distribution at the Galapagos Rift. *Ecology*, **66**: 1067–1080.

Southward, A.J. and Southward, E.C. 1968. Uptake and incorporation of labelled glycine by pogonophores. *Nature, London*, **218**: 875–876.

Southward, A.J. and Southward, E.C. 1980. On the value of dissolved organic matter as food for *Siboglinum ekmani* and other small pogonophores. *Journal of the Marine Biological Association of the United Kingdom*, **60**: 1005–1034.

Southward, A.J. and Southward, E.C. 1981. Dissolved organic matter and the nutrition of the Pogonophora: a reassessment based on recent studies of their morphology and biology. *Kieler Meeresforschungen*, **5**: 445–453.

Southward, A.J. and Southward, E.C. 1987. Pogonophora. In *Animal energetics* (eds. T.J. Pandian and F.J. Vernberg). Academic Press, New York.

Southward, A.J., Southward, E.C., Brattegard, T. and Bakke, T. 1979. Further experiments on the value of dissolved organic matter as food for *Siboglinum fiordicum* (Pogonophora). *Journal of the Marine Biological Association of the United Kingdom*, **59**: 133–148.

Southward, A.J., Southward, E.C., Dando, P.R., Rau, G.H., Felbeck, H. and Flügel, H. 1981. Bacterial symbionts and low $^{13}C/^{12}C$ ratios in tissues of Pogonophora indicate unusual nutrition and metabolism. *Nature, London*, **293**: 616–620.

Southward, A.J., Southward, E.C., Dando, P.R., Barrett, R.L. and Ling, R. 1986. Chemoautotrophic function of bacterial symbionts in small Pogonophora. *Journal of the Marine Biological Association of the United Kingdom*, **66**: 415–437.

Southward, E.C. 1971. Recent researchers on the Pogonophora. *Oceanography and Marine Biology, an Annual Review*, **9**: 193–200.

Southward, E.C. 1982. Bacterial symbionts in Pogonophora. *Journal of the Marine Biological Association of the United Kingdom*, **62**: 889–906.

Southward, E.C. 1986. Gill symbionts in thyasirids and other bivalve molluscs. *Journal of the Marine Biological Association of the United Kingdom*, **66**: 889–914.

Spies, R.B. and DesMarais, D.J. 1983. Natural isotope study of trophic enrichment of marine benthic communities by petroleum seepage. *Marine Biology*, **73**: 67–71.

Spiro, B., Greenwood, P.R., Southward, A.J. and Dando, P.R. 1986. $^{13}C/^{12}C$ ratios in marine invertebrates from reducing sediments: confirmation of nutritional importance of chemoautotrophic endosymbiotic bacteria. *Marine Ecology —Progress Series*, **28**: 233–240.

Stahl, D.A., Lane, D.J., Olsen, G.J. and Pace, N.R. 1984. Analysis of hydrothermal vent-associated symbionts by ribosomal RNA sequences. *Science*, **224**: 409–411.

Strohl, W.R. and Tuovinen, O.H. (eds) 1984. *Microbial chemoautotrophy*. Ohio State University Press, Columbus.

Strohl, W.R., Geffers, I. and Larkin, J.M. 1981. Structure of the sulphur inclusion envelopes from four beggiatoas. *Current Microbiology*, **6**: 75–79.

Terwilliger, R.C. and Terwilliger, N.B. 1985. Respiratory proteins of hydrothermal vent animals. *Bulletin of the Biological Society of Washington*, **6**: 273–287.

Terwilliger, R.C., Terwilliger, N.B., Hughes, G.M., Southward, A.J. and Southward, E.C. 1987. Studies on the haemoglobins of the small Pogonophora. *Journal of the Marine Biological Association of the United Kingdom*, **67**: 219–234.

Trüper, H.G. and Fischer, U. 1982. Anaerobic oxidation of sulphur compounds as electron donors for bacterial photosynthesis. *Philosophical Transactions of the Royal Society B*, **298**: 529–542.

Turekain, K.K., Cochran, J.K. and Nozaki, Y. 1979. Growth rate of a clam from the Galapagos Rise hot spring field using natural radionuclide ratios. *Nature, London*, **280**: 385–387.

Tuttle, J.H. 1985. The role of sulfur-oxidizing bacteria at deep-sea hydrothermal vents. *Bulletin of the Biological Society of Washington*, **6**: 335–343.

Van Dover, C.L. 1986. A comparison of stable isotope ratios ($^{18}O/^{16}O$ and $^{13}C/^{12}C$) between two species of hydrothermal vent decapods (*Alvinocaris lusca* and *Munidopsis subsquamosa*. *Marine Ecology — Progress Series*, **31**: 295–299.

Vetter, R.D. 1985. Elemental sulfur in the gills of three species of clams containing chemoautotrophic bacteria: a possible inorganic energy storage compound. *Marine Biology*, **88**: 33–42.

Welhan, J.A. and Craig, H. 1983. Methane, hydrogen and helium in hydrothermal fluids at 21° N on the East Pacific Rise. In *Hydrothermal processes at seafloor spreading centers* (eds. P.A. Rona, K. Bostrom, L. Laubier and K.L. Smith) pp. 391–409. Plenum Press, New York.

Williams, P.M., Smith, K.L., Druffel, E.M. and Linick, P.W. 1981. Dietary carbon sources of mussels and tubeworms from Galapagos hydrothermal vents determined from tissue ^{14}C activity. *Nature, London*, **292**: 448–449.

Wittenberg, J. 1985. Oxygen supply to intracellular bacterial symbionts. *Bulletin of the Biological Society of Washington*, **6**: 301–310.

Wright, S.H. 1982. A nutritional role for amino acid transport in filter-feeding invertebrates. *American Zoologist*, **22**: 621–634.

5

Lipid biomarkers in marine ecology

J.R. Sargent[1], **R.J. Parkes**[2], **I. Mueller-Harvey**[2] **and R.J. Henderson**[1], [1]NERC Unit of Aquatic Biochemistry, Department of Biological Science, University of Stirling, Stirling FK9 4LA, UK,
[2]Scottish Marine Biological Association, Dunstaffnage Marine Research Laboratory, Oban, Argyll, UK

The advantages of using lipids, particularly fatty acids, as static and dynamic biomarkers of marine microorganisms are discussed and relevant lipid analytical methodology is outlined. The distribution of fatty acids in different classes of algae is described, with emphasis on $(n-3)$ polyunsaturated fatty acids, and various examples of the use of these fatty acids as biomarkers of algae in their natural environment are discussed. The distribution of fatty acids in different groups of bacteria and individual bacterial species is discussed, emphasizing various *cis* and *trans* isomers of monoenoic fatty acids, branched chain fatty acids and cyclopropyl fatty acids. Specific examples of the use of these fatty acids as biomarkers of bacteria in their natural environments are discussed.

INTRODUCTION

Lipids from marine organisms, especially marine animals, have featured prominently in the development of analytical lipid chemistry and an extensive data base has been built up. In the last few decades pure strains of single-celled marine algae and bacteria have been isolated and our understanding of their lipid composition has developed rapidly. Although much remains to be discovered about the lipids of marine microorganisms, current knowledge is sufficient to enable lipids to be used as 'biomarkers' in marine ecosystems. Biomarkers are chemical components of organisms which can be analysed directly from the environment and ideally can be interpreted both quantitatively and qualitatively in terms of *in situ* biomass.

Lipids are particularly useful biomarkers since they are relatively easily extracted, identified and quantified as compared with other major biochemical constituents (protein and carbohydrate). They can be used as biomarkers at two levels. First, as major biochemical constituents accounting for circa 20% of an organism's dry weight, they can indicate biomass. Second, because specific lipids can be located relatively specifically in a given microorganism, they can serve as biomarkers for that organism. Such biomarkers can define the distribution and abundance of microorganisms in ecosystems and, provided that the compounds in question are relatively stable metabolically, can shed light on predator–prey relationships and also the origins and fates of organic matter in various locations including sediments (Bressell and Eglinton, 1984). Moreover, since the biochemical roles of lipids as metabolic energy reserves and structural components of biomembranes are well understood, information concerning the physiological status of the organisms studied can be obtained.

Lipids can be easily radiolabelled especially in single-celled organisms, e.g. from ^3H_2O, $^{14}CO_2$, or [^{14}C]glucose, so that information on their metabolic turnover can be readily obtained. In this way they can be used as dynamic markers in ecosystems. This is particularly important in studies of production rates in natural ecosystems, e.g. primary production where the conventional approach has been to equate production with the amount of radioactivity incorporated from $^{14}CO_2$ into total cellular material. Such an approach is obviously unsatisfactory since it ignores metabolic turnover of individual molecules and such turnover cannot be assessed by studying bulk cellular material alone. However, by selecting particular molecular components which can be readily radiolabelled, e.g. phospholipids or triacylglycerols labelled from ^3H_2O, it is possible to measure both the mass and the rate of turnover of these components. In this way accurate production rates of specified components can be defined and related to the physiological status of a given organism. For example, the results can be related to the extent to which an alga is producing triacylglycerols as energy reserves towards the end of its growth cycle, or the extent to which it is producing polar lipids for membranes during the active stage of growth and cell division early in its growth cycle.

This review is concerned with lipids in single-celled marine algae and marine bacteria. We concentrate on the fatty acid components of lipids (Figs. 1 and 2) since these are particularly useful biomarkers, being essential components of every living cell and having great structural diversity coupled with high biological specificity.

METHODOLOGY

Comprehensive accounts of lipid analytical methodology can be found in standard texts such as Christie (1982) and the following is intended only to outline the general approach with emphasis on fatty acids analysis. Many lipids are labile towards oxidation and hydrolysis so that precautions are required such as routinely storing lipid extracts at −70 °C under an inert gas (N_2 or Ar) in solvents containing an anti-oxidant (e.g. butylated hydroxytoluene).

A variety of organic solvents have been used to extract total lipids from a variety of organisms, but chloroform:methanol (2:1 by volume) probably remains the solvent of choice (Folch et al., 1957; Bligh and Dyer, 1959), if only on the basis of the wealth of accumulated experience using this solvent. A typical extraction scheme used in bacterial lipid analysis (White, 1983) is given in Fig. 3.

Saturated fatty acids

$$CH_3(CH_2)_nCOOH$$

Monounsaturated fatty acids

$$CH_3(CH_2)_xCH=CH(CH_2)_yCOOH$$

PUFA

$$CH_3(CH_2)_x(CH=CH-CH_2-CH=CH)_y(CH_2)_zCOOH$$

Iso fatty acids

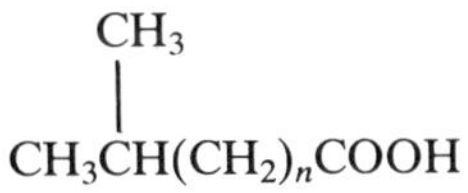

$$\begin{matrix} CH_3 \\ | \\ CH_3CH(CH_2)_nCOOH \end{matrix}$$

Anteiso fatty acids

$$\begin{matrix} CH_3 \\ | \\ CH_3CH_2CH(CH_2)_nCOOH \end{matrix}$$

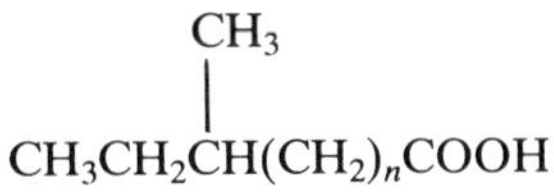

Cyclopropyl fatty acids

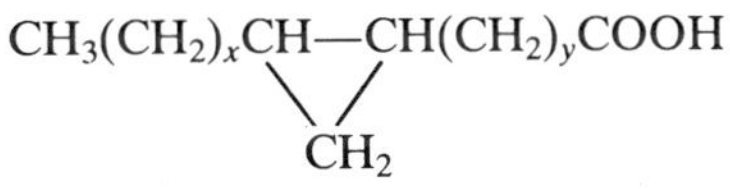

$$\begin{matrix} CH_3(CH_2)_xCH-CH(CH_2)_yCOOH \\ \diagdown \diagup \\ CH_2 \end{matrix}$$

Fig. 1—Fatty acid structures.

Emphasis should be placed on quantitative procedures, including the use of internal standards, during such separations so that the analyses can be related to the wet or dry weights of the original organisms. The most reliable method for quantifying total lipids remains direct weighing. There are, however, other methods available such as total carbon analysis, colorimetric determination of phospholipid–phosphate or analysis of the fatty acids in total lipids by gas–liquid chromatography (GLC).

While information can be obtained by analysing fatty acid methyl esters isolated from total lipid, more penetrating analyses require at least the prior separation of total lipid into polar and neutral classes (Fig. 3), and preferably the complete separation of total lipid into its component classes by thin layer or column

Triacylglycerols

(triglycerides)

CH_2OCOR^1

$CHOCOR^2$

CH_2OCOR^3

R^1, R^2, R^3 = fatty acyl units

Phospholipids

CH_2OCOR^1

$CHOCOR^2$

CH_2OPO_3X

R^1, R^2 = fatty acyl units; X = base (ethanolamine, serine, choline, etc.)

Glycolipids

CH_2OCOR^1

$CHOCOR^2$

CH_2OX

R^1, R^2 = fatty acyl units; X = carbohydrate (galactose, galactosylgalactose, sulphoquinovose etc.)

Fig. 2—Lipid class structures.

chromatography including high performance chromatography. This is because the neutral lipid content and composition of cells is more variable and more dependent on the physiological status of the organism than is the polar lipid content and composition.

To determine the composition of fatty acid methyl esters prepared from total lipid or individual lipid classes, GLC on fused-silica capillary columns is virtually mandatory. A variety of column phases is available commercially for different applications. Our experience (e.g. Sargent *et al.*, 1985) is that polyethylene glycol (Carbowax 20 m) chemically bonded on the fused silica is well capable of routinely resolving up to 30–40 fatty acids from C12 to C24 and containing from zero to six

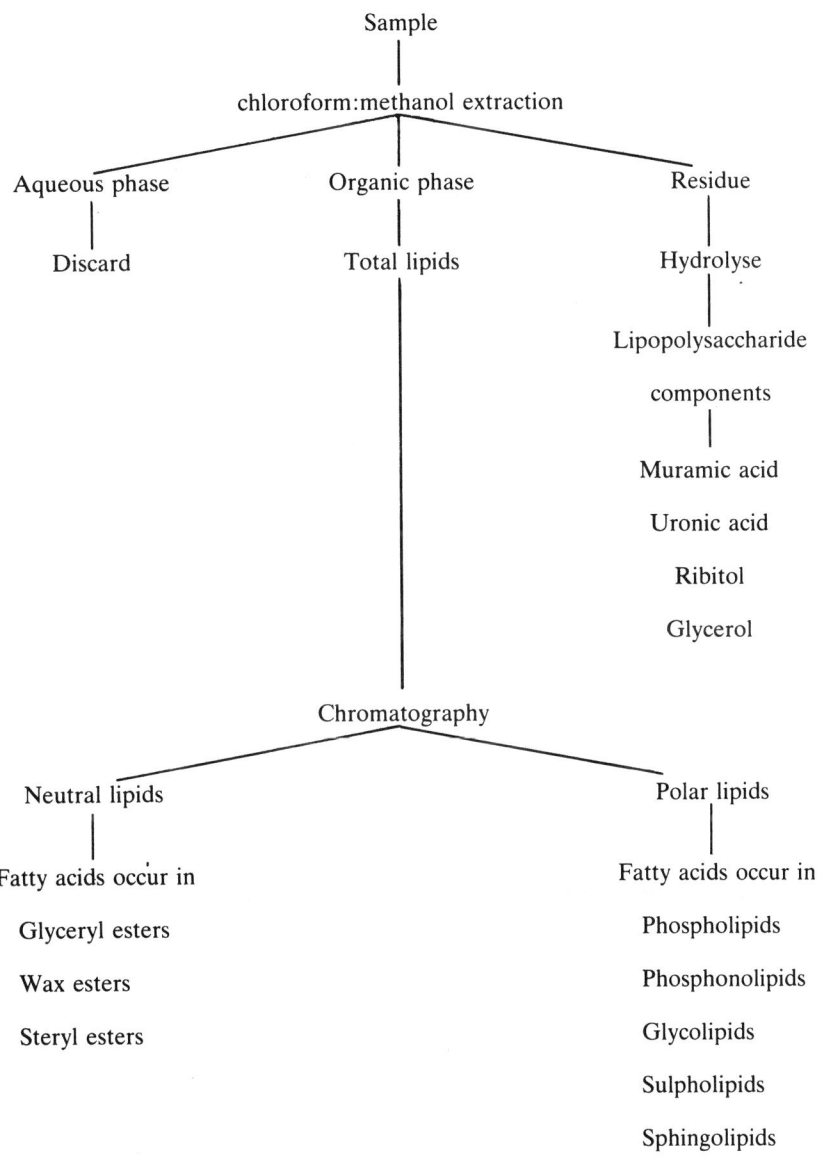

Fig. 3—Extraction and separation scheme for lipids.

double bonds. These include virtually all of the fatty acids likely to be encountered routinely in marine organisms. Intermittent access to GLC–MS facilities is invaluable for precisely identifying the fatty acids.

It must be stressed that much of the older analytical data is incomplete, particularly in terms of isomeric identification of fatty acids, and extreme care should be exercised in comparing data from the same or similar species if long time intervals or different gas chromatography methods are involved.

NOMENCLATURE

A shorthand nomenclature for fatty acids is used based on the system described by Christie (1982). Fatty acids are described in the form of numbers separated by a colon. The number before the colon specifies the carbon chain length and the following number specifies the number of double bonds. The position of the double bond is defined by (n-x), where n is the total number of carbon atoms and x is the number of carbon atoms between the terminal methyl group and the first double bond encountered, e.g. palmitoleic acid is specified as 16:1 (n-7) which is equivalent to 16:1w7, the nomenclature used in some references. The position of the double bonds in fatty acids can also be specified from the carboxyl terminus using the Δ system. Thus 16:1(n-7) is equivalent to Δ9 16:1, and 18:3(n-3) is equivalent to Δ9,12,15 18:3. This illustrates that, unless indicated otherwise, double bonds in biological polyunsaturated fatty acids (PUFA) are invariably separated by one —CH_2 group so that only the position of the double bond nearest to either the methyl or the carboxyl terminus need be indicated. Unless otherwise stated, the (n-x) system is used throughout this article.

The geometry of the double bond is indicated by 'c' for *cis* or 't' for *trans*. The prefixes 'i', 'a', and 'br' refer to *iso*, *anteiso* and methyl-branching of unconfirmed positions respectively. Other methyl-branching is indicated in terms of its position from the carboxylic end of the chain, e.g. 10Me 16:0. Methoxy fatty acids are represented by MeO and cyclopropyl fatty acids are indicated by the symbol $|\nabla|$. Representative structures of fatty acids are reproduced in Fig. 1.

FATTY ACID COMPOSITION OF ALGAL CLASSES

Recent reviews have described the structures, distribution and biosynthetic pathways of lipids in algae (Erwin, 1973; Wood, 1974; Pohl and Zurheide, 1979; Harwood and Russell, 1984). Lipids of actively growing and dividing algae are mainly polar glycolipids (Fig. 2) located in thylakoid membranes (e.g. chloroplasts). These glycolipids are characteristically rich in (n-3)PUFA. Algae also have variable amounts of neutral lipids, mainly triacylglycerols (Fig. 2), present as membrane-bound oil droplets in their cytoplasm. Algal triacylglycerols generally contain lower concentrations of (n-3)PUFA and higher concentrations of saturated and monounsaturated fatty acids, than the corresponding glycolipids.

The biosynthesis of lipids in algae can be markedly influenced by the physiological state of the organism and by environmental conditions, such as light, temperature and nutrient availability. Algae grown at high light intensities tend to have increased concentrations of (n-3)PUFA in their total lipids (Pohl and Zurheide, 1979). Decreasing temperatures tend to increase the concentration of (n-3)PUFA in cellular total lipid (Ackman *et al.*, 1968; Seto *et al.*, 1984). Algae grown under conditions of nitrogen limitation continue to photosynthesize and may accumulate their end products of photosynthesis as lipids, mostly triacylglycerols (Ben-Amotz *et al.*, 1985). This phenomenon also manifests itself as an accumulation of triacylglycerols in the stationary phase of algal cultures. Therefore, the fatty acid composition of total lipid extracted from algae can vary quite markedly throughout the growth cycle.

However, this is not a major problem in practice, and indeed can be an advantage since neutral lipids when present in abundance, i.e. in lipid-rich nutrient-limited algae, can simply be separated from polar lipids by absorption chromatography. The percentage of neutral lipids in total lipid can then be used to assess the physiological status of the organisms. The polar lipids associated with algal biomembranes have a relatively constant fatty acid composition and are particularly useful for the more specific fatty acid biomarkers.

Biomarker	Measures	Reference
Phospholipid 16:0	Total biomass, prokaryotes and eukaryotes	White *et al.* (1979) Harwood and Russell (1984) Gehron and White (1983)
Br fatty acids	Biomass of some bacteria	Leo and Parker (1966)
Cyclopropyl fatty acids	Biomass of some bacteria	Harwood and Russell (1984)
Lipopolysaccharide	Gram-negative bacteria	Rogers (1983)
Teichoic acid	Most Gram-positive bacteria	Rogers (1983)
Muramic acid	Bacteria except Archaebacteria (note that there are different amounts in Gram-positive and Gram-negative bacteria)	Rogers (1983)
Plasmalogens	Some anaerobic bacteria (including rumen bacteria)	Goldfine and Hagen (1972)
Hopanoids	Some bacteria	Rohmer *et al.* (1984)
Phytanyl ether lipid	Archaebacteria	Harwood and Russell (1984)
PUFA	Eukaryotes, gliding bacteria, some deep sea bacteria	Harwood and Russell (1984) Johns and Perry (1977) DeLong and Yayanos (1986)
Phytopigments	Photoautotrophs	Gillan and Johns (1983)

Table 1   General biomarkers

Lipid components of algae and bacteria can be used as general markers of the biomass of these organisms as exemplified in Table 1. Additionally, certain fatty acids and, more so, groups of fatty acids, are preferentially associated with certain classes of algae and can, therefore, be useful biomarkers for these algal classes. Such associations, however, are by no means unique for a given algal class, far less for single algal species, and should be regarded as general trends rather than as absolute rules. Representative fatty acid compositions of total lipids from different algae classes, using selected species within a given algal class, are reproduced in Table 2 (see also Ackman *et al.*, 1968; Chuecas and Riley, 1969; Harrington *et al.*, 1970; Wood, 1974; Joseph, 1975; Pohl and Zurheide, 1979; Ben Amotz *et al.*, 1985) and discussed below.

Table 2 illustrates that the saturated fatty acids 14:0 and 16:0 are present in all algal classes with the exception of the one species of marine blue–green algae so far analysed, where 14:0 is not a significant component. 16:1(*n*-7) is a major monoenoic fatty acid in most algal classes, except the Haptophyceae and Cryptophyceae. 18:1(*n*-9) is a major fatty acid only in the Haptophyceae, whereas 18:1(*n*-7) is

Fatty acid	Cyanobacteria: *Synechococcus* sp.	Bacillariophyceae: *Biddulphia sinensis*	Dinophyceae: *Prorocentrum cordatum*	Haptophyceae: *Emiliania huxleyi*	Cryptophyceae: *Cryptomonas* sp.	Chlorophyceae: *Dunaliella tertiolecta*
14:0	0.3	11.3	3.9	34.8	6	1.3
14:1	—	—	0.3	—	—	0.1
16:0	38.5	13.4	25.7	5.4	4	13.4
16:1(n-7)	16.5	37.7	9.5	0.3	2	1.5
16:2	0.4	3.1	—	0.3	—	1.2
16:3	—	1.8	—	0.1	2	6.1
16:4	—	2.5	—	—	—	24.4
18:0	0.8	1.1	1.2	0.5	—	1.0
18:1(n-9)	1.6	0.4	2.2	14.5	5	2.3
18:1(n-7)	21.1	0.5	4.5	—	—	1.1
18:2(n-6)	18.1	0.4	1.1	2.3	—	5.6
18:3(n-3)	2.7	—	0.8	7.2	7	28.2
18:4(n-3)	—	0.2	0.1	7.8	44	0.6
18:5(n-3)	—	—	27.8	10.1	—	—
20:1	—	—	0.5	—	—	0.6
20:4(n-6)	—	—	0.8	—	—	—
20:5(n-3)	—	24.2	4.4	—	16	—
Other C20 PUFA	—	—	—	—	—	—
22:5(n-3)	—	0.9	1.0	0.6	—	—
22:6(n-3)	—	—	14.5	11.4	10	—

Data from: *Synechococcus* sp., Goodloe and Light (1982); *Biddulphia sinensis*, Volkman *et al.* (1980); *Prorocentrum cordatum*, Nichols *et al.* (1984); *Emiliania huxleyi*, Volkman *et al.* (1981); *Cryptomonas* sp., Beach *et al.* (1970); *Dunaliella tertiolecta*, Ackman *et al.* (1968).

Table 2   Fatty acid compositions of total lipid from representatives of various marine algal classes

particularly abundant in the Cyanophyceae. 20:1 can be abundant in the Cryptophyceae (Chuecas and Riley, 1969) but is not significant in other algal classes. PUFA can be of considerable help in distinguishing between algal classes. For example, C16 PUFA can be substantial components of diatoms and Chlorophyceae. 18:2(n-6) is a major fatty acid in blue–green algae, whereas neither 18:2(n-6) nor 18:3(n-3) are major constituents of marine diatoms or dinoflagellates. However, other C18 PUFA, specifically 18:4(n-3) and 18:5(n-3), are abundant in Dinophyceae, Haptophyceae and Cryptophyceae. Very long chain PUFA are not significant constituents of blue–green algae, but 20:5(n-3) can be a major PUFA in diatoms, dinoflagellates and Cryptophyceae. 22:6(n-3) is particularly abundant in Dinophyceae, Haptophyceae and Cryptophyceae but is not present in significant amounts in other algal classes.

FATTY ACID COMPOSITION OF BACTERIA

Fatty acids are used extensively in bacterial taxonomy and hence there is a large and growing data base for identifying bacteria from their fatty acid compositions (Shaw, 1974; Lechevalier, 1977; Keddie and Bousfield, 1980; Lechevalier, 1982; Bousfield *et al.*, 1983). As in algae, lipids are general and major constituents of bacteria and hence can be used to assess total bacterial biomass or the biomass of groups of bacteria. Additionally, certain lipid components are specific to particular bacteria and can be used to estimate the biomass of these organisms in the environment. These different types of biomarkers are summarized in Tables 1 and 3. A full discussion of the distribution and specificity of these biomarkers in bacteria can be found in a recent review by Parkes (1987).

EXAMPLES OF FATTY ACIDS AS BIOMARKERS

Algae
Fatty acid analysis has been used to help identify algal species that are difficult to classify by conventional taxonomic criteria (Nichols *et al.*, 1983). It has also been used to indicate species composition of marine phytoplankton and to follow the fate of individual phytoplanktonic fatty acids in the plankton. Thus, Sargent *et al.* (1985) studied phytoplankton in two northern Norwegian fjords during the early stages of the spring bloom. Both fjords were dominated by mixed diatoms and the haptophycean *Phaeocystis poucheti*, with diatoms being more prominent than *Phaeocystis* in Balsfjord and the converse holding for Ullsfjord. In addition, a surface slick from Balsfjord composed entirely of *Phaeocystis* was available. Total phytoplanktonic lipids from the Balsfjord spring bloom were richer in C16 (n-3)PUFA and 20:5(n-3) than the Ullsfjord bloom. The *Phaeocystis* slick material was particularly rich in 18:4(n-3) and 18:5(n-3) and, in line with this, the Ullsfjord phytoplanktonic lipids had substantially more 18:4(n-3) and especially 18:5(n-3) than that from Balsfjord. It was also concluded that *Phaeocystis* was prominent in the diets of two dominent herbivorous zooplankton species (*Calanus finmarchicus* and *Thysanoessa inermis*) in both fjords, since their lipids contained substantial amounts

Bacteria	Characteristic fatty acids	Reference
Sulphate reducing		
Desulfovibrio sp. except *D. gigas*	i17:1(*n*-7)	Taylor and Parkes (1983) Edlund *et al.* (1985)
Desulfobacter sp.	10Me16:0	Taylor and Parkes (1983) Dowling *et al.* (1986)
Desulfobulbus sp.	17:1(*n*-6)	Taylor and Parkes (1983) Taylor and Parkes (1985) Parkes and Calder (1985)
Bacillus sp.	br fatty acids, usually unsaturated	Kaneda (1977)
Methane oxidizing:		
Type 1	C16 monoenoic fatty acids	Nichols *et al.* (1985)
Type 2	C18 monoenoic fatty acids	
Type 1: *Methylomonas* sp.	16:1(*n*-8)c, 16:1(*n*-8)t, 16:1(*n*-7)t, 16:1(*n*-5)c, 16:1(*n*-5)t	
Type 2: *Methylosinnus trichosporium*	18:1(*n*-8)c, 18:1(*n*-8)t, 18:1(*n*-7)c, 18:1(*n*-6)c	
Thiobacillus sp.	10-11-Me18:1(*n*-6), ∇19:0 (8,9), 10-11-MeO18:0, 12-13-MeO20:0, 20-OH∇16:0, 2-OH∇18:0, 11-OH and 13-OH19:0	Kerger *et al.* (1986)
Chemotypes	Nine different types of un- specified bacteria defined by specific fatty acids	Gillan and Hogg (1984)
Clostridia	∇15:1	Chan *et al.* (1971)

Table 3   Specific bacterial biomarkers

of 18:4(*n*-3). These deductions were supported by direct visual examination of the phytoplankton and the animals' gut contents.

18:4(*n*-3) can be a prominent component of the fatty acids in calanoid copepod wax esters in Norwegian waters. In addition, it is often a prominent fatty acid in the triacylglycerols of zooplanktonivorous fish such as herring and sand eels (Sargent and Henderson, 1987). Recent work has supported the concept that *Phaeocystis*, traditionally regarded as being of little significance in the diet of zooplankton, can in fact be a major dietary constituent of these animals in the German Wadden Sea (Weiss, 1983). Similarly, Mayzaud *et al.* (1976) have shown that 18:5(*n*-3) may serve as a useful biomarker of food web relationships in water of the Bedford Basin in Canada. This fatty acid, present in substantial amounts in dinoflagellates, can be transmitted to various herbivorous copepods, as well as to carnivorous chaetognaths.

Sargent and Falk-Petersen (1981) and Falk-Petersen *et al.* (1981) used fatty acid analyses to assess the relative importance of phytoplankton in the diets of *Thysanoessa inermis* and *T. rashi* in Balsfjord. The former species was considered to be the more herbivorous of the two since it was richer in 18:4(*n*-3) and contained

18:1(n-7), cis-vaccenic acid, which was considered to be derived from phytoplanktonic 16:1(n-7). This conclusion, of course, assumes the relative absence of bacteria
from the water column since 18:1(n-7) can be a major fatty acid in bacteria (Table 3).

Several studies using lipids and fatty acids as biomarkers have been carried out to
assess the nutritional value, both quantitative and qualitative, of natural phytoplankton at various stages of the bloom. These studies are concerned (a) with lipids being
major sources of energy in marine food webs and (b) with (n-3)PUFA being essential
dietary factors for all marine animals so far studied. Using $^{14}CO_2$, Smith and Morris
(1980) estimated that Antarctic phytoplankton incorporated up to 80% of the carbon
fixed into their total lipids at temperatures below 0 °C and at low light intensities. In
contrast, Li and Platt (1982) estimated that total lipids did not exceed 30% of the
carbon fixed by Arctic phytoplankton under relatively similar conditions. A study by
Sargent et $al.$ (1985) estimated that 20% of the carbon was fixed into phytoplanktonic lipid during the early spring bloom in Balsfjord. However, this study emphasized
that the glycosyl moieties of algal lipids can turn over independently of the fatty acyl
moieties, a situation that is well known for higher plants (see references in Sargent et
$al.$, 1985). As has already been emphasized in the Introduction, there are very
obvious dangers in using $^{14}CO_2$ incorporation as a measure of primary productivity
without a detailed knowledge of molecular turnover in the organisms studied (cf.
Jensen, 1985; Rivkin, 1985). Lancelot (1985) estimated that total lipid accounted for
a relatively constant part, about 20%, of the total carbon fixed by $Phaeocystis$
$poucheti$. We have recently found a similar value for mixed phytoplankton at various
stages of a bloom in a sea water enclosure in Loch Ewe, Scotland (Fraser, Gamble
and Sargent, unpublished data).

The fatty acid composition of total phytoplankton can vary during a bloom as the
algal species composition changes and as the physiological states of individual
organisms change. Diatom-dominated populations developing during the early
stages of spring blooms had total lipid relatively deficient in PUFA (Kattner et $al.$,
1983; Morris, 1984) and were presumably rich in neutral lipid. Our experience has
been that, during the early and middle stages of phytoplankton blooms, algal total
lipids consistently have (n-3)PUFA accounting for 40–50% of the total fatty acids
(Sargent et $al.$, 1985; Fraser, Gamble and Sargent, unpublished data) and are,
therefore, rich in polar lipid. Moreover, under these circumstances the (n-3)PUFA
content of the total lipid of herbivorous zooplankton approximates to 50% of the
total fatty acids and its composition, e.g. the percentage present as C16 (n-3)PUFA,
18:4(n-3) and 18:5(n-3), closely reflects phytoplanktonic total lipid.

Fatty acids have been used as markers in benthic studies. Falk-Petersen and
Sargent (1982) analysed the benthic, filter-feeding mollusc $Chlamys$ $islandica$ which
ingests phytoplankton and the mud-ingesting starfish $Ctenodiscus$ $crispatus$. $Chlamys$
had total lipids dominated by (n-3)PUFA characteristic of the phytoplankton in
Balsfjord. In contrast, $Ctenodiscus$ lipids had less (n-3)PUFA but more (n-6)PUFA,
especially 20:4(n-6), and more iso and $anteiso$ C15 and C17 fatty acids. The latter
two fatty acids are abundant in total lipids extracted from surface sediments in
Balsfjord and are characteristic of bacteria (Table 3). These results indicated that
sediment bacteria contribute to the nutrition of $Ctenodiscus$. The source of the
20:4(n-6) in $Ctenodiscus$ was not identified although it is known that a variety of
marine benthic animals can be rich in this fatty acid (references in Falk-Petersen and

Sargent, 1982). Recent work has suggested that (n-6)PUFA are essential fatty acids for marine animals in addition to (n-3)PUFA (Bell *et al.*, 1986). It is possible that the (n-6)PUFA originate from benthic eukaryotic organisms such as protozoa, whose role in benthic food webs remains to be investigated.

Bacteria

The concentration of bacteria in sediments is much higher than in the water column. Consequently most bacterial biomarker research has been conducted in sediments and examples of the use of bacterial biomarkers will be mainly from this environment.

As previously noted, *iso* and *anteiso* branched chain fatty acids are characteristic of bacteria, and these fatty acids have been found both in sediments and in bacteria cultured from these sediments (Leo and Parker, 1966; Cranwell, 1976; Johns *et al.*, 1977; White *et al.*, 1980). In comparison with bacteria isolated from sediments and the fatty acid profile of the sediment itself, other fatty acids have been considered characteristic of bacteria. These include the following: ∇10 Me16:0; Δ17:0 and 19:0; 18:1(n-7); the 15:1, 17:1(n-6) and 17:1(n-8) isomers (especially when these occur in pairs); certain monoenoic branched chain fatty acids and possibly some *trans* fatty acids. It is important to note that, although the individual fatty acids are characteristic of bacteria, they are not unique to these organisms and hence it is the combination of these fatty acids which can be considered as a general marker for bacteria (Perry *et al.*, 1979). The relative proportions of these markers can also be used to analyse community structure (Parkes and Taylor, 1983; Gillan and Hogg, 1984) since they have different distributions within different bacterial types. The concentration of typical bacterial fatty acids within sediments indicates that bacteria make a significant contribution to the total biomass and often represent the largest proportion of the sediment biomass (Perry *et al.*, 1979; Van Vleet and Quinn, 1979; Gillan *et al.*, 1981; Parkes and Taylor, 1983; Federle *et al.*, 1983; Gillan and Hogg, 1984).

The study of deep-sea environments by the biomarker approach seems to be particularly appropriate as more conventional methods of bacterial analysis require growth of bacteria or metabolism of radiolabelled compounds and these are difficult to conduct under the appropriate conditions of temperature and pressure that exist in the deep sea (Jannasch and Taylor, 1984). The use of this approach on deep-sea sediments in the Venezuelan Basin (water depth 3500–8400 m) has demonstrated that these sediments are dominated by bacteria but, rather surprisingly, fatty acids believed to be markers for anaerobic bacteria were also present (Baird and White, 1985). In contrast, an estuarine sediment had a much higher biomass but was less dominated by bacteria. A further unexpected finding in the deep-sea environment is the presence of high concentrations of 20:5(n-3) and 22:6(n-3) in deep sea bacteria (Delong and Yayanos, 1986). Thus far, with the exception of the marine gliding bacterium *Flexibacter polymorphus* (Johns and Perry, 1977), long chain PUFA have been considered to be absent from prokaryotes and, therefore, to be characteristics of eukaryotes.

An important aspect of this approach is the need to obtain bacterial cultures that are representative of the bacteria *in situ*. It is also important that these bacteria are grown under conditions which enable their characteristic lipids to be produced because, in a manner similar to that already discussed for algae, the fatty acids of bacteria can vary considerably with different culture conditions (e.g. temperature, growth substrates, growth state (Lechevalier, 1982)). Parkes and Taylor (1983) used a multiple chemostat system to enrich bacterial communities at discrete stages along a redox gradient in an attempt to obtain a bacterial community characteristic of the sediment. The bacterial community isolated in the first vessel of this system was then grown under different conditions within a single-stage chemostat so as to select for three distinct types of communities: aerobes, facultative aerobes and facultative anaerobes. The fatty acids of the lipids of these communities were compared with the mixed culture of anaerobic sulphate-reducing bacteria which developed in the end vessel of the multiple chemostat system. All four communities could be distinguished in terms of their fatty acids, one of the most marked differences being the presence of significant amounts of cyclopropyl fatty acids in the aerobic communities and their absence in the anaerobic communities. These results suggested that cyclopropyl fatty acids are characteristic of aerobic sediment bacteria. This suggestion was supported by the presence of the highest concentrations of cyclopropyl fatty acids being found in the aerobic surface layers of a sediment, and their decrease with depth as anaerobic conditions prevailed.

The isolation of microorganisms is not, however, an absolute requirement for the biomarker approach as there are other ways of 'calibrating' biomarkers. One such approach is the direct correlation of a particular microbial activity with the presence of a specific biomarker. This approach was used by Martz et al. (1983) to correlate the concentration of the biomarker for methane-producing bacteria (phytanyl ether lipids, Table 1) with concentrations in, and efflux of methane from, sediments. Direct manipulation of natural microbial communities to enhance a specific group of microorganisms, followed by analysis of the resulting changes in fatty acids, provides another approach. Bobbie and White (1980) used this technique to stimulate the development of either prokaryotes or fungi in natural populations of marine organisms attached to Teflon squares. Scanning electron micrographs confirmed that the desired populations had been obtained and fatty acid analysis demonstrated the expected differences between the two communities (see Table 1). The bacterial community was enriched in branched chain, ∇17:0 and 18:1(n-7) fatty acids whilst the fungus community was enriched in 18:2, and C18 and C20 PUFA.

Taylor and Parkes (1985) used anaerobic sediment slurries, supplemented with growth substrates for specific types of sulphate-reducing bacteria, to determine whether biomarkers previously found in pure cultures of these organisms (Taylor and Parkes, 1983) could be applied to natural sediments. They found that the biomarkers for various types of sulphate-reducing bacteria were indeed stimulated. Interestingly, the biomarker for *Desulfovibrio*, i17:1(n-7), the most commonly isolated sulphate-reducing bacterium, was not stimulated and this tends to cast doubt on the environmental importance of this bacterium. This work also demonstrates the potential of lipid biomarkers for studying competition between sediment microbes under realistic incubation conditions, since some of the added substrates could potentially be used by different types of sulphate-reducing bacteria. There is an obvious need to work towards more subtle manipulations of environmental samples in this type of work, so that the results become increasingly relevant to *in situ* conditions. This should be possible with a combination of labelled isotopes (both ^{14}C and ^{13}C) and the improvements that are taking place in lipid analysis.

The use of labelled substrates for biomarker studies greatly extends the usefulness of the technique as it allows the productivity as well as the biomass of different bacterial types to be measured. By judicious selection of labelled substrates and the lipid fraction to be analysed, a spectrum of microbial activities can be measured comparable with the spectrum of microbial biomasses that can be estimated using biomarkers (Tables 1 and 3). For example, a method has been developed by Moriarty *et al.* (1985) to measure bacterial productivity within sediments using the rate of $^{32}PO_4$ incorporation into phospholipids. This may have an advantage in sediments over the more commonly used procedure of measuring productivity using [^{3}H]thymidine incorporation into DNA, since the thymidine method does not measure the productivity of anaerobic sulphate-reducing bacteria, which can play a dominant role within sediments (Jørgensen, 1982; Moriarty *et al.*, 1985). The same basic techniques can be used to measure the productivity of other groups of organisms, e.g. $^{35}SO_4$ incorporation into sulpholipids can potentially measure eukaryotic activity (Moriarty *et al.*, 1985).

The relative concentrations of biomarkers or the relative rate of their production can also provide an indication of the physiological state of microbial groups. Sediment disturbance can be assessed by the ratio of [^{3}H]thymidine incorporation to $^{32}PO_4$ incorporation, since phospholipid synthesis seems to be rapidly enhanced by sediment disturbance, whilst DNA production is not immediately affected (Moriarty *et al.*, 1985). A similar estimate of sediment disturbance can be obtained from the relative rates of incorporation of [^{14}C]acetate into the fatty acids of phospholipids and the endogenous storage lipid, poly-β-hydroxy-alkanoate (PHA) (Findlay *et al.*, 1985). The biosynthesis of phospholipid fatty acids measures cellular growth whereas PHA biosynthesis measures carbon accumulation (unbalanced growth). Findlay *et al.* (1985) found that disturbing a sediment with a garden rake 30 min before analysis could be detected by an increase in the phospholipid fatty acid:PHA ratio, but only when the [^{14}C]acetate was injected into the sediment. When the label was introduced by the slurry technique the effect of raking the sediment surface could not be detected, since slurrying the sediment caused a bigger disturbance to the microbial community. This observation is obviously pertinent to how labelled compounds in other techniques should be added to sediment. Bioturbation caused by sand dollar feeding in an estuarine sediment could be detected by an increased phospholipid fatty acid:PHA ratio (Findlay *et al.*, 1985). The relative concentration of PHA can be used to indicate the recent nutritional status of a prokaryotic community (Nickels *et al.*, 1979). In addition, recent work has suggested that the *trans:cis* ratio of monoenoic fatty acids may be used as a 'starvation' or 'stress' index for determining the nutritional status of bacteria and, as a consequence, may be used to address the question of bacterial dormancy in natural aquatic environments (Guckert *et al.*, 1986).

CONCLUSIONS

Although it is arguable that no single organic compond *per se* can ever specifically identify a given organism, there is now little doubt that individual lipids can be extremely helpful in identifying microorganisms. Certainly relatively small groups of organic compounds can be highly characteristic of quite narrow groups of microorganisms or even an individual microorganism. Thus, lipid biomarker methodology can greatly augment more traditional and laborious methods of identifying organisms in their natural environment. It is the case that, so far, the method has mainly been applied qualitatively. Its real potential lies as a quantitative method which, when combined with radioactive tracers, can provide insight into the dynamic state of specific groups of microorganisms in their natural environment. As GLC and GLC–MS methodology becomes, on the one hand, more technologically advanced and, on the other, easier to apply routinely, it is likely that the real potential of the method will be realized, especially when allied to computerized data processing and statistical analysis systems. Already a microbial identification system of computerized high resolution GLC of fatty acids, based on an extensive reference library of known compositions of individual organisms, is being marketed commercially by one company. The ecology of marine microorganisms can benefit greatly from such technological advances and also provide an excellent testing ground for their further development.

REFERENCES

Ackman, R.G., Tocher, C.S. and McLachlan, J. 1968. Marine phytoplankter fatty acids. *Journal of the Fisheries Research Board of Canada*, **25**: 1603–1620.

Baird, B.H. and White, D.C. 1985. Biomass and community structure of the abyssal microbiota determined from the ester-linked phospholipids recovered from Venezuela basin and Puerto Rico trench sediments. *Marine Geology*, **68**: 217–231.

Beach, D.H., Harrington, G.W. and Holz, G.G. 1970. The polyunsaturated fatty acids of marine and freshwater cryptomonads. *Journal of Protozoology*, **17**: 501–510.

Bell, M.V., Henderson, R.J. and Sargent, J.R. 1986. The role of polyunsaturated fatty acids in fish. *Comparative Biochemistry and Physiology*, **83B**: 711–719.

Ben-Amotz, B., Tornabene, T.G. and Thomas, W.H. 1985. Chemical profile of selected species of microalgae with emphasis on lipids. *Journal of Phycology*, **21**: 72–81.

Bligh, E.G. and Dyer, W.J. 1959. A rapid method for total lipid extraction and purification. *Canadian Journal of Biochemistry and Physiology*, **37**: 911–917.

Bobbie, R.J. and White, D.C. 1980. Characterisation of benthic microbial community structure by high resolution gas chromatography of fatty acid methyl esters. *Applied and Environmental Microbiology*, **39**: 1212–1222.

Bousfield, I.J., Smith, G.L., Dando, T.R. and Hobbs, G. 1983. Numerical analysis of total fatty acid profiles in the identification of coryneform, nocardioform and some other bacteria. *Journal of General Microbiology*, **129**: 375–394.

Brassell, S.C. and Eglinton, G. 1984. Lipid indicators of microbial activity in marine sediments. In *Heterotrophic activity in the sea* (eds. J.E. Hobbie and P.J.LeB. Williams), pp. 105–136. Academic Press, London.

Chan, M., Himes, R.H. and Akagi, J.M. 1971. Fatty acid composition of thermophilic, mesophilic and psychrophilic clostridia. *Journal of Bacteriology*, **106**: 876–881.

Christie, W.W. 1982. *Lipid analysis*, 2nd edition. Pergamon Press, Oxford.

Chuecas, L. and Riley, J. 1969. The component fatty acids of the lipids of some marine phytoplankton species. *Journal of the Marine Biological Association of the UK*, **49**: 97–116.

Cranwell, P.A. 1976. Decomposition of aquatic biota and sediment formation: organic compounds in detritus resulting from microbial attack on the alga, *Ceratium hirundinella. Freshwater Biology*, **6**: 41–48.

DeLong, E.F. and Yayanos, A.A. 1986. Biochemical function and ecological significance of novel bacterial lipids in deep-sea procaryotes. *Applied and Environmental Microbiology*, **51**: 730–737.

Dowling, N.J., Widdel, F. and White, D.C. 1986. Comparison of the phospholipid-linked fatty acid biomarkers of acetate-oxidising sulphate-reducers and other sulphide-forming bacteria. *Journal of General Microbiology*, **132**: 1815–1825.

Edlund, A., Nichols, P.D., Roffey, R. and White, D.C. 1985. Extractable and lipopolysaccharide fatty acid and hydroxy acid profiles from *Desulfovibrio* species. *Journal of Lipid Research*, **26**: 982–988.

Erwin, J.A. 1973. Comparative biochemistry of fatty acids in eukaryotic organisms. In *Lipids and biomembranes of eukaryotic organisms* (ed. J.A. Erwin), pp. 42–143. Academic Press, New York.

Falk-Petersen, I.-B. and Sargent, J.R. 1982. Reproduction of asteroids from Balsfjorden, northern Norway: analysis of lipids in the gonads of *Ctenodiscus crispatus, Asterias lincki* and *Pteraster militaris. Marine Biology*, **69**: 291–298.

Falk-Petersen, S., Gatten, R.R., Sargent, J.R. and Hopkins, C.C.E. 1981. Ecological investigations on the zooplankton community in Balsfjorden northern Norway: seasonal changes in the lipid class composition of *Meganyctiphanes norvegica* (M. Sars), *Thysanoessa raschii* (M. Sars) and *T. inermis* (Kroyer). *Journal of Experimental Marine Biology and Ecology*, **54**: 209–224.

Federle, T.W., Livingston, R.J., Meeter, D.A. and White, D.C. 1983. Modifications of estuarine sedimentary microbiota by exclusion of epibenthic predators. *Journal of Experimental Marine Biology and Ecology*, **73**: 81–94.

Findlay, R.H., Pollard, P.C., Moriarty, D.J.W. and White, D.C. 1985. Quantitative determination of microbial activity and community nutritional status in estuarine sediments: evidence for a disturbance artifact. *Canadian Journal of Microbiology*, **31**: 493–498.

Folch, J., Lees, M. and Sloane Stanley, G.H. 1957. A simple method for the isolation and purification of total lipid from animal tissues. *Journal of Biological Chemistry*, **226**: 497–509.

Gehron, M.J. and White, D.C. 1983. Sensitive assay of phospholipid glycerol in environmental samples. *Journal of Microbiological Methods*, **1**: 23–32.

Gillan, F.T. and Hogg, R.W. 1984. A method for the estimation of bacterial biomass and community structure in mangrove-associated sediments. *Journal of Microbiological Methods*, **2**: 275–293.

Gillan, F.T. and Johns, R.B. 1983. Normal-phase HPLC analysis of microbial carotenoids and neutral lipids. *Journal of Chromatographic Science*, **21**: 34–38.

Gillan, F.T., Johns, R.B., Verheyen, T.V., Volkman, J.K. and Bavor, H. 1981. Trans-monounsaturated acids in a marine bacterial isolate. *Applied and Environmental Microbiology*, **41**: 849–856.

Goldfine, H. and Hagen, P.O. 1972. Bacterial plasmalogens. In *Ether lipids: chemistry and biology* (ed. F. Snyder), pp. 329–350. Academic Press, New York.

Goodloe, R.S. and Light, R.J. 1982. Structure and composition of hydrocarbons and fatty acids from a marine blue–green *Synechococcus* sp. *Biochimica et Biophysica Acta*, **710**: 485–492.

Guckert, J.B., Hood, M.A. and White, D.C. 1986. Phospholipid, ester-linked fatty acid profile changes during nutrient deprivation of *Vibrio cholerae*: increase in the trans/cis ratio and proportions of cyclopropyl fatty acids. *Applied and Environmental Microbiology*, **52**: 794–801.

Harrington, G.W., Beach, D.H., Dunham, J.E. and Holz, G. 1970. The polyunsaturated fatty acids of marine dinoflagellates. *Journal of Protozoology*, **17**: 213–219.

Harwood, J.L. and Russell, N.J. 1984. *Lipids in plants and microbes*. 162 pp. George Allen and Unwin, London.

Jannasch, H.W. and Taylor, G.D. 1984. Deep sea microbiology. *Annual Reviews of Microbiology*, **38**: 487–514.

Jensen, L.M. 1985. ^{14}C-labelling patterns of phytoplankton: specific activity of different product pools. *Journal of Plankton Research*, **7**: 643–652.

Johns, R.B. and Perry, G.J. 1977. Lipids of the marine bacterium *Flexibacter polymorphus*. *Archives of Microbiology*, **114**: 267–271.

Johns, R.B., Perry, G.J. and Jackson, K.S. 1977. Contribution of bacterial lipids to recent marine sediments. *Estuarine and Coastal Marine Science*, **5**: 521–529.

Jørgensen, B.B. 1982. Mineralisation of organic matter in the sea bed — the role of sulphate reduction. *Nature*, **296**: 643–645.

Joseph, J. 1975. Identification of 3,6,9,12,15-octadecapentaenoic acid in laboratory cultured photosynthetic dinoflagellates. *Lipids*, **19**: 395–430.

Kaneda, T. 1977. Fatty acids of the genus *Bacillus*: an example of branched chain preference. *Bacteriological Reviews*, **41**: 391–418.

Kattner, G.G., Gercken, G. and Ebelein, K. 1983. Development of lipids during a spring phytoplankton bloom in the northern North Sea. I. Particulate fatty acids. *Marine Chemistry*, **14**: 149–162.

Keddie, R.M. and Bousfield, I.J. 1980. Cell wall composition in the classification and identification of Coryneform bacteria. In *Microbiological classification and identification: SAB Symposium Series No. 8* (eds. M. Goodfellow and R.G. Board), pp. 167–188. Academic Press, New York and London.

Kerger, B.D., Nichols, P.D., Antworth, C.P., Sand, W., Bock, E., Cox, J.C., Langworthy, T.A. and White, D.C. 1986. Signature fatty acids in the polar lipids of acid producing *Thiobacillus* sp.: Methoxy, cyclopropyl, alpha-hydroxy-cyclopropyl and branched and normal monoenoic fatty acids. *FEMS Microbiological Ecology*, **32**: in press.

Lancelot, C. 1985. Metabolic changes in *Phaeocystis poucheti* (Hariot) Lagerheim during the spring bloom in Belgian coastal waters. *Estuarine, Coastal and Shelf Science*, **18**: 593–600.

Lechevalier, M.P. 1977. Lipids in bacterial taxonomy — a taxonomist's view. *Critical Reviews of Microbiology*, **5**: 109–210.

Lechevalier, M.P. 1982. Lipids in bacterial taxonomy. In *Handbook of Microbiology* (eds. A.I. Laskin and H.A. Chevalier), Vol. 4, 2nd edition, pp. 435–541. CRC Press, Boca Raton, FL.

Leo, R.G. and Parker, P.L. 1966. Branched chain fatty acids in sediments. *Science*, **152**: 649–650.

Li, W.K.W. and Platt, T. 1982. Distribution of carbon among photosynthetic end-products in phytoplankton of the eastern Canadian arctic. *Journal of Phycology*, **18**: 466–471.

Martz, R.F., Sebacher, D.I. and White, D.C. 1983. Biomass measurement of methane forming bacteria in environmental samples. *Journal of Microbiological Methods*, **1**: 53–61.

Mayzaud, P., Eaton, C.A. and Ackman, R.G. 1976. The occurrence and distribution of octadecapentaenoic acid in a natural plankton population. A possible food chain index. *Lipids*, **11**: 858–862.

Moriarty, D.J.W., White, D.C. and Wassenberg, T.J. 1985. A convenient method for measuring rates of phospholipid synthesis in seawater and sediments: its relevance to the determination of bacterial productivity and the disturbance artifacts introduced by measurements. *Journal of Microbiological Methods*, **3**: 321–330.

Morris, R.J. 1984. Studies of a spring phytoplankton bloom in an enclosed experimental ecosystem. II. Changes in the component fatty acids and sterols. *Journal of Experimental Marine Biology and Ecology*, **75**: 59–70.

Nichols, P.D., Volkman, J.K. and Johns, R.B. 1983. Sterols and fatty acids of the marine unicellular alga, FCRG 51. *Phytochemistry*, **22**: 1447–1452.

Nichols, P.D., Jones, G.J., DeLeeuw, J.W. and Johns, R.B. 1984. The fatty acid and sterol composition of two marine dinoflagellates. *Phytochemistry*, **23**: 1043–1047.

Nichols, P.D., Smith, G.A., Antworth, C.P., Hanson, R.S. and White, D.V. 1985. Phospholipid and polysaccharide normal and hydroxy fatty acids as potential signatures for methane-oxidising bacteria. *FEMS Microbiological Ecology*, **31**: 327–335.

Nickels, J.S., King, J.D. and White, D.C. 1979. Poly-beta-hydroxybutyrate metabolism as a measure of unbalanced growth and estuarine detrital microbiota. *Applied and Environmental Microbiology*, **37**: 459–465.

Parkes, R.J. 1987. Analysis of microbial communities within sediments using biomarkers. In *Ecology of microbial communities* (eds. M. Fletcher, T.R.G. Gray and J.G. Jones), pp. 147–177. Cambridge University Press, Cambridge.

Parkes, R.J. and Calder, A.G. 1985. The cellular fatty acids of three strains of *Desulfobulbus*, a propionate utilising sulphate-reducing bacterium. *FEMS Microbiological Ecology*, **31**: 361–363.

Parkes, R.J. and Taylor, J. 1983. The relationships between fatty acid distributions and bacterial respiratory types in contemporary sediments. *Estuarine, Coastal and Shelf Science*, **16**: 173–189.

Perry, G.J., Volkman, J.K., Johns, R.B. and Bavor, H.J. 1979. Fatty acids of bacterial origin in contemporary marine sediments. *Geochimica et Cosmochimica Acta*, **43**: 1715–1725.

Pohl, P. and Zurheide, S. 1979. Fatty acids and lipids in marine algae and the control of their biosynthesis by environmental factors. In *Marine algae in pharmaceutical science* (eds. H.A. Hoppe, T. Levring and Y. Tanaka), pp. 473–523. Walter de Gruyter, Berlin.

Rivkin, R.B. 1985. Carbon-14 labelling patterns of individual marine phytoplankton from natural populations. *Marine Biology*, **89**: 135–142.

Rogers, H.J. 1983. Bacterial cell structure. In *Aspects of microbiology* (eds. J.A. Cole, C.J. Knowles and D. Schlessinger), Vol. 6. Van Nostrand Reinhold (UK), Wokingham.

Rohmer, M., Bouvier-Nave, P. and Ourisson, G. 1984. Distribution of hopanoid triterpenes in prokaryotes. *Journal of General Microbiology*, **130**: 1137–1150.

Sargent, J.R. and Falk-Petersen, S. 1981. Ecological investigations on the zooplankton community in Balsfjorden, northern Norway: lipids and fatty acids in *Meganyctiphanes norvegica. Thysanoessa raschi* and *T. inermis* during mid-winter. *Marine Biology*, **62**: 131–137.

Sargent, J.R. and Henderson, R.J. 1987. Marine (*n*-3) polyunsaturated fatty acids. In *Developments in oils and fats* (ed. R.J. Hamilton), in the press. Elsevier, Amsterdam.

Sargent, J.R., Eilertsen, H.C., Falk-Petersen, S. and Taasen, J.P. 1985. Carbon assimilation and lipid production in phytoplankton in northern Norwegian fjords. *Marine Biology*, **85**: 109–116.

Seto, A., Wang, H.L. and Hesseltine, C.W. 1984. Culture conditions affect eicosapentaenoic acid content of *Chlorella minutissima. Journal of the American Oil Chemists' Society*, **61**: 892–894.

Shaw, N. 1974. Lipid composition as a guide to the classification of bacteria. *Advances in Applied Microbiology*, **17**: 63–108.

Smith, A.E. and Morris, I. 1980. Synthesis of lipid during photosynthesis by phytoplankton of the southern ocean. *Science*, **207**: 197–199.

Taylor, J. and Parkes, R.J. 1983. The cellular fatty acids of the sulphate-reducing bacteria, *Desulfobacter* sp., *Desulfobulbus* sp. and *Desulfovibrio desulfuricans. Journal of General Microbiology*, **129**: 3303–3309.

Taylor, J. and Parkes, R.J. 1985. Identifying different populations of sulphate reducing bacteria within marine sediment systems using fatty acid biomarkers. *Journal of General Microbiology*, **131**: 631–642.

Van Vleet, E.S. and Quinn, T.G. 1979. Early diagenesis of fatty acids and isoprenoid alcohols in estuarine and coastal sediments. *Geochimica et Cosmochimica Acta*, **43**: 289–303.

Volkman, J.K., Eglinton, G. and Corner, E.D.S. 1980. Sterols and fatty acids of the marine diatom *Biddulphia sinensis. Phytochemistry*, **19**: 1809–1813.

Volkman, J.K., Smith, D.J., Eglinton, G., Forsberg, T.E.V. and Corner, E.D.S. 1981. Sterol and fatty acid composition of four marine haptophycean algae. *Journal of the Marine Biological Association of the UK*, **61**: 509–527.

Weiss, T. 1983. Feeding of calanoid copepods in relation to *Phaeocystis* blooms in the German Wadden Sea area off Sylt. *Marine Biology*, **74**: 87–94.

White, D.C. 1983. Analysis of microorganisms in terms of quantity and activity in natural environments. In *Microbes in their natural environment* (eds. J.H. Slater, R. Whittenbury and J.W.T. Wimpenny), pp. 37–66. Cambridge University Press, Cambridge.

White, D.C., Davis, W.M., Nickels, J.S., King, J.D. and Bobbie, R.J. 1979. Determination of the sedimentary biomass by extractable lipid phosphate. *Oecologia*, **40**: 51–62.

White, D.C., Bobbie, R.J., Nickels, J.S., Fazio, S.D. and Davis, W.M. 1980. Nonselective biochemical methods for the determination of fungal mass and community structure in estuarine detrital microflora. *Botanica Marina*, **23**: 239–250.

Wood, B.J.B. 1974. Fatty acids and saponifiable lipids. In *Algal physiology and biochemistry* (ed. W.D.P. Stewart), pp. 236–265. Blackwell Scientific Publications, Edinburgh.

6

Analytical flow cytometry and its application to marine microbial ecology

P.H. Burkill, NERC Institute for Marine Environmental Research, Prospect Place, Plymouth PL1 3DH, UK

Analytical flow cytometry is a novel biomedical technique for the rapid ($> 10^3$ s^{-1}) characterization, quantification and sorting of particles based on simultaneous, multiple measurements of cellular light scatter and fluorescence. Light scatter provides particulate micromorphometric characterization, while fluorescence is related to biochemical properties of particles. Monodisperse particles at concentrations up to 5×10^6 ml^{-1} held in aqueous suspension may be directly analysed. Particles within the size range 0.5–150 µm may be measured by light scatter, while the fluorescence detection limit is typically equivalent to 10^3–10^4 fluorescein molecules per particle. Microbial cytometric applications are diverse, reflecting the multidisciplinary nature of the technique, but have been centred principally on the autofluorescent photosynthetic pigments of cyanobacteria and other phytoplankton. Although the technique has been applied to freshwater bacteria, it has yet to be routinely applied to marine bacteria. Direct applications to protozoa are hampered because of low numerical concentration in natural waters, although experimental studies on protozoan grazing have been carried out. Cytometric analysis and sorting of natural microparticulate matter has been successfully carried out at sea. These applications are reviewed together with those fringe areas relevant to cytometry. The unique nature of this powerful analytical technique, together with commercial developments in cytometer design and technology, will accelerate the future application of flow cytometry in marine microbial ecology.

INTRODUCTION

With the growing public awareness of environmental problems and the scientific realization of the importance of microbes in biogeochemical cycling, environmental microbiology has recently become a fast-developing field. While advances in concepts of biogeochemical cycling have been made (Laws, 1983), progress in environmental microbiology also depends on technological innovation. Hypotheses can only be tested with the production of data of appropriate quality, and our understanding can only advance with a suitable marriage between concept and technology. The history of the development of oceanography testifies to the number of techniques adapted from other, often better funded, scientific fields. The application of analytical flow cytometry (AFC) to oceanography is another example of this; developed in the biomedical sciences for cancer research, it is now beginning to be applied in environmental research as a powerful technique for the rapid, accurate and sensitive analysis and sorting of particles. In this review, the technique of AFC is presented together with current applications and future aspirations in marine phytoplankton, bacterial and protozoan ecology.

ANALYTICAL FLOW CYTOMETRY AS A RESEARCH TECHNIQUE

Flow cytometry is now a vast research field with around 1500 instruments world-wide operating mainly with biomedical research and clinical applications. This field has been recently reviewed by Steinkamp (1984) and Van Dilla *et al.* (1985), while Shapiro (1985) provides a practical and charismatic reference source for all aspects of flow cytometry.

The cytometer
Flow cytometers vary enormously in their design and research capability; they range from relatively simple microscope-based systems through to complex research-mode flow sorters. The capabilities and costs of the main categories are shown in Table 1. All categories can give excellent analytical precision, but the simpler systems are typically less stable and less sensitive than the laser-based cytometers and are incapable of sorting particles. Although cytometers can be laboratory built, the commonest instrument in marine microbial applications is Coulter Electronics EPICS 741. Cytometer descriptions in thie review refer to this instrument unless stated otherwise.

The principles of cytometry
AFC is a biomedical research technique routinely used for the characterization, enumeration and sorting of cells or other particles from heterogeneous populations (Horan and Wheeless, 1977; Melamed *et al.*, 1979; Kruth, 1982). The analytical and sorting procedures are based on simultaneous measurements of multiple biochemical parameters such as chlorophyll, protein, nucleic acid and immuno-fluorescent properties, as well as light scattering, of cells up to 150 μm in size, at rates exceeding 10^3 cells s^{-1}. The technique, which is depicted in Fig. 1, uses a laser-based flow cytometer (Fig. 2) that measures colour-differentiated fluorescence and light scattering to analyse and sort cells. Cells in a continuous stream of electrolyte, such as seawater, are presented one at a time to the beam of a laser. As they pass through the beam, cells will scatter light and may also fluoresce. Light scatter detectors are situated in the low (typically 2°–20°) forward angle for the measurement of cell size and at 90° for assessing topographic and internal cell structure. Fluorescence may be

Design	Irradiation	Parameters	Price range (£000)	Manufacturer or reference	Model
Simple, analysis only					
Microscope based	Hg lamp	1	<30	Fukuda et al. (1982)	
				Olson et al. (1983)	
				Leitz	MPV
Intermediate, analysis only					
Frame	Hg lamp	4	50–75	Becton Dickinson	FACS Analyser
	25 mW laser	5		Becton Dickinson	FACScan
	150 mW laser	4		Coulter	Profile
	100 mW laser	4		Ortho	Spectrum
Research mode, sorting analysers					
Frame	3 W laser	4	90–200	Coulter	EPICS C
	5 W laser	5		Coulter	EPICS 741
	3×3–5 W laser	6		Coulter	EPICS 753
	4 W laser	5		Becton Dickinson	400 series
	Various	Up to 8		Ortho	Cytofluorograf

Table 1   Research capability and price of different designs of flow cytometer

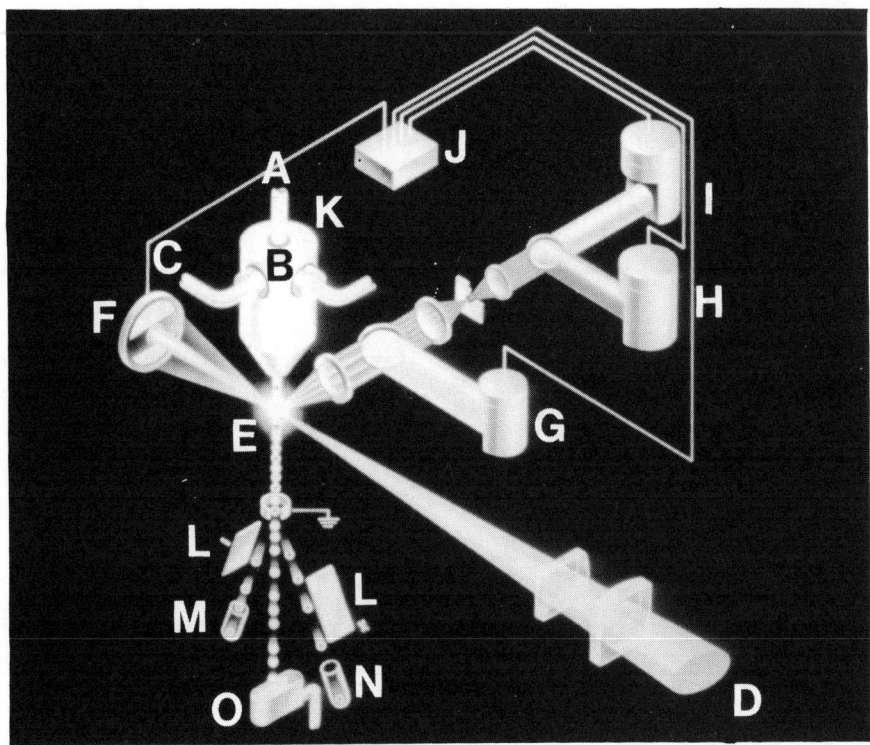

Fig. 1—A schematic diagram of particle analysis and sorting by AFC. Cells or other suspended particles are introduced (A) to the flow chamber (B) where they are hydrodynamically focused by sheath fluid (C). Monochromatic light from an argon-ion laser (D) irradiates each particle as it passes singly through the beam. Particles scatter and may fluoresce light from the analysis point (E). A photodiode (F) quantifies the forward-angle light scatter to measure particle size, while photomultiplier tubes (G,H,I) pick-up the 90° light scatter for particle refractive index, and two-colour, wavelength-selective fluorescence to measure biochemical properties of the particle. Signals from these sensors are processed electronically (J) and held in a computer file. The sample stream is vibrated by a piezoelectric transducer (K), producing uniform liquid droplets at $35\,000\,s^{-1}$ travelling at $10\,m\,s^{-1}$. Droplets containing particles exhibiting the pre-set scatter and fluorescence characteristics are electrostatically charged at the droplet formation break-off point. As these droplets pass charged plates (L), they are deflected into sort containers (M,N). Particles not meeting the required sort criteria pass undeflected into waste (O). Particles may be analysed and sorted at up to $5000\,s^{-1}$, with collection purities of $>98\%$. (Courtesy of Coulter Electronics Ltd., Luton.)

due either to autofluorescence, such as chlorophyll in algae, or to applied, induced or immunologically coupled fluorescence from specific fluorochromes, such as fluorescein for cell protein. The scatter and fluorescence characteristics of cell size, granularity, fluorescence intensity and spectral properties are collected by refined optical sensors and stored in computer memory. Sorting is achieved by ultrasonically vibrating the sample stream with a piezo-electric crystal to yield a series of regularly spaced droplets of uniform size each containing no more than one cell, as shown in Fig. 1. Those droplets exhibiting the desired scatter and/or fluorescence properties

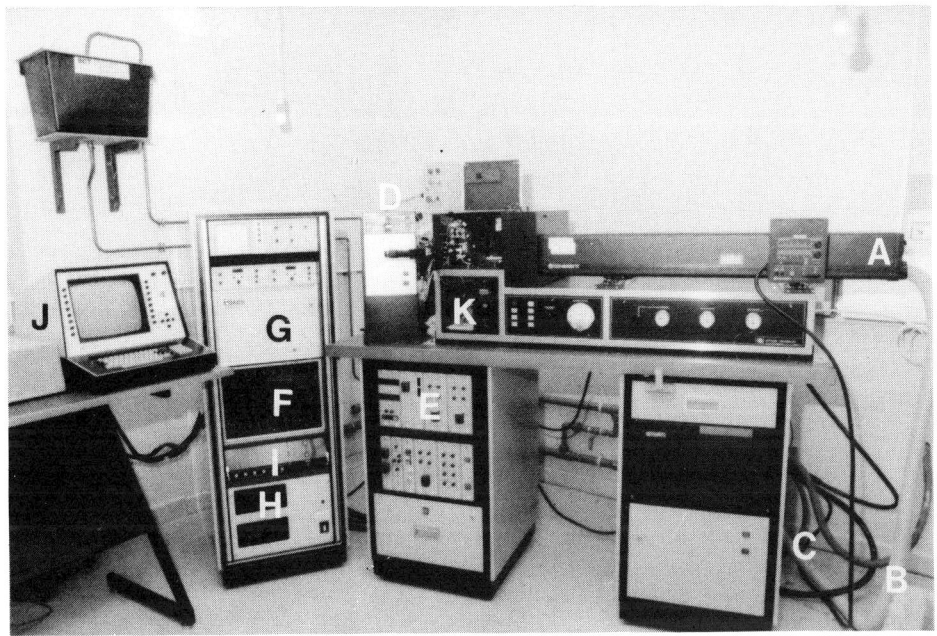

Fig. 2—The research-mode sorting analytical flow cytometer (EPICS 741) at IMER, Plymouth. A 5 W argon-ion laser (A) powered with three-phase electricity (B), and water cooled (C), irradiates particles from a sample (D) as they pass the analysis zone. Optical sensors are controlled (E) as analog signals are converted to digital signals (F) before being held in a computer file in memory (G) or disc (H). Sort-logic boards (I) pre-programmed by computer (J) control particle sorting into containers (K).

are electrostatically charged and, as they pass at a velocity of $10\,\mathrm{m\,s}^{-1}$ between charged plates, are deflected into appropriate sort-collection vessels.

Particle sorting is one of the unique characteristics of AFC; another is that the particles are analysed singly. This has two powerful consequences; firstly, measurements of the intra-population variability of individuals are made. This is in contrast to conventional bulk measurements when the distribution of properties of individuals within a population cannot be determined. Intra-population variability provides a powerful descriptor of population. Consider the situation depicted in Fig. 3, where histograms of the cellular DNA content of three different hypothetical microbial cultures are shown. Fig. 3a shows a culture in balanced exponential growth, while Fig. 3b depicts a highly synchronized culture with cells at an identical stage of the mitotic cycle. In contrast, Fig. 3c reveals three sub-populations. Analysed by bulk measurements, all three cultures would show the same average cellular DNA concentration, yet it is clear from analysis of individual cells that the populations have distinctly different DNA, and hence growth, characteristics.

The second consequence of individual particle analysis is that by simultaneous measurement of multiple parameters, a multidimensional (or multivariate) description of the particles can be obtained. Fig. 4 shows one-dimensional histograms of intensity of light scatter (Fig. 4a) and of fluorescence (Fig. 4b) obtained from a

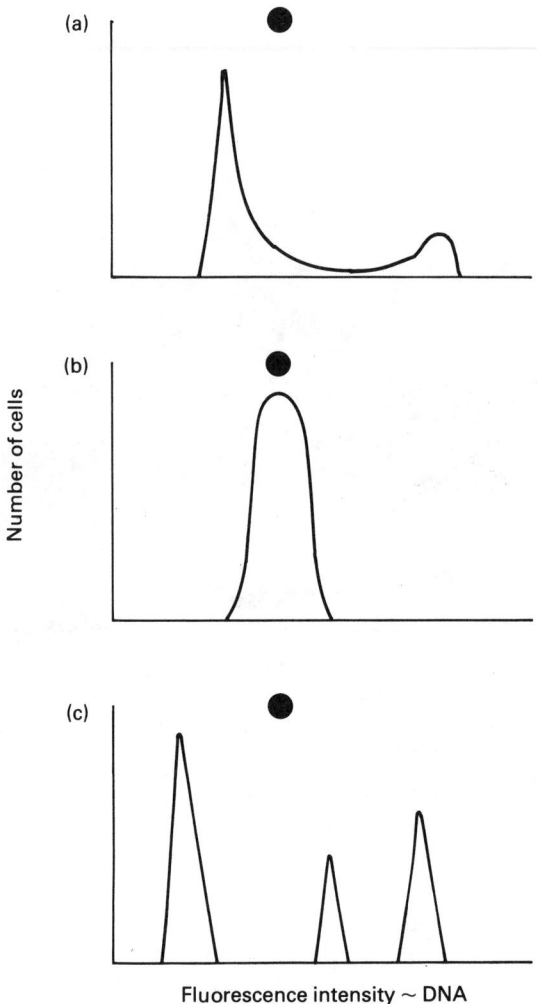

Fig. 3—Analysis of individual particles, a characteristic of AFC, provides a more powerful description than conventional bulk measurements. The first population a is in balanced exponential growth, while the second population b is in a highly synchronized growth state with all cells at an identical stage of the mitotic cycle. Population c comprises three discrete sub-populations. Bulk measurements would only suggest that hypothetical populations a, b and c have the same mean cellular DNA content (●), yet cytometric analysis of individual particles clearly reveals intra-population differences.

hypothetical sample. In the light scatter and fluorescence plots, a broad histogram is revealed in the former while in the latter a much narrower histogram is present together with a suggestion of another population with negligible fluorescence. By themselves, these two one-dimensional histograms provide inadequate discrimination of the populations which are revealed in the two-dimensional light scatter and fluorescence histogram (Fig. 4c) as having three distinct components. Components 1

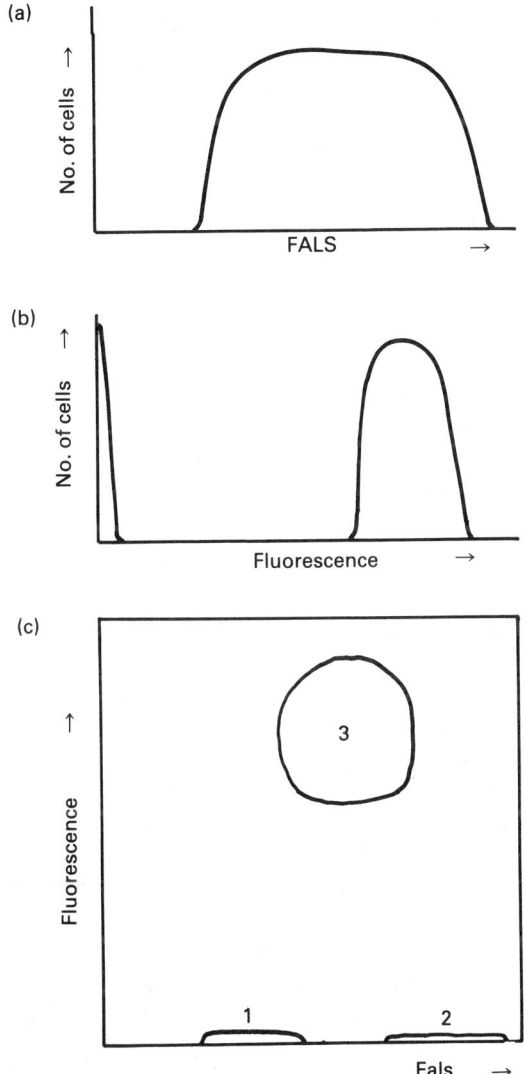

Fig. 4—Simultaneous measurements of multiple parameters allows a more precise characterization of the cells than measurement of single parameters. A histogram of cellular FALS reveals a disperse population a. Measurements of fluorescence in the same population reveal two distinct components b. Simultaneous bivariate measurements of dual parameters c reveal clearly the presence of three distinct populations.

and 3 are characterized by the absence of fluorescence and differ from each other in terms of their light scatter characteristics. Component 2 is highly fluorescent and has light scatter characteristics intermediate between the other two components. Thus, whereas one-dimensional analyses of multi-component populations may give inadequate resolution, such populations may be better defined by the multidimensional analytical capability of AFC. Whereas it is difficult to display simultaneously more than three dimensions, sophisticated multivariate statistical techniques (Mardia *et al.*, 1979), such as multidimensional scaling, allow the full potential offered from the simultaneous collection of up to six parameters from each particle by AFC analysis (Dean, 1985).

Particle characterization

Particle characterization is restricted typically to six of the following parameters:

- forward angle light scatter (FALS)
- 90° light scatter (90°LS)
- log 90° light scatter (L90°LS)
- integral 'green' fluorescence intensity (IGFL)
- log integral 'green' fluorescence intensity (LIGFL)
- peak 'red' fluorescence intensity (PRFL)
- log peak 'red' fluorescence intensity (LPRF)
- peak 'green' fluorescence intensity (PGFL)
- log peak 'green' fluorescence intensity (LPGFL)
- time of flight (TOF)
- time

FALS is used to size particles (Mullaney *et al.*, 1969), while 90°LS is primarily sensitive to particle refractive index (Salzman, 1982). Particle fluorescence is used to characterize the cellular composition and/or biochemistry. The fluorescence-detection electronics may record 'integral' and 'peak' signals. Here the 'integral' signal signifies the total quantity of light received by a sensor, while the 'peak' signal records the maximum amplitude of the signal as the particle passes through the laser beam. There may be a striking difference between integral and peak signals. Imagine two phytoplankton cells, one spherical, the other rod shaped, each containing the same quantity of homogeneously dispersed chlorophyll. Each cell will generate identical integral signals, but the spherical cell will generate a higher peak signal due to a greater cross sectional concentration of chlorophyll than the thinner rod. Thus peak and integral signals may be used to determine the internal distribution of cellular constituents and allow geometrical characterization of cells. TOF measures the passage time of a particle or particle component through the laser beam and provides another morphometric characteristic (Cram *et al.*, 1985). In addition to these parameters it is possible to generate ratios of parameters. In short, the instrument produces a multitude of multidimensional signals with which cellular micromorphometry and chemical constitution may be determined. Steinkamp (1984) and Shapiro (1985) have reviewed cytometric particle characterization. The signals most commonly used in microbial ecology are FALS and integral fluorescence, and these will be briefly described below.

FALS and particle sizing

FALS is used commonly to measure particle size. Although size is the main variable affecting the magnitude of the FALS signal, it is also affected by shape, density, pigmentation, granularity and refractive index (Mullaney and Dean, 1969; Salzman, 1982). As a result, absolute cell sizing by FALS can be difficult to achieve. The scattering of light can be described by Maxwell's electromagnetic field equations

(Stratton, 1941) and the theoretical solutions for homogeneous spheres are well known (Kerker, 1969). In this type of particle, the magnitude of FALS has been shown to be linearly related to cell diameter (Phinney *et al.*, 1987b), to cross sectional area (Salzman, 1982) and to volume (Mullaney and Dean, 1970). For non-spheroidal particles, there is no comparable model, although light scatter signals have been considered both conceptually (Wang and Barber, 1979) and for marine microbes (Yentsch and Yentsch, 1984). In spite of its imprecision as a means of sizing microbes, FALS has a major advantage — that it is a non-destructive probe and therefore can be used in the analysis and sorting of unstained viable cells.

Fluorescence and biochemical properties

Quantitative cytochemical staining techniques based on fluorochromes, coupled with AFC, provide a rapid and accurate method for determining biochemical, cytological and functional properties of microbes and other particles. As a quantitative technique, AFC permits three levels of characterization of cell properties. These are

(i)   semi-quantitation
(ii) relative quantitation
(iii) absolute quantitation

The semi-quantitative approach is used to characterize cells without concern for the interpretation of the characterization parameters. Examples are found in feeding experiments of Stoecker *et al.* (1986) and Cucci *et al.* (1985). The relative quantitation approach is important in cell cycle analysis where G1, S and G2+M phases of the mitotic cycle must be discerned by their relative DNA concentrations, as in Olson *et al.* (1983). An example of the absolute quantitation of a cellular constituent is shown in Fig. 5, where the concentration of cellular chlorophyll-a content in different phytoplankton is shown to be linearly related to AFC fluorescence intensity (Burkhill and Mantoura, in preparation). While semi- and relative quantitation of cellular constituents is common, absolute quantitation, although offering tremendous potential, is rarely carried out.

As a particle characterization technique, fluorescence has three major advantages. These are as follows:

(1) Fluorescence intensity is directly proportional to specific cellular concentration (Kerker *et al.*, 1982; and Fig. 5).
(2) It is highly sensitive, with typical detection limits of about 1000 to 10 000 fluorochrome molecules in commercial instruments and down to 150 molecules in purpose-built cytometers (Watson, 1987).
(3) Some non-fluorescent cellular molecules can be chemically converted to a fluorescent form (Yentsch, 1981).

A great multiplicity of fluorochromes are used in biomedical AFC; these and their uses have been recently reviewed elsewhere (Steinkamp, 1984; Shapiro, 1985) and will not be considered here.

In the microbial field, fluorescent signals may be derived from the following:

(1) the autofluorescence of photosynthetic pigments present in phytoplankton (Yentsch and Yentsch, 1979);

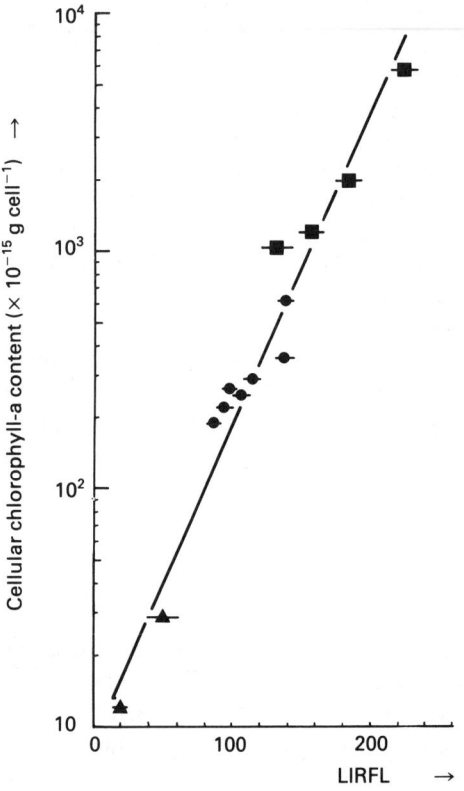

Fig. 5—Plot showing the calibration of LIRFL against cellular chlorophyll-a content of sorted sub-sets of three species of phytoplankton (▲, *Agmanellum quadruplicatum*, a cyanobacterium; ●, *Phaeodactylum tricornutum*, a diatom; ■, *Amphidinium carterae*, a dinoflagellate). Irradiation was with the 488 nm laser line set at 200 mW; fluorescence was measured at > 630 nm. Chlorophyll-a was determined by high-performance liquid chromatographic analysis of a known number of sorted cells. From Burkill and Mantoura (in preparation).

(2) induced fluorescence derived from the interaction of a non-fluorescent stain with a cellular constituent to yield a fluorescent product (Rotman and Papermaster, 1966);

(3) applied fluorochromes such as Hoechst 33342 labelling cellular DNA (Olsen *et al.*, 1983);

(4) immuno-fluorescently coupled antibodies for specific binding to cell markers (Campbell *et al.*, 1983).

Each of these will be discussed in the application section below.

Practical considerations

The power of multidimensional analysis and sorting by AFC can only be achieved when samples comprise a uni-disperse cell preparation. Other practical criteria for AFC are the following: the optimum concentration range of particles is 5×10^4 to

5×10^6 cells ml^{-1}; the upper size limit of particles for analysis is about $150\,\mu m$, and for sorting is about $75\,\mu m$. The smallest resolvable particle size, which depends on the detection limit of the sensor and the type of fluorochrome, is about $0.5\,\mu m$ for light scatter or requires typically 10^4 molecules of fluorescein or 10^6 molecules of chlorophyll-a per particle (Burkhill and Mantoura, in preparation).

Fortunately, micro-organisms fulfill many of these prerequisites. Fig. 6a shows the good correspondence between the sizes of microorganisms in different microbial taxa and the optimal size range for analysis of particles by AFC. Protozoa, most cyanobacteria and other phytoplankton, together with the larger bacteria, can be resolved using light scatter down to a threshold of $0.5\,\mu m$. Bacteria smaller than this can only be detected with applied or induced fluorescence. The photosynthetic pigments of phytoplankton allow their quantitation by fluorescence.

While cell concentration can be manipulated in culture work to ensure a satisfactory concentration of particles, this cannot be easily achieved when working with marine natural water samples. Fig. 6b shows typical Coulter counter size spectra of particle concentrations measured in the euphotic zone of eutrophic near-shore (North Sea), intermediate coastal (Celtic Sea) and oligotrophic off-shore (Celtic Sea shelf break) waters. These data, which range down to a particle diameter of $1.6\,\mu m$, have been supplemented with typical concentrations of cyanobacteria and bacteria found in the Celtic Sea (Burkhill, unpublished data) to extend the size spectrum downwards. As the Coulter counter does not differentiate between microorganisms and detritus, the concentrations of biotic components $> 1.6\,\mu m$ in size may be considerably lower than those figured. It can be seen that the concentration of particles in natural water samples is sub-optimal for AFC analysis. The time taken to analyse 1000 particles at different concentrations is shown in Fig. 6c, which demonstrates that picoplankton populations are amenable to AFC. To analyse particles $\gtrsim 2\,\mu m$ at concentrations shown in Fig. 6b requires analysis times of > 30 minutes with standard AFC settings. These times can be reduced by sample concentration (Olsen *et al.*, 1985) or by increasing sample throughput either by increasing sample flow rate of by increasing flow-tip size. Analytical precision is likely to be reduced in these situations.

The determination of absolute particle concentration by cytometry is often difficult. Sorting instruments and some analysers use pressure-feed fluidics to drive the sample through the cytometer at constant velocity; this design does not readily allow measurements of the sample volume analysed, and particle concentrations are expressed relatively. In biomedical research, this is often satisfactory. When absolute particle concentrations are required, an internal calibrant such as reference fluorospheres may be used. It is important that the reference beads are cytometrically distinct from the other particles. Particle concentration may be determined as follows: let A be the cytometric cell count, B be the initial concentration of reference beads, C the volume of beads added, D the cytometric count of reference beads and E the initial sample volume. It follows that the particle concentration, F, is:

$$F = A \times B \times C / D \times E$$

This approach, described by Stewart and Steinkamp (1982), has been used to determine concentrations of cyanobacteria in natural waters (Olson *et al.*, 1985).

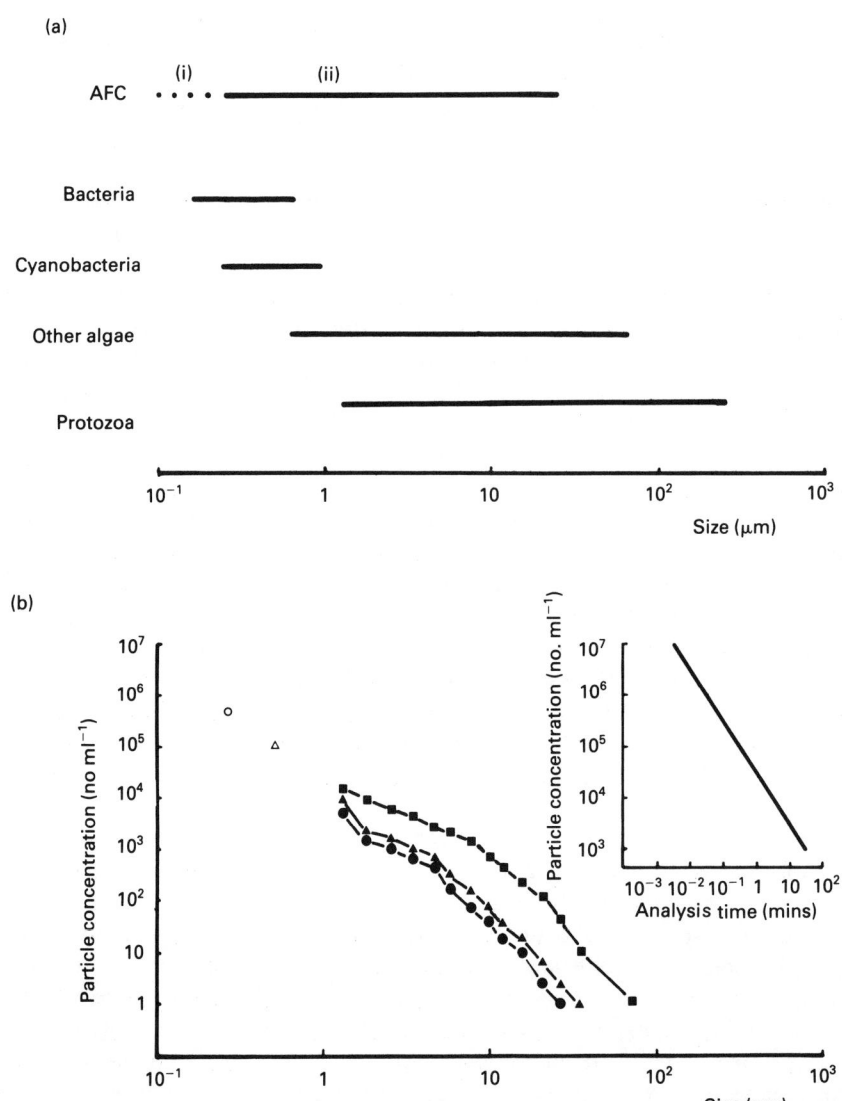

Fig. 6—a, Size spectrum of different marine microbial taxa and its inter-relationship with the size of particles suitable for cytometric analysis by fluorescence (i) and by light scatter and fluorescence (ii). b, Concentration of different size particles measured by Coulter counter in the euphotic zone of eutrophic, mesotrophic and oligotrophic marine waters (■: North Sea; ▲: Celtic Sea; ●: Celtic Sea shelf break, and by microscopy for bacteria (○) and cyanobacteria (△) in the Celtic Sea. c, Time taken to analyse 1000 particles of different concentrations, using standard instrument settings.

While the internal standard approach has the advantage of simplicity, some caution is urged in its use, as the detergents commonly used to ensure adequate bead dispersion have been shown to cause cell lysis in some flagellate species (Burkhill, unpublished). The new models of analysers that are now becoming available can aspirate a pre-programmed volume of sample, thereby avoiding this indirect method of determining absolute particle concentration.

A simple volumetric sample injector has been described for cytometers (Steen, in Visser and Tanke, 1985). However, for reasons given above, this injection system may be restricted to analysis-only cytometers. To allow volumetric injection with sorting instruments, the constant sample pressure must be precisely maintained. This could be carried out by introduction of a quantitative sample into the analysis stream using a Rheodyne injector system as routinely used in high-performance liquid chromatography (Mantoura, personal communication).

THE APPLICATION OF ANALYTICAL FLOW CYTOMETRY TO MICROBIAL ECOLOGY

Although the application of AFC to marine microbial ecology is relatively new, it is a fast expanding field with diverse applications. These range from lab studies of cell toxins (Yentsch, 1981) to shipboard research on picoplankton (Olson *et al.*, 1985), and involve many different cellular biochemical properties. A summary of microbial constituents that can be characterized by cytometric fluorescence and light scatter is shown in Table 2. Illustrative examples of laboratory-based AFC measurements on eucaryote and procaryote phytoplankton, bacteria and protozoa as well as shipboard analyses of natural particulates are given below.

Phytoplankton
Phytoplankton may be cytometrically characterized by the presence of chlorophyll-a autofluorescence with an emission maximum at 680 nm (Table 3). The unique excitation and emission spectral signatures that characterize the cyanobacteria allow their differentiation from all other phytoplankton except cryptophytes. Although these taxa contain both chlorophyll and phycobiliprotein, the relative proportions of these two photopigments in the two groups differ; the ratio of phycobiliprotein to chlorophyll-a is higher in cyanobacteria than in cryptophytes under culture conditions. The two taxa may be distinguished by differences in size, since cryptophytes are typically $>2\,\mu m$ in diameter (Stockner and Antia, 1986). It is not yet known whether this is reflected in differences in FALS signals. The spectral differences and the relative photopigment proportions between phycoerythrin-containing cyanobacteria and other algae can be enhanced by using the 514 nm laser line in AFC, since this wavelength preferentially excites phycoerythrin (Wood *et al.*, 1985).

Eucaryotic phytoplankton
Cytometric chlorophyll-a analysis was first used by Paau *et al.* (1978) to study fresh water algal cells, and to differentiate them from bacteria in mixed culture (Paau *et al.*, 1979). The magnitude of chlorophyll fluorescence was noted to vary greatly

Biochemical property or particle type	Characterization
Light scatter	
Size	FALS
Refractive index	90°LS
Cytoplasm granularity	90°LS
Nucleus–cytoplasm ratio	90°LS
Bio-optical properties	FALS and 90°LS
Autofluorescence	
Autotrophic phytoplankton	Chlorophyll-a
Autotrophic cyanobacteria	Phycoerythrin
Bioluminescent microbes	Luciferin–luciferase
Applied, induced and immuno-labelled fluorochromes	
Protein	FITC
	Fluorescamine
	OPA
Live–dead cells	FDA
Membrane potential	Cyanine dye
DNA × RNA	AO
	PI
DNA	DAPI
	Mithramycin
	Hoechst 33342
Nitrifying bacteria	Fluorescent antibodies
Pollutant microbial assay	Pollutant conjugate fluorescent antibodies
Enzyme activity	Membrane permeable substrate specific stain
Intracellular metabolite	Membrane permeable metabolite specific stain

FITC, fluorescein isothiocyanate; OPA, ortho-phthaldehyde; FDA, fluorescein diacetate; AO, acridine orange; PI, propidium iodide; DAPI, 4,6-diamidino-2-phenylindole-2-HCl.

Table 2   A summary of different biochemical and micromorphometric properties of marine microbes that may be characterized by fluorescence and light-scattering techniques

among different algal species by Trask *et al.* (1982). Light scatter has also been used to discriminate between different algal species, either alone (Price *et al.*, 1978), or in conjunction with chlorophyll fluorescence, to determine cellular size and shape (Trask *et al.*, 1982).

The effects of different environmental factors on cellular chlorophyll content under culture conditions have been investigated in several studies. The effects of light and nutrient variation on chloropigments have been studied in cultures of a diatom, a dinoflagellate and a coccolithophore (Olson *et al.*, 1983), using a purpose-built microscope-based cytometer. They demonstrated a decrease in cellular chlorophyll content as the nitrogen supply becomes limiting, and found this becomes evident well before changes are detectable in the population growth curve. This testifies to the sensitivity of even a simple cytometer and to the resolving power of single cell analysis.

Taxon	Wavelength		Pigment		Reference
	Excitation	Emission	Principal	Accessory	
Chlorophytes	450	680	chl-a	Lutein	1
Diatoms	525–530	680	chl-a	Fucoxanthin	1
Dinoflagellates	525–530	680	chl-a	Peridinin	1
Cryptophytes	490	560–570	chl-a	Phycobiliproteins	1
Cryptophytes	544–568	680	chl-a	Phycoerythrin	2
Cyanobacteria	490	560–570	chl-a	Phycobiliproteins	1
Cyanobacteria	565	575	chl-a	Phycoerythrin	2
Cyanobacteria	615–620	635–645	chl-a	Phycocyanin	2
Cyanobacteria	498, 540	578	chl-a	Phycoerythrin	2
	565			Phycocyanin	

References: 1, Yentsch and Yentsch (1979); 2, Prezelin and Boczar (1986).
Note: other chlorophylls and carotenoids, in addition to the above, may be present in phytoplankton.

Table 3 Examples of autofluorescence spectral signatures and photopigment complements of different phytoplankton taxa

Studies of the effects of light on *Protogonyaulax tamarensis* showed that this dinoflagellate responded to reduced irradiance by increasing chlorophyll fluorescence, with a corresponding increase in cell volume when reproduction was maintained, and a decrease in cell volume when division ceased (Yentsch *et al.*, 1985). Interestingly, the adaptation to low light was not accompanied by changes in the activities of photosynthetic carboxylating enzymes such as RUBPCase.

Satisfactory preservation of chlorophyll-a fluorescence is critical to a wider scale application of AFC to phytoplankton ecology. This important topic has been considered by Yentsch and Pomponi (1986), who report that Lugol's iodine, glutaraldehyde (with and without cacodylic acid) and liquid nitrogen are unsatisfactory as preservatives of cellular chlorophyll in phytoplankton.

The presence of chlorophyll fluorescence can complicate measurement of other cellular constituents labelled with applied fluorochromes in algae. Ideally, the excitation spectrum of the fluorochrome should be sufficiently different from that of chlorophyll. However, the presence of accessory pigments greatly extends the spectrum of light harvesting in algae. As a result, undisturbed fluoresence cannot be obtained. Two choices remain; to use fluorochromes with emission spectra that differ from chlorophyll ($\lambda_{em} = 680$ nm), or to remove the chloropigment. The latter option, which precludes use of living cells, has been achieved by photo-oxidation (Yentsch *et al.*, 1983b) or solvent extraction (Olson *et al.*, 1983) in fixed algae.

Other cellular constituents investigated in algae by AFC include nucleic acids (Trask *et al.*, 1982; Olson *et al.*, 1983; Yentsch *et al.*, 1983b) and toxins (Yentsch, 1981), while physiological processes in algae have also been studied (Rivkin *et al.*, 1986).

Labelling of algal DNA was carried out using Hoechst 33342 and DAPI by Trask *et al.* (1982), who used nucleic acid content in combination with chlorophyll content to characterize different algae. They considered the fluorescence distributions were too broad to allow quantitative analysis of DNA content. Relative quantitation of

DNA and its variation through the cell cycle have been achieved by AFC in a study that demonstrated clearly the G1, G2 + M and S phases of the mitotic cycle in three algal taxa (Olson *et al.*, 1983).

Saxitoxin, one of the major toxins found in dinoflagellates responsible for paralytic shellfish poisoning, can be assayed by AFC. This assay depends on the chemical conversion of the toxin to a fluorescent form by the addition of hydrogen peroxide. Yentsch (1981) found the cellular saxitoxin levels highly variable in six *Gonyaulax* spp. She attributed this to variability in cell age and life history.

The sorting capability of AFC allows it to be used as a preparative step in analytical procedures. This capability has been assessed (Rivkin *et al.*, 1986) using sensitive physiological indices in phytoplankton. On passage through the cytometer, cells experience physical shear within the sample–sheath stream, exposure to high intensity irradiation and, when sorted, an electrostatic charge. The effects of each of these factors has been assessed using the uptake of ^{14}C in algae (Rivkin *et al.*, 1986). The results are shown in Table 4. In both a diatom and a dinoflagellate, photosynthetic rates were significantly lower after AFC owing to the effect of exposure to the laser beam rather than the effects of shear or electrostatic charge. However, the cellular content of radiolabel accumulated prior to AFC was not affected by cytometric analysis. These results suggest that AFC may be used as a preparative technique for living cells providing that care is taken and a quantitative recovery is not required. With fixed cells, AFC may be used quantitatively to concentrate and collect cells of similar types.

	Carbon uptake ($pg\,C\,cell^{-1}\,h^{-1}$) at indicated incubation irradiance			
	P. tamarensis		*D. brightwellii*	
Treatment	$208\,\mu E$ $m^{-2}s^{-1}$	$20\,\mu E$ $m^{-2}s^{-1}$	$208\,\mu E$ $m^{-2}s^{-1}$	$20\,\mu E$ $m^{-2}s^{-1}$
Control	105 ± 7	30 ± 5	210 ± 18	35 ± 3
Sorted	50 ± 8	15 ± 2	150 ± 15	25 ± 3
Shear only	98 ± 4	28 ± 4	190 ± 13	32 ± 2
Shear + EC	87 ± 8	24 ± 2	180 ± 12	28 ± 3
Shear + 25 mW laser	55 ± 4	18 ± 3	160 ± 12	22 ± 2
Shear + 500 mW laser	12 ± 4	3 ± 1	29 ± 9	5 ± 3

Values represent means and standard deviations of four replicates. EC, electrostatic charge. Samples were incubated with photosynthesis-saturating ($208\,\mu E\,m^{-2}\,s^{-1}$) and photosynthesis-limiting ($20\,\mu E\,m^{-2}$ s^{-1}) irradiances for 1–2 h. (After Rivkin *et al.*, 1986.)

Table 4　The influence of different components of AFC analysis and sorting on the carbon uptake of a dinoflagellate (*Protogonyaulax tamarensis*) and a diatom (*Ditylum brightwellii*)

Other aspects of algal physiology which are beginning to be addressed include cell enzymes which may be immunologically assayed by AFC. Successful production of polyclonal antibodies to intracellular enzymes has been reported (Perry in Yentsch and Pomponi, 1986), including those to enzymes involved in CO_2 uptake

(RUBPCase), nitrogen transport and reduction (nitrate reductase) and nitrogen incorporation (glutamine synthetase and glutamine synthesis dehydrogenase). These exciting new approaches offer tremendous potential for investigating, at the cellular level of discrimination, physiological mechanisms by which ecologically relevant processes occur. The field of immuno-assay by AFC is now set to become a major research area in microbial ecology.

The application of antibodies that cross-react with the different biochemical components of phytoplankton cell surfaces provides a potential means for characterizing different algal taxa (Yentsch and Pomponi, 1986). The differing cell wall constituents of phytoplankton taxa are shown in Table 5. Monoclonal antibodies have been used as molecular probes for cell wall antigens in higher algae (*Fucus*) by Vreeland *et al.* (1984).

Taxon	Cell wall composition		
	Cellulose	Pectin	Other
Chlorophyte	+	+	Occasionally calcium carbonate
Diatoms	+	+	Silicon
Cyanobacteria	+	+	
Euglenoids	−	−	−
Dinoflagellate	+	+	

Table 5   Cell wall composition in various algal taxa (Yentsch and Pomponi, 1986)

The cell wall materials identified in Table 5 may also be characterized by fluorochromes such as Calcofluor White M2R, which binds to cellulose and other β-linked glucans (Hughs and McCully, 1975). This fluorochrome has been used to label the thecal plates of dinoflagellates (Fritz and Triemer, 1985).

Procaryotic phytoplankton

Unicellular chroococcoid cyanobacteria may be characterized by their small (0.4–1 μm) size and the presence of chlorophyll-a and phycobiliproteins (Wood *et al.*, 1985; Stockner and Antia, 1986). The presence of phycobiliproteins allowed Wood *et al.* (1985) to use the AFC configuration shown in Fig. 7 to differentiate phycoerythrin-containing *Synechococcus* from other phytoplankton and to analyse them.

While cyanobacteria are generally treated as a single taxonomic and ecological group, they show distinct variation in pigment composition (Wood, 1983; Wood *et al.*, 1985), DNA base composition (Waterbury in Wood *et al.*, 1985) and serological groups (Campbell *et al.*, 1983). Aspects of these differences may be investigated by AFC.

Serological groups of different strains of cyanobacteria have been identified and enumerated by epifluorescence microscopy coupled with immuno-fluorescent techniques (Campbell *et al.*, 1983). In this novel work, antisera were produced against five strains of *Synechococcus* and *Synechocystis*. Application to formalin-fixed samples from a temperate estuary revealed that phycoerythrin-containing

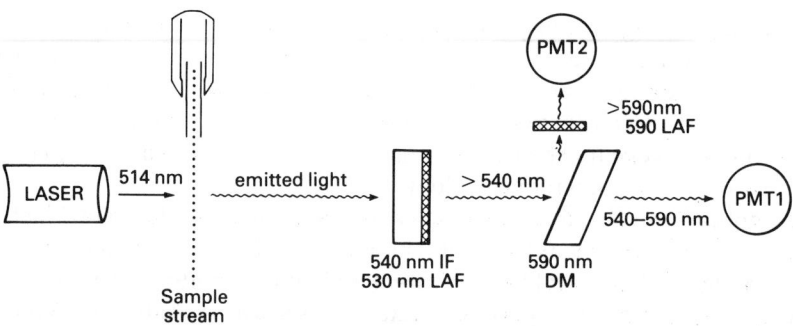

Fig. 7—Optical configuration for differentiating phycoerythrin-containing cyanobacteria from other phytoplankton by AFC. IF, long pass interference filter; LAF, long pass absorbance filter; DM, dichroic mirror; PMT 1, photomultiplier collecting phycoerythrin fluorescence; PMT 2, photomultiplier collecting chlorophyll fluorescence. From Wood *et al.* (1985). Reproduced with permission from the American Society of Limnology and Oceanography, Inc.

Synechococcus populations dominated the cyanobacteria in late summer and autumn; phycocyanin-containing *Synechococcus* and *Synechocystis* were minor components at this time. The population numbers varied, reaching levels of 3×10^5 cells ml^{-1}, suggesting that cyanobacteria may be significant primary producers in estuaries, as well as in oceanic regions (Platt *et al.*, 1983).

Different pigment types in *Synechococcus* have been studied using AFC by Wood *et al.* (1985). Three chromatic groups of *Synechococcus* spp. were found. Two of these were identified within the phycoerythrin-containing *Synechococcus* spp.: type 1 containing both phycoerythrobilin and phycourobilin chromophores, and type 2 with only phycoerythrobilin chromophores. A third group of *Synechococcus* lacked phycoerythrin but contained phycocyanin. Although the structural differences in the principal phycobiliproteins of these groups are slight, the spectral properties differ sufficiently to allow their successful discrimination by AFC.

Bacteria

Bacteria have been classified according to their light scatter patterns (Wyatt, 1968; Koch, 1968), and although light scatter and fluorescence have been used to characterize bacteria cytometrically, their DNA base pairing and growth (Mullaney and Dean, 1970; Van Dilla *et al.*, 1983; Boye *et al.*, 1983), these features have yet to be exploited in marine ecology.

Bacteria have been differentiated from algae using ethidium bromide and chlorophyll fluorescence (Paau *et al.*, 1979). Simultaneous measurements of the relative concentration of protein and nucleic acid have been investigated using FITC, DAPI and PI in freshwater bacteria (Bailey *et al.*, 1978; Hutter *et al.*, 1980).

The use of fluorochromes, such as AO, routinely employed both to enumerate and size marine bacteria by epifluorescence microscopy and as a common nucleic acid stain in biomedical AFC (Crissman *et al.*, 1979), has not yet been developed for marine microbial AFC. This approach has great potential, particularly for characterizing properties such as cell size and DNA content which relate to bacterial productivity. One of the problems that will need addressing concerns AFC discrimination between 'free' bacteria and others that are attached to particles. Paul (1982) has evaluated this problem for epifluorescent microscopic analysis, and concluded that the Hoechst bisbenzimide fluorochromes 33342 and 33258 allowed satisfactory resolution between free and attached bacteria.

Another epifluorescent microscopy technique amenable to adaptation for AFC analysis is the detection and differentiation of actively metabolizing bacteria from dormant or dead bacteria. The technique depends on the uptake of a non-fluorescent, non-polar derivative (fluorescein diacetate, FDA) which readily penetrates cell membranes and is hydrolysed intracellularly to form a fluorescent product (fluorescein). A microscope-based FDA hydrolysis technique has been evaluated for detecting active bacteria in freshwater (Chrzanowski *et al.*, 1984).

An immuno-fluorescent assay of the marine ammonium-oxidizing bacterium *Nitrosococcus oceanus* has been developed by Ward and Perry (1980). This rapid and direct means of quantifying ammonium-oxidizing bacteria, developed as a poly-clonal antibody to cell surface protein antigens, allows the ecologically important process of marine nitrification to be investigated. Concentrations of nitrifying bacteria (*N. oceanus* and *Nitrosomonas marina*) ranged from 10^3 to 10^7 ml^{-1} in oceanic and estuarine waters respectively (Ward, 1982). Although amenable to AFC, this powerful assay remains to be developed for use in AFC.

Protozoa
In spite of the common use of epifluorescence microscopy to identify and quantitate marine protozoa (Haas, 1982; Caron, 1983; Sherr and Sherr, 1983a, b, 1984; Sherr *et al.*, 1986), this group is yet to be studied by AFC. Although population levels *in situ* are typically sub-optimal for AFC analysis (see Fig. 6), cultured protozoa should be amenable to AFC.

The ability to characterize microalgae by AFC has been used to investigate food selection by a marine ciliate under controlled conditions (Stoecker *et al.*, 1986). When offered a mixture of algae (dinoflagellates, cryptophytes and green algae), *Belanion* sp. feeds preferentially on the dinoflagellates. The selection for dinoflagel-lates (*Heterocapsa triquetra*) predominated even when this group was less abundant (Fig. 8).

Shipboard applications and the analysis of natural particulates
The problems associated with adequate fixation of autofluorescence in phytoplank-ton, and the strong trend in marine microbial ecology towards experimental investigations which require immediate feed-back, makes shipboard AFC highly desirable. However, AFC operation at sea presents various problems. The most important of these can be best summarized as follows: a cytometer is, in reality, an optical bench on which the precise alignment of the optical components is paramount; the ship, in contrast, is a highly mobile platform which vibrates at a range of frequencies. Operation of an optical bench on a lively platform is likely to be a frustrating experience. To minimize the effect of movement and vibration on the cytometer is important. The more heavily damped rolling and pitching movements to be found on large (> 50 m) ships makes these types of vessel more suitable as an AFC platform. Minimizing vibration transmission requires careful fixing. Fortun-ately these problems can be overcome, and EPICS 741 (Burkhill, in preparation;

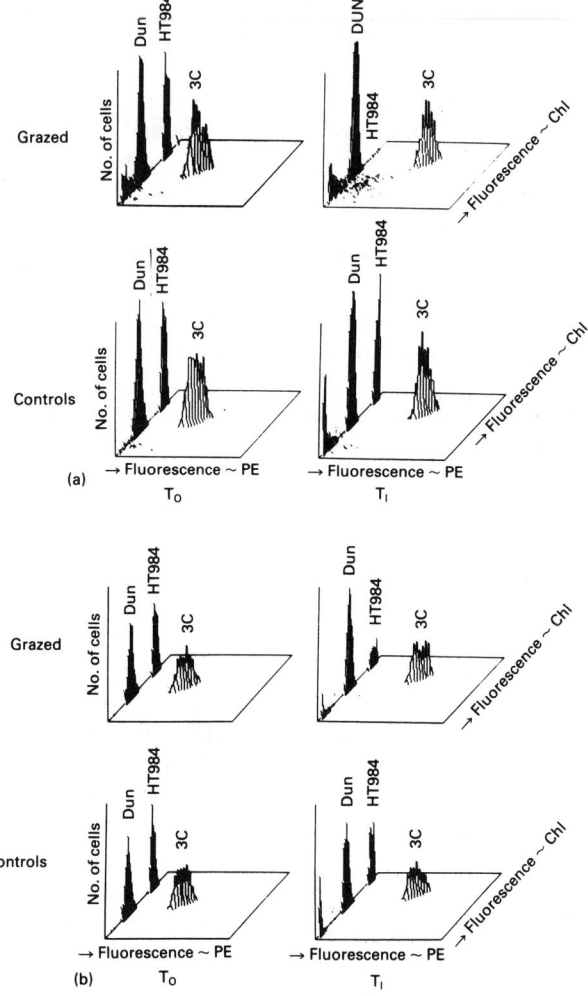

Fig. 8—Food selection by *Balanion* determined by AFC. The ciliate was given a mixture of *Dunaliella tertiolecta* (Dun), *Heterocapsa triquetra* (HT984) and *Chroomonas salina* (3C) at mixtures of about 10^3 (plots a) and 10^4 (plots b) cells ml^{-1}. Log phycoerythrin fluorescence (PE) and log chlorophyll-a fluorescence (Chl) are shown on the x and y axes respectively. Peak height is relative to number of cells of each type. Figures show controls (no *Balanion*) and experimental (*Balanion* present) treatments at the beginning (T_0) and end (T_1) of a grazing experiment. From Stoecker *et al.* (1986). Reproduced with permission from Elsevier Science Publishers B.V.

Olson *et al.*, 1985) and FACS analysers (Cucci, personal communication; Yentsch *et al.*, 1986) have been successfully operated at sea.

A conceptual scheme has been proposed (Fig. 9) for analysing heterogeneous particles in natural waters (Yentsch *et al.*, 1983a). In this scheme, particles may be characterized as organic or inorganic, living or dead, autotrophic or heterotrophic, and separated into different pigment groups (Yentsch and Phinney, 1985). Few of the components in this characterization scheme have been systematically addressed

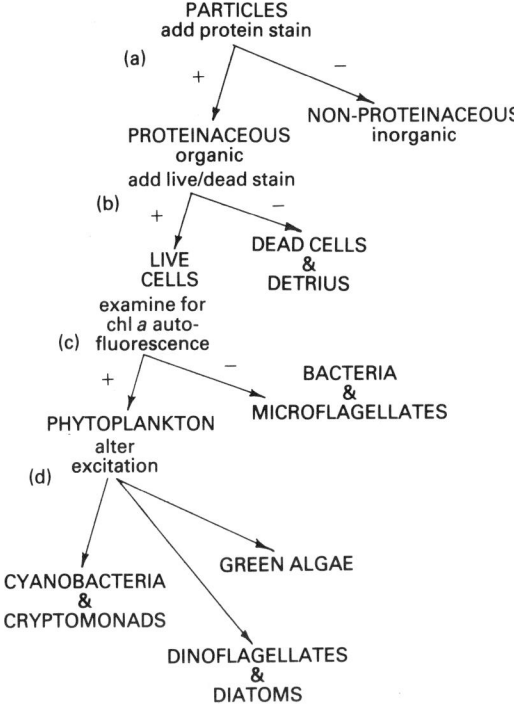

Fig. 9—A conceptual scheme for the characterization of natural particulates on the basis of applied, induced and autofluorescent criteria. A sample of natural particulates may be characterized into the following: a, organic and inorganic components by a protein-binding fluorochrome such as FITC; b, living and dead components using a viability stain such as FDA; c, autotrophs and heterotrophs using the presence of chlorophyll autofluorescence; d, autotrophic components using spectral differences. The scheme remains speculative and only the last type of characterization has been attempted. From Yentsch *et al.* (1983a). Reproduced with permission from the American Society of Limnology and Oceanography, Inc.

in natural waters, although protocols for allometric and ataxonomic phytoplankton analyses have been determined (Yentsch *et al.*, 1984, 1986).

The vertical depth distributions of picoplankton have been studied in the Gulf Stream (Olson *et al.*, 1985) and Celtic Sea (Burkhill, in preparation). The picoplankton at both sites were shown by AFC to be dominated by cyanobacteria of the *Synechococcus* type. The vertical distribution patterns of these cyanobacteria differed from other picoplankton in the Celtic Sea (Fig. 10). Maximum cyanobacteria abundance was found at 21 m depth at a time when the chlorophyll maximum occurred at 31 m. The picoplankton showed an increase in pigment content per cell (as indicated by elevated LIRFL) with depth, down to 41 m, as the community photoadapted to decreased light levels.

THE FUTURE FOR MARINE MICROBIAL ANALYTICAL FLOW CYTOMETRY

AFC offers tremendous potential to further our understanding of many areas in marine microbial ecology. At the moment, this potential is limited by a number of

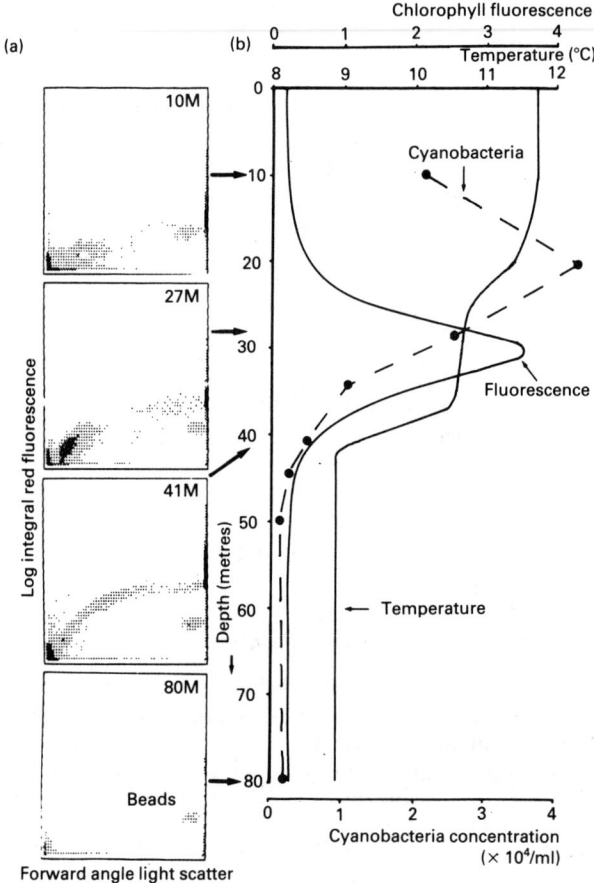

Fig. 10—Vertical profile obtained by rosette sampler in Celtic Sea on 7 June 1986, showing shipboard analysis of picoplankton (scattergrams in a) and cyanobacteria concentration, chlorophyll fluorescence and temperature profiles (plots in b). AFC data are presented as bivariate scattergrams with log cellular chlorophyll concentration expressed by LIRFL as the ordinate, and cell size as measured by FALS as the abscissa. The three slice levels operate at 5, 50 and 100 particles per pixel for dots, crosses and black hatching respectively on the scattergrams. The 2.2 µm fluorosphere beads, labelled in the 80 m sample and clearly identifiable in samples from other depths, were added as an internal standard to calibrate particle size and concentration. Cyanobacteria concentrations in b were determined by epifluorescence microscopy while temperature and chlorophyll fluorescence were measured by standard Neil Brown sensors. From Burkill (in preparation).

factors. These include an unawareness by marine scientists of the current capability of the technique, and the lack of rigorously developed analytical protocols for marine research either *de novo* or by adaptation from the biomedical field, as well as the complexity of the instrumentation and its adaptation to the requirements of marine science.

AFC has an instant appeal to many who feel it will immediately revolutionize their research. This is often unrealistic, since many marine cytometric techniques await development, and techniques that are being used are often empirical. For instance, micromorphometric characterization, and particularly cytometric sizing, of marine particles is under-developed; with few cytometers capable of Coulter impedance-volume measurements, there is a pressing need for a more rigorous understanding of the factors that influence light scatter in marine particles. Although the autofluorescence properties of chlorophyll and phycoerythrin have been used successfully to study phytoplankton, the full capability of AFC in this field will only be harnessed with the development of an analytical protocol for the quantitative analysis of the absolute concentration of cellular photopigments. One of the hurdles in the development of this protocol, and this is also true of much marine AFC work, is the lack of suitable standards and reference materials, as has been discussed by Phinney *et al.* (1987b). Although fluorosphere reference beads, quantitatively labelled over a range of fluorescein levels, are available, there is nothing equivalent for chlorophyll or phycoerythrin. The quantitative stabilization of photosynthetic pigments on reference beads to act as cytometric standards would be of enormous benefit for phytoplankton research.

The stabilization of photosynthetic pigments in fixed material is another area which is critical for the future, wider-scale application of marine AFC. Until this can be achieved, field deployment of cytometers will remain mandatory for many aspects of phytoplankton community ecology.

The aspect of AFC research that probably offers the greatest challenge for microbial ecologists lies in the application of fluorochrome-coupled monoclonal or polyclonal antibodies. These have been used extensively in biomedical sciences where the production of mammalian antibodies is commercially viable. This is not yet the case for antibodies to marine antigens. Since production of highly purified antibodies is demanding, it is likely that the tremendous potential offered by immuno-assay by AFC in the marine sciences will take time and nurturing to be developed. A good foundation for the use of immuno-fluorescence in microbial ecology is Bohlool and Schmidt (1977).

The effects of cytometric sorting, as a preparative step in the purification of samples required for other analyses, need to be further assessed. The development of protocols such as cellular elemental analysis of sorted material will require calibrations against analytical techniques of comparable sensitivity. In the case of cellular carbon and nitrogen measurements, the sensitivity for calibration, based on sub-nanogram quantities of sorted material, is not readily available; obtaining it will provide an exacting challenge.

Many of the present and future technical developments in AFC will be of benefit to microbial ecology. Significant improvements in optical configuration are now allowing the use of air-cooled lasers, thereby reducing size and installation requirements. As the equipment becomes more portable, shipboard and field operation will become easier. As cytometers become cheaper and more 'user friendly' with implementation of automatic analytical protocols, multi-user operation will be facilitated. At that stage, we should be looking forward to AFC as an important tool at the cutting-edge of many major marine microbial research programmes.

ACKNOWLEDGEMENTS

It is a pleasure to acknowledge first, the advice of Clarice Yentsch, Terry Cucci and Dave Phinney (Bigelow Laboratory of Ocean Sciences) who catalysed setting up the marine AFC facility at IMER, Plymouth, and second, my IMER colleagues, Fauzi Mantoura for many stimulating and fruitful cytometry discussions, and Dave Robins who also provided the Coulter counter data in Fig. 6b. This work forms part of the Biogeochemistry Research Programme at the Institute for Marine Environmental Research, a component of the UK Natural Environment Research Council.

REFERENCES

Bailey, J.E., McQuitty, D.M., Lee, L.Y. and Oro, J.A. 1978. Measurement of structured microbial population dynamics by flow microfluorometry. *American Institute of Chemical Engineers Journal*, **24**(4): 570–577.

Bohlool, B.B. and Schmidt, E.L. 1977. The immunofluorescence approach in microbial ecology. In *Advances in microbial ecology*, Vol. 4 (ed. M. Alexander), pp. 203–241. Plenum Press, New York and London.

Boye, E., Steen, H.B. and Skarstad, K. 1983. Flow cytometry: a promising tool in experimental and clinical microbiology. *Journal of General Microbiology*, **129**: 973–980.

Campbell, L., Carpenter, E.J. and Iacono, V.M. 1983. Identification and enumeration of marine chroococcoid cyanobacteria by immunofluorescence. *Applied and Environmental Microbiology*, **46**(3): 553–559.

Caron, D.A. 1983. Technique for enumeration of heterotrophic and phototrophic nanoplankton, using epifluorescence microscopy, and comparison with other procedures. *Applied and Environmental Microbiology*, **46**(2): 491–498.

Chrzanowski, T.H., Crotty, R.D., Hubbard, J.G. and Welch, R.P. 1984. Applicability of the fluorescein diacetate method of detecting active bacteria in freshwater. *Microbial Ecology*, **10**: 179–185.

Cram, L.S., Bartholdi, M.F., Wheeless, J.J. and Gray, J.W. 1985. Morphological analysis by scanning flow cytometry. In *Flow cytometry: instrumentation and data analysis* (eds. M.A. van Dilla, P.N. Dean, O.D. Laerum and M.R. Melamed), pp. 163–194. Academic Press, London.

Crissman, H.A., Stevenson, A.P., Kissane, R.J. and Tobey, R.A. 1979. Techniques for quantitative staining of cellular DNA for flow cytometric analysis. In *Flow cytometry and sorting* (eds. M.R. Melamed, P.F. Mullaney and M.L. Mendelsohn), pp. 243–284. Wiley, New York.

Cucci, T.L., Shumway, S.E., Newell, R.C., Selvin, R., Guillard, R.R.L. and Yentsch, C.M. 1985. Flow cytometry: a new method for characterisation of differential ingestion, digestion and egestion by suspension feeders. *Marine Ecology Progress Series*, **24**: 201–204.

Dean, P.N. 1985. Methods of data analysis in flow cytometry. In *Flow cytometry: instrumentation and data analysis* (eds. M.A. Van Dilla, P.N. Dean, O.D. Laerum and M.R. Melamed), pp. 195–223. Academic Press, London.

Fritz, L. and Triemer, R.E. 1985. A rapid simple technique utilising Calcifluor White M2R for the visualisation of dinoflagellate thecal plates. *Journal of Phycology*, **21**: 662–664.

Fukuda, M., Hoshino, K., Naito, M. and Morita, T. 1982. A fluorescence cytophotometer operated under computer control for multi-parameter cell analysis. *Histochemistry*, **76**: 1–13.

Haas, L.W. 1982. Improved epifluorescence microscopy for observing planktonic micro-organisms. *Annales de l'Institute Oceanographique*, **58S**: 261–266.

Horan, P.K. and Wheeless, L.L. 1977. Quantitative single cell analysis and sorting. *Science*, **198**: 149–157.

Hughs, J. and McCully, M.E. 1975. The use of an optical brightener in the study of plant structure. *Stain Technology*, **50**: 319–329.

Hutter, K.-J. and Eipel, H.E. 1978. Flow cytometric determinations of cellular substances in algae, bacteria, moulds and yeasts. *Antonie van Leeuwenhoek*, **44**: 269–282.

Hutter, K.-J., Stohr, M. and Eipel, H.E. 1980. Simultaneous DNA and protein measurements of microorganisms. In *Flow Cytometry IV* (eds. O.D. Laerum, T. Lindmo and E. Thorud), pp. 100–102. Universitetforlaget, Bergen.

Kerker, M. 1969. *The scattering of light and other electromagnetic radiation.* Academic Press, New York.

Kerker, M., Van Dilla, M.A., Brunsting, A., Kratohvil, J.P., Hsu, P., Wang, D.S., Gray, J.W. and Langlois, R.G. 1982. Is the central dogma of flow cytometry true: that fluorescence intensity is proportional to cellular dye content? *Cytometry*, **3**: 71–78.

Koch, A.L. 1968. Theory of angular dependence of light scatter by bacteria and similar sized biological objects. *Journal of Theoretical Biology*, **18**: 133.

Kruth, H.S. 1982. Flow cytometry: rapid biochemical analysis of single cells. *Analytical Biochemistry*, **125**: 225–242.

Laws, E.A. 1983. Man's impact on marine nitrogen cycling. In *Nitrogen in the marine environment* (eds. E.J. Carpenter and D.G. Capone), pp. 459–485. Academic Press, London.

Mardia, K.V., Kent, J.T. and Bibby, J.M. 1979. Multivariate analysis. Academic Press, London.

Melamed, M.R., Mullaney, P.F. and Mendelsohn, M.L. 1979. *Flow cytometry and sorting.* Wiley, New York.

Mullaney, P.F. and Dean, P.N. 1969. Cell sizing: a small-angle light scattering method sizing particles of low refractive index. *Applied Optics*, **8**: 2361.

Mullaney, P.F. and Dean, P.N. 1970. The small angle scattering of biological cells. *Biophysical Journal*, **10**: 764–772.

Mullaney, P.F., Van Dilla, M.A., Coulter, J.R. and Dean, P.N. 1969. Cell sizing: a light scattering photometer for rapid volume determination. *Reviews of Scientific Instruments*, **40**: 1029–1032.

Olson, R.J., Frankel, S.L., Chisholm, S.W. and Shapiro, H.M. 1983. An inexpensive flow cytometer for the analysis of fluorescence signals in phytoplankton: chlorophyll and DNA distributions. *Journal of Experimental Marine Biology and Ecology*, **68**: 129–144.

Olson, R.J., Vaulot, D. and Chisholm, S.W. 1985. Marine phytoplankton distributions measured using shipboard flow cytometry. *Deep-Sea Research*, **32**: 1273–1280.

Paau, A.S., Oro, J. and Cowles, J.R. 1978. Application of flow microfluorometry to the study of algal cells and isolated chloroplasts. *Journal of Experimental Botany*, **29**: 1011–1020.

Paau, A.S., Cowles, J.A., Oro, J., Bartel, A. and Hungerford, E. 1979. Separation of algal mixtures and bacterial mixtures with flow-microfluorometer using chlorophyll and ethidium bromide fluorescence. *Archives of Microbiology*, **120**: 271–273.

Paul, J.H. 1982. Use of Hoechst dyes 33258 and 33342 for enumeration of attached and planktonic bacteria. *Applied and Environmental Microbiology*, **43**: 939–944.

Phinney, D.A., Cucci, T.L. and Yentsch, C.M. 1987a. Perspectives on aquatic flow cytometry I: instrumentation and analysis. *Journal of Plankton Research*, submitted.

Phinney, D.A., Cucci, T.L. and Yentsch, C.M. 1987b. Perspectives on aquatic flow cytometry II: importance of standards and controls. *Journal of Plankton Research*, submitted.

Platt, T., Subba Rao, D.V. and Irwin, B. 1983. Photosynthesis of picoplankton in the oligotrophic ocean. *Nature*, **301**: 702–704.

Prezelin, B.B. and Boczar, B.A. 1986. Molecular bases of cell absorption and fluorescence in phytoplankton: potential applications to studies in optical oceanography. *Progress in Phycological Research*, **4**: 349–464.

Price, B.J., Kollman, V.H. and Salzman, G.C. 1978. Light scatter analysis of micro-algae. Correlation of scatter patterns from pure and mixed asynchronous cultures. *Biophysical Journal*, **22**: 29.

Rivkin, R.B., Phinney, D.A. and Yentsch, C.M. 1986. Effects of flow cytometric analysis of and cell sorting on photosynthetic carbon uptake by phytoplankton in cultures and from natural populations. *Applied and Environmental Microbiology*, **52**: 935–938.

Rotman, B. and Papermaster, B.W. 1966. Membrane properties of living mammalian cells studied by enzymatic hydrolysis of fluorogenic esters. *Proceedings of the National Academy of Sciences*, **55**: 134–141.

Salzman, G.C. 1982. Light scattering analysis of single cells. In *Cell analysis* (ed. N. Catsimpoolas), pp. 111–143. Plenum, New York and London.

Shapiro, H.M. 1985. *Practical flow cytometry*. Alan Liss, New York.

Sherr, B.F. and Sherr, E.B. 1983a. Enumeration of heterotrophic microprotozoa by epifluorescence microscopy. *Estuarine and Coastal Shelf Science*. **16**: 1–7.

Sherr, E.B. and Sherr, B.F. 1983b. Double-staining epifluorescence technique to assess frequency of dividing cells and bacteriovory in natural populations of heterotrophic microprotozoa. *Applied Environmental Microbiology*, **46**: 1388–1393.

Sherr, B.F. and Sherr, E.B. 1984. Role of heterotrophic protozoa in carbon and energy flow in aquatic ecosystems. In *Current perspectives in microbial ecology* (eds. M.J. Klug and C.A. Reddy), pp. 412–423. American Society of Microbiology, New York.

Sherr, E.B., Sherr, B.F., Fallon, R.D. and Newell, S.Y. 1986. Small, aloricate ciliates as a major component of the marine heterotrophic nanoplankton. *Limnology and Oceanography*, **31**: 177–183.

Steinkamp, J.A. 1984. Flow cytometry. *Reviews of Scientific Instruments*, **55**: 1375–1400.

Stewart, C.C. and Steinkamp, J.A. 1982. Quantitation of cell concentration using the flow cytometer. *Cytometry*, **2**(4): 238–243.

Stockner, J.G. and Antia, N.J. 1986. Algal picoplankton from marine and freshwater ecosystems: a multidisciplinary perspective. *Canadian Journal of Fisheries and Aquatic Sciences*, **43**: 2472–2503.

Stoecker, D.K., Cucci, T.L., Hulburt, E.M. and Yentsch, C.M. 1986. Selective feeding by *Balanion* sp. (Ciliata: Balanionidae) on phytoplankton that best support its growth. *Journal of Experimental Marine Biology and Ecology*, **95**: 113–130.

Stratton, J.A. 1941. *Electromagnetic theory*. McGraw-Hill, New York.

Trask, B.J., van den Engh, G.J. and Elgershuizen, J.H.B.W. 1982. Analysis of phytoplankton by flow cytometry. *Cytometry*, **2**(4): 258–264.

Van Dilla, M.A., Langlois, R.G., Pinkel, D., Yajko, D. and Hadley, W.K. 1983. Bacterial characterisation by flow cytometry. *Science*, **220**: 620–622.

Van Dilla, M.A., Dean, P.N., Laerum, O.D. and Melamed, M.R. 1985. *Flow cytometry: instrumentation and data analysis*. Academic Press, London.

Visser, J.W.M. and Tanke, H.J. 1985. Local modifications to commercial instruments. In *Flow cytometry: instrumentation and data analysis* (eds. M.A. Van Dilla, P.N. Dean, O.D. Laerum and M.R. Melamed), pp. 224–259. Academic Press, London.

Vreeland, V., Slomich, M. and Laetsch, W.M. 1984. Monoclonal antibodies as molecular probes for cell wall antigens of the brown alga, *Fucus*. *Planta*, **162**: 506–517.

Wang, D.S. and Barber, P.W. 1979. Scattering by inhomogeneous non-spherical objects. *Applied Optics*, **18**: 1190.

Ward, B.B. 1982. Ocean distribution of ammonium-oxidizing bacteria determined by immunofluorescent assay. *Journal of Marine Research*, **40**: 1155–1172.

Ward, B.B. and Perry, M.J. 1980. Immunofluorescent assay for the marine ammonium-oxidizing bacterium *Nitrosococcus oceanus*. *Applied and Environmental Microbiology*, **39**(4): 913–918.

Watson, J.V. 1987. Flow cytometry in biomedical science. *Nature*, **325**: 741–742.

Wood, A.M. 1983. Occurrence and ecological significance of different pigment types of marine *Synechococcus*. *Eos*, **63**: 960.

Wood, A.M., Horan, P.K., Muirhead, K., Phinney, D.A., Yentsch, C.M. and Waterbury, J.B. 1985. Discrimination between different types of pigments in marine *Synechococcus* spp. by scanning spectroscopy, epifluorescence microscopy and flow cytometry. *Limnology and Oceanography*, **30**: 1303–1315.

Wyatt, P.F. 1968. Differential light scattering: a physical method for identifying living bacteria. *Applied Optics*, **7**: 1879.

Yentsch, C.M. 1981. Flow cytometric analysis of cellular saxitoxin in the dinoflagellate *Gonyaulax tamarensis* var *excavata*. *Toxicon*, **19**(5): 611–621.

Yentsch, C.S. and Phinney, D.A. 1985. Spectral fluorescence: an ataxonomic tool for studying the structure of phytoplankton populations. *Journal of Plankton Research*, **7**: 617–632.

Yentsch, C.M. and Pomponi, S.A. 1986. Automated individual cell analysis in aquatic research. *International Review of Cytology*, **105**: 183–243.

Yentsch, C.S. and Yentsch, C.M. 1979. Fluorescence spectral signatures: the characterization of phytoplankton populations by the use of excitation and emission spectra. *Journal of Marine Research*, **37**: 471–483.

Yentsch, C.M. and Yentsch, C.S. 1984. Emergence of optical instrumentation for measuring biological properties. *Oceanography and Marine Biology, an Annual Review*, **22**: 55–98.

Yentsch, C.M., Horan, P.K., Muirhead, K., Dortch, Q., Haugen, E., Legendre, L., Murphy, L.S., Perry, M.J., Phinney, D.A., Pomponi, S.A., Spinrad, R.W., Wood, M., Yentsch, C.S. and Zahuranec, B.J. 1983a. Flow cytometry and cell sorting: a technique for analysis and sorting of aquatic particles. *Limnology and Oceanography*, **28**: 1275–1280.

Yentsch, C.M., Mague, F.M., Horan, P.K. and Muirhead, K. 1983b. Flow cytometric DNA determinations on individual cells of the dinoflagellate, *Gonyanlax tamarensis* var. *excavata*. *Journal of Experimental Marine Biology and Ecology*, **67**: 175–183.

Yentsch, C.M., Cucci, L. and Phinney, D.A. 1984. Flow cytometry and cell sorting: problems and promises for biological ocean science research. In *Lecture notes on coastal and estuarine studies*, No. 8, *Marine phytoplankton and productivity* (eds. O. Holm Hansen, L. Bolis and R. Gilles), pp. 141–155. Springer-Verlag, Berlin.

Yentsch, C.M., Cucci, T.L., Phinney, D.A., Selvin, R. and Glover, H.E. 1985. Adaptation to low photon flux densities in *Protogonyaulax tamarensis* var. *excavata*, with reference to chloroplast photomorphogenesis. *Marine Biology*, **89**: 9–20.

Yentsch, C.M., Cucci, T.L., Phinney, D.A. and Topinka, J.A. 1986. Real-time characterization of individual marine particles at sea: flow cytometry. In *Lecture notes on coastal and estuarine studies*, No. 17, *Tidal mixing and plankton dynamics* (eds. J. Bowman, M. Yentsch and W.T. Peterson), pp. 414–448. Springer-Verlag, Berlin.

7

Microcalorimetry as a tool in microbiology and microbial ecology

Lena Gustafsson, Department of Marine Microbiology, University of Göteborg, Carl Skottsbergs Gata 22, S-413 19 Göteborg, Sweden

All biological processes are accompanied by alterations in heat. The quantity of heat that is evolved by a culture during growth, excluding chemical side reactions, is equal to the change in enthalpy resulting from the metabolism of the organisms. This change in enthalpy can be measured by microcalorimetry and represents the sum of the contributions from all the reactions that occur during a metabolic process. In this context, microcalorimetry can serve as a general analytical tool to detect and monitor biological processes as well as a tool to perform energy balance calculations for such processes. The most exact interpretation of heat measurements is evidently given by utilization of defined metabolic systems. Nevertheless, as calorimetric methods are entirely non-specific and all types of metabolism are followed by heat changes, calorimetry can be a useful tool in studying complex ecosystems.

INTRODUCTION

Metabolism is always accompanied by evolution of heat, which can be viewed as a prerequisite for life. This assertion is explained by the fact that the second law of thermodynamics states that the degree of order in the entire system, i.e. the cells plus their environment, must always decrease. Biosynthesis is an order-producing process that must be coupled to reactions which to a higher degree disorder the environment. This occurs by a continual evolution of heat to the environment by the overall metabolism.

In a system where a constant pressure and temperature prevail, the change in heat (Q_p) can be directly related to the change in enthalpy (ΔH). Thus, if a cell culture is defined as being the system under examination, the quantity of heat that is evolved by the culture during growth, excluding chemical side reactions, is equal to the change in enthalpy of the metabolism of the cells (ΔH_{met}). In this type of experimental work one often deals with liquid systems, in which a constant pressure and temperature as well as an approximately constant volume are assumed to prevail. This results in a negligible pressure–volume work ($P \Delta V$). Consequently, for all practical purposes, the change in enthalpy in solution corresponds to the change in internal energy (ΔE), as $\Delta H = \Delta E + P \Delta V$. (For an introduction to the application of thermodynamics on biological systems see for example Wilkie (1960), Forrest (1972), Chang (1977), Lamprecht and Zotin (1978) and Jones (1979). The change in enthalpy of a metabolic process that can be measured by calorimetry, is the sum of the contributions from all the reactions that occur during cellular metabolism, i.e. the catabolic and anabolic reactions, as well as the reactions due to macromolecular organization, transport processes, movement and maintenance.

In biology, calorimetry can be used in two contexts. It serves as a general analytical tool to detect and monitor specific biological processes as well as a tool to perform energy balance calculations for such processes. As all biological processes are accompanied by alterations in heat, calorimetric methods are general and thus entirely non-specific. This limits the application of these methods for many defined analytical purposes, but also represents an advantage in the detection of unknown and undefined phenomena. Another characteristic of great importance is that calorimetry allows for the measurement of the continuous alteration in heat of a biological system without disturbing the system by, for example, the addition of reagents. Furthermore, in contrast to spectrophotometric methods, calorimetric methods do not require optically clear objects.

The most exact interpretation of heat measurements by calorimetry is evidently given by utilizing defined metabolic systems. Nevertheless, calorimetry can be a useful tool in studying complex ecosystems. Calorimetry and its use in biochemistry and biology have been reviewed by Brown (1969), Forrest (1972), Spink and Wadsö (1976), Lamprecht and Schaarschmidt (1977), Jones (1979) and Beezer (1980).

INSTRUMENTATION

A diversity of calorimetric principles and different instrumental designs have been used in biological calorimetric studies (for review see Spink and Wadsö (1976); for fundamentals and practice of calorimetry, see also Hemminger and Höhne (1984)). The term 'direct calorimetry' is reserved for experiments that involve the use of calorimeters. The term microcalorimetry has been applied when the instruments are highly sensitive and require only a small quantity of the sample. Calorimeters can be classified as being of the adiabatic or the heat-conduction type, or to lie in between these two extreme calorimetric principles. Ideally no heat is exchanged between the calorimetric vessel and the surroundings when using adiabatic instruments, while heat is quantitatively transferred from the reaction vessel to a heat sink (exothermic reaction) or in the opposite direction (endothermic reaction). For isothermal heat-conduction instruments, the heat sink is kept at a constant temperature. Because microcalorimeters of the heat-conduction type have been used in most of the recent work with microbial systems, this presentation will be focused on this calorimetry principle. In heat-conduction calorimeters, the heat evolved is passed from the measuring cell, through the surrounding thermopile wall and then absorbed by a heat sink. For steady state processes and approximately for slow processes such as microbial growth, the produced thermopile voltage (V) is directly proportional to the thermal power or the heat effect (dQ/dt) evolved:

$$V = \text{constant} \times dQ/dt$$

Under such conditions the calorimeter clearly acts as a wattmeter, giving the rate of the process, and can be used for kinetic studies.

The quantity of heat evolved (Q) is proportional to the area (A) under the power–time curve:

$$Q = \text{constant} \times A$$

The integral of the power–time curve is proportional to the extent of the reaction. Thus, by integration of the power–time curve obtained, the enthalpy change (ΔH) of a process can be calculated.

For microbial growth studies using liquid cell cultures, different kinds of heat-conduction calorimeters have been used (for review see Spink and Wadsö, 1976), mainly of the batch or the flow-through type. As the name implies, batch calorimetry refers to a method where the reaction is generally performed in a closed vessel. A major disadvantage of most batch calorimeters is the inaccessibility of the calorimetric vessel, also rendering it almost impossible to achieve true aerobic conditions during measurements. These problems can be avoided by the use of flow calorimeters (Stoesser and Gill, 1967; Monk and Wadsö, 1968) in which measurements of the heat of metabolism of liquid cell cultures are performed in a steady state flow through the instruments. The connection of a flow calorimeter to an external fermenter facilitates simultaneous measurements of many parameters, e.g. determinations of cell density, pH, oxygen tension, as well as metabolite, substrate and product concentrations or uptake–production rates (Eriksson and Wadsö, 1971; Brettel et al., 1972, 1980, 1981a, b; Gustafsson, 1979b; Nichols et al., 1979; Bowden and James, 1985a; Larsson and Gustafsson, 1987). Heat production in continuous cultures can also be measured by flow microcalorimetry. In this case corrections have to be made, however, because of the time lag of the flow between the fermenter and the flow-through cell resulting in reduction of the substrate concentration (Brettel et al., 1981a).

The basis calorimetric design of a batch calorimeter, followed by the flow version (Monk and Wadsö, 1968), was practically identical in an ampoule version (Wadsö, 1974). The development of the ampoule calorimeter, in which the samples are loaded in simple closed ampoules (1–4 ml, glass or steel), made it possible to measure the heat production of such environmental particulate material as soil and sediment samples (Ljungholm et al., 1979a, b, 1980; Gustafsson and Gustafsson, 1983). Recently, a flexible multichannel microcalorimeter system has been developed that, in one instrument, includes different types of calorimetry, e.g. flow and ampoule calorimetry (Suurkuusk and Wadsö, 1982). By further improvements of this multichannel system, measurements of environmental samples can be carried out using a perfusion vessel (Nordmark et al., 1984) which consists of a vessel (1–4 ml, steel) in which stirring and aeration during the measuring period is possible, instead of the hermetically closed ampoules (Gustafsson and Gustafsson, 1985). Dermoun et al. (1985) have recently developed a batch type instrument that is a modification of the classical Tian Calvet (heat-conduction type) microcalorimeter (for detailed explanation see Hemminger and Höhne, 1984). Their design allows for aeration and stirring during measurement, thereby enabling aerobic growth studies on homogenous and heterogeneous samples. Using a different instrumental approach, Lock and Ford (1983) developed an inexpensive split-flow microcalorimeter of the semiadiabatic type. This instrument is useful for experiments where it is desirable to measure the rate of heat production or total heat evolved from attached or sedimentary aquatic microorganisms, over or through which a small flow of water can be passed.

MICROBIAL GROWTH STUDIES

Studies of microbial systems using calorimetric methods have increased the knowledge of the energy balance of microbial growth and of microbial growth kinetics, largely due to the development of refined instruments (reviewed by Forrest, 1969; Belaich, 1980; Lamprecht, 1980).

Energy balance studies

To perform energy balance calculations of microbial growth initial and final states have to be defined. The initial state can be achieved by a defined substrate, inoculated with a known amount of microbial cells. The final state may be chosen as the point at which the growth reaction stops owing to exhaustion of one growth-limiting component of the substrate. At the final state, the products of the growth reaction must also be known. The elemental composition of the microbial cells can be analysed in order to construct a growth reaction equation. Battley (1960a) proposed the following growth reaction equation for the aerobic growth of *Saccharomyces cerevisiae* on glucose:

$$C_6H_{12}O_{6(aq)} + 3.84O_{2(aq)} + 0.29NH_{3(aq)} \rightarrow 4.09CO_{2(aq)}$$

$$+ 4.73H_2O_{(l)} + 1.95CH_{1.72}O_{0.44}N_{0.15}$$

Subsequently, Battley (1960b) measured calorimetrically the enthalpy change (ΔH_{exp}) that accompanied this growth reaction at 30°C. The experimentally obtained value (ΔH_{exp}) was corrected for chemical side-reactions, such as the solubilization of gases and the neutralization of bases, by using tabulated standard values. This resulted in a value of $\Delta H_{exp,corr}$ of $-2004\,kJ$ per mole of glucose consumed. In this case, the correction for side-reactions amounted to less than 1% of the experimentally obtained value. It is important to evaluate the value of side-reactions, however, as they may greatly influence the experimentally obtained value (Spink and Wadsö, 1976).

A growth reaction summarizes all metabolic reactions. The experimentally obtained enthalpy charge that accompanies a growth reaction at constant pressure and temperature, corrected for side reactions ($\Delta H_{exp,corr}$), is equal to the change in enthalpy of the total metabolism (ΔH_{met}). This growth reaction is composed of one catabolic reaction and one anabolic reaction, and may be described as follows (Belaich, 1980):

energy source → products of catabolism

$$\Delta H = \Delta H_{cat}$$
$$C(H_2O) + NH_4OH \rightarrow C_XH_YO_ZN + nH_2O$$
$$\Delta H = \Delta H_{an}$$

Using standard enthalpies of formation, a theoretical enthalpy change value for the catabolic reaction ($\Delta H_{cat,theor}$) is easily obtained from the difference between the sum of enthalpies of formation of the products and enthalpies of formation of the reactants. Battley (1960b) calculated this to be -2833 kJ per mole of glucose oxidized aerobically to carbon dioxide and water.

The enthalpy change of the anabolic reaction ($\Delta H_{an,theor}$) can also be calculated from standard enthalpies of formation. The heat of formation of the cell material can be calculated from values of the elemental analysis and the measured heat of combustion of the cells. Great differences, however, have been reported for the heat of formation of different cell material (Belaich, 1980). A discrepancy between reported heats of formation of cell material can be expected, however, as the chemical composition of cells varies for each particular type of organism, environmental condition and growth phase. Using published thermodynamic data, Belaich (1980) obtained variations in heats of formation of cell material from different microorganisms within the extremes of -1.21 to -17.02 kJ per gram of biomass. In these calculations, Belaich used two different reported values for the heat of combustion, i.e. -22.59 and -15.47 kJ per gram of biomass, while Lamprecht (1980) reported a mean value of -19.33 ± 2.09 kJ per gram of biomass for the heat of combustion of different yeasts grown under different conditions by calculating from various literature values. Belaich calculated the resulting enthalpy change for the production of 1 g dry mass of cells from glucose plus NH_4OH to vary from -9.11 to $+4.39$ kJ g^{-1}. This can be compared with an estimate of the enthalpy change based on the calculation of the contribution of all the different steps of the anabolic reaction. Such an estimate yielded a value of $+0.39$ kJ per gram of cell dry mass produced and, as pointed out by Belaich (1980), a value of that size is in agreement with the assumption that the enthalpy change of the anabolic reaction (ΔH_{an}) is negligible, provided that the energy source is a sugar.

Such an assumption is also in agreement with the values reported by Battley (1960a, b, c), using a theoretical calculation based on the fact that 73.4% of the glucose available was oxidized by the catabolic reaction. This should result in a heat evolution (Q_{cat}) of -2079 kJ per mole of glucose consumed. The corrected experimental enthalpy change value ($\Delta H_{exp,corr}$) of -2004 kJ per mole of glucose as reported by Battley (1960b), which should be equal to ΔH_{met}, agrees with the value of Q_{cat} within 4%. When the anabolic reaction was considered and the value of $+0.39$ kJ per gram of cell dry mass produced was used, the calculated $\Delta H_{met,theor}$ agrees with the experimental $\Delta H_{met,exp}$ within 5% (Belaich, 1980). The seemingly small contribution of the anabolic reaction to ΔH_{met} during growth on sugars has been found both for yeasts and bacteria (reviewed by Belaich, 1980; Gustafsson, 1979a; cf. Brettel et al., 1980).

Analytical studies

Methods that monitor microbial growth, turbidometrically or by direct measurements of the cell concentration and biomass, yield the cumulative status of growth at the time point of measurement. Calorimetry gives a measure of the instantaneous metabolic rate (dQ/dt), however, and therefore relates to the first derivative of growth. Consequently, a higher degree of resolution of the growth can be achieved using calorimetric measurements and this enables observations of fine temporal structures of the growth process.

A simultaneous exponential increase in both the biosynthesis of new cellular material and the amount of heat produced has been shown during the main part of the growth period for bacteria and yeasts (Forrest, 1969; Eriksson and Wadsö, 1971; Beezer et al., 1978; Belaich, 1980). An inconstancy has also been reported (Delin et al., 1969; Schaarschmidt et al., 1975; Gustafsson and Norkrans, 1976; Gustafsson, 1979b; Bowden and James, 1985a; Larsson and Gustafsson, 1987); however, that may generally be attributed to the age of and the incubation conditions used for the inoculum, the media composition and/or environmental changes, e.g. in pH or oxygen tension. Such a non-constant specific rate of heat production (dQ/dt per unit of biomass) may reflect a change in metabolism during growth in batch conditions. For example, when growing S. cerevisiae aerobically with glucose as the carbon and energy source, a respiro-fermentative metabolism has been shown (Fiechter et al., 1981; van Dijken and Scheffers, 1986). This metabolism results in a varying fermentative activity accompanied by a changing respiratory quotient and specific rate of production of heat during the presumably fermentative phase (Schaarschmidt et al., 1975; Brettel et al., 1980; Larsson and Gustafsson, 1987). Thus, to evaluate the process of metabolism, it is essential to combine heat measurements with measurements of cell concentration, fluxes of substrate, products, and intracellular metabolites. This is easily performed using flow microcalorimetry.

Because of the subsequential utilization of different substrates, power–time curves with more than one phase may be obtained. Organisms may produce some intermediary products, e.g. ethanol is produced when S. cerevisiae is grown aerobically on glucose (Schaarschmidt et al., 1975; Lamprecht, 1980; Larsson and Gustafsson, 1987) or acetic acid is produced by Escherichia coli and Klebsiella aerogenes (Eriksson and Wadsö, 1971; Bowden and James, 1985a). Separate phases of metabolic activity have also been shown for dissimilatory nitrate reducers owing to the sequential use of different electron acceptors (Reiling and Zuber, 1983; Samuelsson et al., 1986). Samuelsson and coworkers performed these experiments using a flow microcalorimeter under strict anaerobic conditions. The microcalorimeter used, Bioactivity Monitor LKB 2277 (Suurkuusk and Wadsö, 1982), was modified to use stainless steel tubing throughout and the flow of culture was drawn through the calorimeter using a pump that was placed in a hood and was flushed with N_2 gas. The calorimetric experiments were combined with analyses of nitrate, nitrite and nitrous oxide. The continuous reduction of the electron acceptor, nitrate, by a denitrifier, Pseudomonas fluorescens, was shown by the detection of just one phase in the power–time curve, while a dissimilatory ammonium producer, Pseudomonas putrefaciens, gave two phases because of the subsequent consumption of nitrate and the accumulated nitrite. In contrast to P. fluorescens, the denitrifying thermophilic Bacillus stearothermophilus gave a similar pattern to P. putrefaciens (Reiling and Zuber, 1983).

Calorimetry may be useful as an analytical tool for biotechnical, medical, or environmental protection and control purposes. Microcalorimetry may be a convenient way to ensure consistency of industrial strains and the composition of complex industrial media. When using power–time curves as 'fingerprints', particular attention must be paid to the preparation of the inocula. Methods are presently available that yield consistent uniform inocula by freezing in liquid nitrogen (Beezer *et al.*, 1976; Nichols *et al.*, 1979). The physiological state can be defined by microcalorimetry and it has been shown to be extremely important for the degree of tolerance of yeast cells when exposed to a stress situation (Blomberg *et al.*, 1987). This illustrates how microcalorimetry may serve as a tool in industry to enable the correct choice of an inoculum and time for harvesting. Microcalorimetry may also be valuable as a tool to measure the effectiveness of antimicrobial agents in industry (Harju-Jeanty, 1982; Beaubien and Jolicoeur, 1985), medicine and for environmental protection and control (Binford *et al.*, 1973; Beezer, 1977; Beezer and Chowdhry, 1980; Chowdhry *et al.*, 1983; Redl and Tiefenbrunner, 1981; Beezer *et al.*, 1983; Semenitz *et al.*, 1983; Gustafsson and Gustafsson, 1983). Power–time curves of zero-order kinetics were used by Beezer and coworkers to determine the effect of different antifungal drugs as a more rapid and simple method for measuring antimicrobial activity. The assay was more sensitive and reproducible, but lacked the capacity of the classical agar plate technique. Bowden and James (1985b) reported a quantitative study of energy changes occurring during growth of *K. aerogenes* in the presence of nalidixic acid (NA). Increasing concentrations of NA gave rise to decreasing yields and increasing amounts of heat and carbon dioxide released.

MICROBIAL ECOLOGY STUDIES

Pamatmat (1982) stated that 'total heat flux is an indication of the rate of degradation of potential chemical energy originally fixed by photosynthesis and represents benthic energy flow'. This proposal was stated for the heat production rate by sediments, but can be extended to encompass heat fluxes in many environmental samples. As such, the total rate of heat production from an environmental sample equals the rate of the combined processes, including metabolic processes and chemical side reactions. The latter may, however, also be coupled to the metabolic reactions (Pamatmat, 1982). The great advantage of direct calorimetry, as compared with other techniques used for studying metabolism, is the non-specificity of the measurement because contributions from all types of metabolism will be included. An important consideration, however, because of this non-specificity of heat measurements, is that the interpretation of the complex signal from a heterogenous environmental sample may only be feasible in combination with data from more specific measurements. Techniques that can be useful in combination with heat measurements are among others gas analyses, rate measurements of turnover of elements, measurements of specific enzyme activities and levels of energy related metabolites such as nucleotides. A combination of different techniques with calorimetry would contribute to the understanding of the structure and activity of the population and to which kinds of metabolism dominate the ecosystem under examination. More specific techniques run the risk of overestimating or underestimating the value of some specific process, and it is therefore of invaluable importance to be able to correlate those to a measurement of the total activity. Today, the only technique that gives a reliable measurement of the total activity in a natural sample is microcalorimetry.

To date, the application of microcalorimetry in the field of microbial ecology has been used to a limited extent. Consequently, in the following discussion I will briefly describe different approaches of the technique in the field of microbial ecology.

Several authors (Hesselink van Suchtelen, 1931; Mortensen, *et al.*, 1973; Konno, 1976; Ljungholm *et al.*, 1979a, b, 1980; Sparling, 1981a, b, 1983) have proposed the usage of direct calorimetry for measuring metabolic activity in soil. Mortensen and Ljungholm with their coworkers have used an ampoule microcalorimeter (Wadsö, 1974) for their soil studies. Cylindrical steel 1 ml ampoules were loaded with sieved soil of compost or mor type (Mortensen *et al.*, 1973). The soils were tested for biological activity. In these experiments, sterile soils, attained by repeated autoclaving or irradiation, generated heat that amounted to about 10% of that obtained in untreated samples. Consequently, it seems that non-biological reaction in the soil contributed to some extent to the rate of heat production. The heat production rate was measurable in approximately 1 g of untreated soil. Fertilization resulted in an increase in the amount of heat released, while storage of the soils, even at a relatively low temperature (4 °C), reduced the rate of heat production. Addition of water to soil samples (2–5 g dry mass enclosed in ampoules; volume 10 ml) gave rise to a heat production of non-biological origin, probably because of changes of the hydration equilibria in the soil (Ljungholm *et al.*, 1979a). In long term experiments (2–3 months), soil samples were enclosed in plastic (polyethene) ampoules sealed at both ends by silicone rubber membranes (Ljungholm *et al.*, 1979a, 1980). These ampoules were highly permeable to oxygen and carbon dioxide and during measurements were hermetically enclosed in steel ampoules. In between the calorimetric measurements, the plastic ampoules were exposed to a controlled atmosphere of specific temperature and humidity. The rate of heat production decreased with time in untreated and glucose-fertilized soils. From studies of the effects of the gas phase composition in the ampoules, this decline was suggested to depened upon accumulation of carbon dioxide rather than an oxygen deficiency. A reduction in microbial activity was also obtained after leaching with dilute sulphuric acid, compared with the activity in water-leached soil (Ljungholm *et al.*, 1980).

Sparling (1983) investigated 25 soils from northern Britain by measuring the rate of heat (100 ml stainless steel flow-through cell, Calvet microcalorimeter) and carbon dioxide production in these samples. These parameters were correlated (Sparling 1981a, b) and the fact that an average amount of 21.1 J of heat was released per cubic centimetre of gas evolved suggests that the metabolism was largely aerobic. The addition of glucose (amendment) greatly increased the heat output and a linear relationship between the biomass and heat production rate was obtained, i.e. 180.05 ± 34.61 (SD) mW g^{-1} biomass C. The author reported that 'the rate of heat output from a range of soils showed large variations when expressed in terms of soil dry weight, but the variation was much reduced when the rate of heat output was expressed in terms of the biomass. Within treatments, the rate of heat output from the soil biomass was remarkably uniform, and soil treatments such as storage or amendment had very much greater effects on heat output than did the type of soil sampled. The findings suggest that microcalorimetry gave reasonable indication of the overall rates of biomass catabolism in the soils, and that the biomass of different soils behaved in a generally similar manner.'

Pamatmat (1982) presented data for the heat production rate of sediment collected from a San Francisco Bay mud flat. Samples were taken, by inverting Petri dishes, from different depths, i.e. from 0 to 7 cm. For measurements during exposure to aerobic conditions, samples were flooded with water. Measurements of the anaerobic metabolism were performed by filling dishes with sediment to minimize available oxygen. Anaerobic layers from a depth of 1–2 cm and deeper evolved heat at fairly steady rates, and these rates decreased with increasing depth. The heat production rate of the uppermost aerobic layer (0–1 cm) decreased with time of measurement owing to insufficient aeration. The anaerobic metabolism (with chemical side reactions included) resulted in a heat production rate of -1600 J h^{-1} m^{-2}, while in the aerated sample, which included anaerobic and aerobic metabolism as well as chemical side reactions, the heat production rate was -1960 J h^{-1} m^{-2}. These kinds of heat measurements are useful for mechanistic studies of the functions of various ecosystems, especially in combination with measurements of other parameters, e.g. ATP concentration and electron transport activity (Pamatmat et al., 1981).

In order to obtain a reliable value of the *in situ* metabolic activity in different environments, precautions must be taken to minimize the disturbance to the ecosystems. Lasserre and Tournier (1984), however, reported identical microcalorimetric responses with undisturbed and disturbed sediment in their microcosm systems in eutrophication studies. Microcosms were established from the superficial layer of sediments and water, representing different ecosystems located on the Atlantic coast of France. The sediments were gently homogenized and sieved (mesh size 2 mm). Glass chambers (35 ml) were loaded with 10 ml sediment and 25 ml of the sea water. The water 5 mm above the sediment surface was pumped via a filter (100 µm) at a flow rate of 23 ml h^{-1}, the so-called circulating interface, through an LKB flow microcalorimeter. During experiments, the temperature was maintained at 18 ± 1 °C in the microcosms and these were placed in darkness to eliminate photosynthesis. Peptone was used to imitate an acute eutrophication process. During the exponential transitory phase that followed peptone addition, a linear correlation between heat production rate and the rate of oxygen utilization was observed and the pattern of the power–time curve was closely correlated with the $^{14}CO_2$ turnover. Conversely, different changes were not correlated when attaining the steady state subsequent to the transitory phase.

Three stations, i.e. one lagoonal enclosure, one intertidal beach, and one estuarine mud flat, were compared with respect to seasonal variations (Tournier and Lasserre, 1984). Characteristic differences did not exist between the three systems as a result of the heat of response upon eutrophication (peptone 4 mg ml^{-1}). In contrast, the characteristics of the power–time and oxygen tension curves were related to seasonal trends.

For some purposes, a high degree of reproducibility between subsamples is more important than reliable *in situ* values, e.g. for establishing a system for toxicity testing (Gustafsson and Guftafsson, 1983). The rates of heat production from river sediment samples (1.2–1.5 ml), flooded with 0.8–0.5 ml river water, were measured. The subsamples were enclosed in glass ampoules (3 ml) that allowed aeration of the water phase via injection needles between measurements. A calorimeter of the heat-conduction type (Bioactivity Monitor LKB 227) was used. The sediment samples were treated with different concentrations of a quaternary amine, whereby a qualitatively good agreement was obtained between measurements of the specific heat production rate (dQ/dT per dry sample mass) and measurements of the specific respiratory activity, as measured using the Warburg technique. Increasing concentrations of the quaternary amine decreased both the total metabolic and the respiratory activity. The highest concentration of the quaternary amine used, 1000 mg per litre of sediment–water mixture, gave a burst in activity in both types of measurements after 20 h of exposure, whereafter a low activity level was obtained. In addition, the pool of ATP and the number of microorganisms, as determined using the most probable number technique, decreased with an increased concentration of the quarternary amine. The combination of the methods in this study showed the dominating nature of aerobic respiratory metabolism in these samples and its sensitivity to increasing concentrations of the amine, as well as a sensitivity of the total metabolism.

In these experiments, we obtained a decreasing specific rate of heat production owing to oxygen deprivation during the measuring period (cf. Pamatmat, 1982). In a subsequent study (Gustafsson and Gustafsson, 1985), the effect of oxygen starvation was almost overcome by the usage of a perfusion vessel instead of closed ampoules. The perfussion vessel (LKB; see Nordmark *et al.*, 1984) enabled stirring and aeration of the water phase during the course of the incubation. An expected increased heat production rate owing to aeration of the sample can be masked, however, by a high dry sample weight because diffusion of air through the entire sample may be limited. Increased aeration rate in the perfusion vessel was followed by an increase in the heat production rate.

Using the instrument constructed according to their own design (Lock and Ford, 1983; see the Instrumentation section), Lock and Ford (1985) studied the rate of heat production from river epilithon obtained by utilizing the growth that developed on black glass beads (diameter, 1.5 mm) that remained in a river in North Wales for more than 2 months. Thereafter, the microcalorimeter cells were loaded with the beads. At the start of an experiment, the epilithon was perfused with filtered (0.7 μm) river water at ambient river temperature until equilibrium was reached. By removing fractions of different molecular weights from the river water, the contribution of each fraction to the total metabolic activity as determined by monitoring the rates of heat production. The expected rapid response to changes in the energy supply did not occur. These authors proposed a lag in the availability to the heterotrophic microorganisms of dissolved and colloidal organic matter. In a subsequent study, epilithon were collected from boreal river and stream systems (Quebec, Canada) containing a considerable pool of organic carbon. In comparison with epilithon from the temperate river (North Wales), a greater part of the metabolic activity was derived from fractions containing high organic molecular weight components from the river water. The authors proposed that the epilithon of the boreal rivers was more adapted to high concentrations of humic material. The level of metabolic activity was substantially lower, however, in the boreal epilithon ($< 1.05 \, \mu W \, cm^{-2}$) than in the temperate ones ($0.8–6.8 \, \mu W \, cm^{-2}$). Epilithon from both river systems produced heat in organic-matter-free medium corresponding to up to 48% of the total metabolic activity. This was assumed to indicate utilization of stored products in the epilithon matrix. In long term perfusions, a relatively large increase in heat production rate was obtained when high molecular weight material was removed. Ford and Lock (1987) suggested 'a competitive access model, whereby slowly metabolized high molecular weight compounds saturate adsorption sites on the epilithon surface'.

CONCLUDING REMARKS

All metabolism will result in heat evolution, and therefore direct calorimetry is generally applicable to every kind of biological system. Microcalorimetry may be used as a precise analytical tool in a defined system or as a tool to measure the total metabolic activity in a complex and heterogenous environmental sample. However, because of the non-specificity of calorimetric measurements, the success in interpreting the experimental results to yield valuable and unique information will generally be dependent on combining the data with other results obtained using more specific techniques. A great advantage of microcalorimetry is that it enables continuous measurements and is a technique that is generally easy to handle.

This presentation is not a complete review of the applications of microcalorimetry in the various fields of microbiology but rather contains selected studies hopefully highlighting the general applicability of microcalorimetry as a tool in basic and applied microbiology.

ACKNOWLEDGEMENTS

My sincere appreciation to my colleagues for constructive criticism of the manuscript and to M. Jehler and A. Renås for secretarial assistance.

REFERENCES

Battley, E.H. 1960a. Growth-reaction equations for *Saccharomyces cerevisiae*. *Physiologia Plantarum*, **13**: 192–203.

Battley, E.H. 1960b. Enthalpy changes accompanying the growth of *Saccharomyces cerevisiae* (Hansen). *Physiologia Plantarum*, **13**: 628–640.

Battley, E.H. 1960c. A theoretical approach to the study of the thermodynamics of growth of *Saccharomyces cerevisiae* (Hansen). *Physiologia Plantarum*, **13**: 674–686.

Beaubien, A. and Jolicoeur, C. 1985. Applications of flow microcalorimetry to process control in biological treatment of industrial wastewater. *Journal of the Water Pollution Control Federation*, **57**: 95–100.

Beezer, A.E. 1977. Microcalorimetric studies of micro-organisms. In *Application of calorimetry in life sciences* (eds. I. Lamprecht and B. Schaarschmidt). Walter de Gruyter, Berlin.

Beezer, A.E. 1980. *Biological calorimetry*. Academic Press, London.

Beezer, A.E. and Chowdhry, B.Z. 1980. Microcalorimetric investigations of drugs. In *Biological calorimetry* (ed. A.E. Beezer), pp. 195–246. Academic Press, London.

Beezer, A.E., Newell, R.D. and Tyrrell, H.J.V. 1976. Application of flow microcalorimetry to analytical problems: the preparation, storage and assay of frozen inocula of *Saccharomyces cerevisiae*. *Journal of Applied Bacteriology*, **41**: 197–207.

Beezer, A.E., Newell, R.D. and Tyrrell, H.J.V. 1978. Flow microcalorimetric investigation of yeast growth in a complex medium. *Microbios*, **22**: 73–84.

Beezer, A.E., Hunter, W.H. and Storey, D.E. 1983. A structure–activity correlation for the antibacterial action of a series of *m*-alkoxy phenols against *Escherichia coli*. *Journal of Pharmacy and Pharmacology*, **35**: 406–407.

Belaich, J.P. 1980. Growth and metabolism in bacteria. In *Biological calorimetry* (ed. A.E. Beezer), pp. 1–42. Academic Press, London.

Binford, J.S., Binford, L.F. and Adler, P. 1973. A semiautomated microcalorimetric method of antibiotic sensitivity testing. *American Journal of Clinical Pathology*, **59**: 86–94.

Blomberg, A., Larsson, C. and Gustafsson, L. 1987. Tolerance of *Saccharomyces cerevisiae* to salt stress in relation to physiological state. To be published.

Bowden, C.P.P. and James, A.M. 1985a. Microcalorimetry studies of energy changes during the growth of *Klebsiella aerogenes* in simple salts/glucose media; correlation of specific power and size of the ATP pool. *Microbios*, **43**: 93–105.

Bowden, C.P.P. and James, A.M. 1985b. Microcalorimetry studies of energy changes during the growth of *Klebsiella aerogenes* in simple salts/glucose media: inhibition by nalidixic acid. *Microbios*, **44**: 201–216.

Brettel, R., Corti, L., Lamprecht, I. and Schaarschmidt, B. 1972. Combination of a continuous culture with a flow-microcalorimeter. *Studia Biophysica, Berlin*, **34**: 71–76.

Brettel, R., Lamprecht, I. and Schaarschmidt, B. 1980. Microcalorimetric investigations of the metabolism of yeasts. VII. Flow-calorimetry of aerobic batch cultures. *Radiation and Environmental Biophysics*, **18**: 301–309.

Brettel, R., Lamprecht, I. and Schaarschmidt, B. 1981a. Microcalorimetric investigations of the metabolism of yeasts. Growth in chemostat cultures on glucose. *European Journal of Applied Microbiology and Biotechnology*, **11**: 205–211.

Brettel, R., Lamprecht, I. and Schaarschmidt, B. 1981b. Microcalorimetric investigations of the metabolism of yeasts. Growth in batch and chemostat cultures on ethanol medium. *European Journal of Applied Microbiology and Biotechnology*, **11**: 212–215.

Brown, H.D. 1969. *Biochemical microcalorimetry*. Academic Press, New York.

Chang, R. 1977. *Physical chemistry with applications to biological systems*. Collier Macmillan, London.

Chowdhry, B.Z., Beezer, A.E. and Greenhow, E.J. 1983. Analysis of drugs by microcalorimetry. Isothermal power-conduction calorimetry and thermometric titrimetry. *Talanta*, **30**: 209–243.

Delin, S., Monk, P. and Wadsö, I. 1969. Flow microcalorimetry as an analytical tool in microbiology. *Science Tools*, **16**: 22–26.

Dermoun, Z., Boussand, R., Cotten, D. and Belaich, J.P. 1985. A new batch calorimeter for aerobic growth studies. *Biotechnology and Bioengineering*, **27**: 996–1004.

Eriksson, R. and Wadsö, I. 1971. Design and testing of a flow microcalorimeter for studies of aerobic bacterial growth. *Proceedings 1st European Biophysics Congress*, Vol. IV, pp. 319–327. Medical Academy, Vienna.

Fiechter, A., Fuhrmann, G.F. and Käppeli, O. 1981. Regulation of glucose metabolism in growing yeast cells. *Advances in Microbial Physiology*, **22**: 123–183.

Ford, T.E. and Lock, M.A. 1987. Epilithic metabolism of dissolved organic carbon in boreal forest rivers. *FEMS Microbiology Ecology*, in press.

Forrest, W.W. 1969. Bacterial microcalorimetry. In *Biochemical microcalorimetry* (ed. H.D. Brown), pp. 165–180. Academic Press, New York.

Forrest, W.W. 1972. Microcalorimetry. In *Methods in microbiology* (eds. J.R. Norris and D.W. Ribbons), Vol. 6B, pp. 285–318. Academic Press, London and New York.

Gustafsson, K. and Gustafsson, L. 1983. Heat changes, respiratory activities, ATP pools and metabolic potentialities in natural sediment samples treated with a quaternary amine. *OIKOS*, **4**: 64–72.

Gustafsson, K. and Gustafsson, L. 1985. A microcalorimetric perfusion vessel used for measurements of total activity in sediment samples. *Journal of Microbiological Methods*, **4**: 103–112.

Gustafsson, L. 1979a. On the mechanism of halotolerance. *PhD Thesis*. University of Gothenburg, Sweden.

Gustafsson, L. 1979b. The ATP pool in relation to the production of glycerol and heat during growth of the halotolerant yeast *Debaryomyces hansenii*. *Archives of Microbiology*, **120**: 15–23.

Gustafsson, L. and Norkrans, B. 1976. On the mechanism of salt tolerance. Production of glycerol and heat during growth of *Debaryomyces hansenii*. *Archives of Microbiology*, **110**: 117–183.

Harju-Jeanty, P. 1982. Microcalorimetry, a new technique for studying slime problems in papermaking waters. *Appita*, **36**: 26–31.

Hemminger, W. and Höhne, G. 1984. *Calorimetry. Fundamentals and practice*. Verlag Chemie, Weinheim.

Hesselink van Suchtelen, F.A. 1931. Energetics and microbiology of the soil V. *Archiv für Pflanzenbau*, **7**: 519–541.

Jones, M.N. 1979. *Biochemical thermodynamics*. Elsevier, Amsterdam.

Konno, T. 1976. Calorimetric application on measurements of activity of soil microorganisms. *Netsu*, **3**: 148–151.

Lamprecht, I. 1980. Growth and metabolism in yeasts. In *Biological calorimetry* (ed. A.E. Beezer), pp. 43–112, Academic Press, London.

Lamprecht, I. and Schaarschmidt, B. 1977. *Application of calorimetry in life sciences*. Walter de Gruyter, Berlin.

Lamprecht, I. and Zotin, A.I. 1978. *Thermodynamics of biological processes*. Walter de Gruyter, Berlin and New York.

Larsson, C. and Gustafsson, L. 1987. Glycerol production in relation to the ATP pool and heat production rate of the yeasts *Debaryomyces hansenii* and

Saccharomyces cerevisiae during salt stress. *Archives of Microbiology*, **147**: 358–363.

Lasserre, P. and Tournier, T. 1984. Use of microcalorimetry for the characterization of marine metabolic activity at the water-sediment interface. *Journal of Experimental Marine Biology and Ecology*, **74**: 123–139.

Ljungholm, K., Norén, B., Sköld, R. and Wadsö, I. 1979a. Use of microcalorimetry for the characterization of microbial activity in soil. *OIKOS*, **33**: 15–23.

Ljungholm, K., Norén, B. and Wadsö, I. 1979b. Microcalorimetric observations of microbial activity in normal and acidified soils. *OIKOS*, **33**: 24–30.

Ljungholm, K., Norén, B. and Odham, G. 1980. Microcalorimetric and gas chromatographic studies of microbial activity in water leached, acid leached and restored soils. *OIKOS*, **34**: 98–102.

Lock, M.A. and Ford, T.E. 1983. Inexpensive flow microcalorimeter for measuring heat production of attached and sedimentary aquatic microorganisms. *Applied and Environmental Microbiology*, **46**: 463–467.

Lock, M.A. and Ford, T.E. 1985. Microcalorimetric approach to determine relationships between energy supply and metabolism in river epilithon, *Applied and Environmental Microbiology*, **49**: 408–412.

Monk, P. and Wadsö, I. 1968. A flow micro-reaction calorimeter. *Acta Chemica Scandinavica*, **22**: 1842–1852.

Mortensen, U., Norén, B. and Wadsö, I. 1973. Microcalorimetry in the study of the activity of microorganisms. *Bulletins from the Ecological Research Committee (Stockholm)*, **17**: 189–197.

Nichols, S.C., Prichard, F.E. and James, A.M. 1979. Microcalorimetry studies of energy changes during the growth of *Klebsiella aerogenes* in simple salts/glucose media. 1 Establishment of standard conditions. *Microbios*, **25**: 187–203.

Nordmark, G., Laynez, J., Schön, A., Suurkuusk, J. and Wadsö, I. 1984. Design and testing of a new microcalorimetric vessel for use with living cellular systems and in titration experiments. *Journal of Biochemical and Biophysical Methods*, **10**: 187–202.

Pamatmat, M.M. 1982. Heat production by sediment. Ecological significance. *Science*, **215**: 395–397.

Pamatmat, M.M., Graf, G., Bengtsson, W. and Novak, C.S. 1981. Heat production, ATP concentration and electron transport activity of marine sediments. *Marine Ecology — Progress Series*, **4**: 135–143.

Redl, B. and Tiefenbrunner, F. 1981. Microcalorimetric measurements of microbial metabolic activity in sewage digesting systems. *European Journal of Applied Microbiology and Biotechnology*, **12**: 234–238.

Reiling, H.E. and Zuber, H. 1983. Heat production and energy balance during growth of a prototrophic denitrifying strain of *Bacillus stearothermophilus*. *Archives of Microbiology*, **136**: 243–253.

Samuelsson, M-O., Cadez, P. and Gustafsson, L. 1986. Heat production by a denitrifying bacterium, *Pseudomonas fluorescens*, and a dissimilatory ammonium producing bacterium, *Pseudomonas putrefaciens*, grown anaerobically with nitrate as electron acceptor. In Samuelsson, M-O. Dissimilatory nitrate reduction in the marine environment, *PhD Thesis*, University of Gothenburg, Sweden.

Schaarschmidt, B., Zotin, A.I., Brettel, R. and Lamprecht, I. 1975. Experimental investigation of the bound dissipation function. Change of the Ψ_u-function during the growth of yeast. *Archives of Microbiology*, **105**: 13–16.

Semenitz, E., Casey, P.A., Pfaller, W. and Gstraunthaler, G. 1983. Microcalorimetric, turbidimetric, phase-contrast microscopic, and electron microscopic investigations of the actions of amoxicillin, clavulanic acid and augmentin on amoxicillin-sensitive and amoxicillin-resistant strains of *Escherichia coli*. *Chemotherapy*, **29**: 192–207.

Sparling, G.P. 1981a. Microcalorimetry and other methods to assess biomass and activity in soil. *Soil Biology and Biochemistry*, **13**: 93–98.

Sparling, G.P. 1981b. Heat output of the soil biomass. *Soil Biology and Biochemistry*, **13**: 373–376.

Sparling, G.P. 1983. Estimation of microbial biomass and activity in soil using microcalorimetry. *Journal of Soil Science*, **34**: 381–390.

Spink, C. and Wadsö, I. 1976. Calorimetry as an analytical tool in biochemistry and biology. In *Methods of biochemical analysis* (ed. D. Glick), Vol. 23, pp. 1–159. Wiley, New York.

Stoesser, P.R. and Gill, S.J. 1967. Precision flow-microcalorimeter. *Reviews of Scientific Instruments*, **38**: 422–425.

Suurkuusk, J. and Wadsö, I. 1982. A multichannel microcalorimetry system. *Chemica Scripta*, **20**: 155–163.

Tournier, T. and Lasserre, P. 1984. Microcalorimetric characterization of seasonal metabolic trends in marine microcosms. *Journal of Experimental Marine Biology and Ecology*, **74**: 111–121.

van Dijken, J.P. and Scheffers, W.A. 1986. Redox balances in the metabolism of sugars by yeasts. *FEMS Microbiology Reviews*, **32**: 199–224.

Wadsö, I. 1974. A microcalorimeter for biological analysis. *Science Tools*, **21**: 18–21.

Wilkie, D.R. 1960. Thermodynamics and the interpretation of biological heat measurements. In *Progress in biophysics and biochemical chemistry*, Vol. 10, p. 259.

8

Genetically engineered organisms in the ocean environment — risks and benefits

Rita R. Colwell, Department of Microbiology, University of Maryland, College Park, MD 20742, USA

Substantial evidence of genetic modification of microorganisms in natural aquatic environments exists. Studies of deep ocean dumpsites show changes occurring in the microbial populations of receiving waters, including altered bacterial community structure. Allochthonous microorganisms can be detected after dumping of wastes at deep sea sites. Initial effects are detectable at the time of dumping, followed by sustained community structural changes and, eventually, genetic modification of the natural population, including increased incidence of plasmids. Evidence of plasmid selection and genetic exchange in aquatic environments, including the ocean, has been obtained, e.g. high incidence of plasmid-containing bacteria in polluted waters, presence of free DNA in marine environments, co-existence of identical plasmids in different co-habiting strains of bacteria commensal to marine animals, and *in situ* plasmid transfer. Genetically engineered marine microorganisms introduce a new technology, offering opportunities for enhancing biodegradation *in situ* and fish production, including more rapid growth, larger growth size, and extended migration. Risks associated with release of genetically engineered organisms to the environment, whether intentional or accidental, are not known, but it has been suggested that adverse effects might include community and structure alteration and exchange of genetic material of pathogens with autochthonous organisms. Microorganisms, once released to the ocean environment, are unlikely to be retrievable. Genetically engineered marine microorganisms explicitly engineered to persist and function in the environment may provide many benefits but also pose new problems and raise issues which must be considered.

INTRODUCTION

Advances in molecular genetics during the past decade ushered in a new technology, now recognized as biotechnology. As a result, the biological sciences have entered the marketplace with all of the drama associated with massive influxes of venture capital to biotechnology. A variety of successful applications have been achieved in medicine, fulfilling early claims of the substantive potential of biotechnology.

In the marine sciences, the potential of biotechnology has only begun to be recognized (Colwell, 1983). Cloning of the growth hormone genes in fish and shellfish has been undertaken and technological advances in aquaculture based on biotechnology can be expected in the near future. Concomitantly, cloning of genes governing the production of pharmacologically active metabolites from marine organisms is also underway, with the enormous potential of yielding new drugs effective in treating leukaemia and other forms of cancer, and combating virus infections, including those associated with *Herpes* and related viruses.

The role of marine bacteria as producers of marine toxins, e.g. tetradotoxin, the Puffer fish toxin, recently shown to be produced by marine and estuarine *Vibrio* spp., including *Vibrio cholerae* (Tamplin *et al.*, 1987), is also being elucidated at the molecular genetic level. The possibilities for biotechnological applications from marine sources are immense (Colwell, 1983; Ahmed and Attaway, 1987).

With the potential benefits come certain risks, including the uncertainties associated with release of genetically engineered microorganisms (GEMs) into the marine environment. These uncertainties include the question of whether GEMs will survive and persist in the oceans, whether lateral transfer of genetic material from GEMs to autochthonous bacteria will occur, whether adverse effects on natural cycles and processes will be caused by the GEMs, and whether diseases of indigenous biota will be associated with GEMs. Ravages of disease inflicted on the fisheries stocks of the world oceans by GEMs could be hypothesized, even though the probability is remote. Questions also may be raised as to whether fish engineered to produce anti-freeze proteins, and thereby becoming capable of migrating further north than previously possible, will upset the natural ecosystem. Additional concerns no doubt can, and will be, formulated. To provide a rational basis for discussion of hypothesized events related to release of GEMs into the marine environment, it is important to review the present status of gene transfer occurring in the marine environment, in order to establish a baseline, or reference point, against which the effects of GEM release can be measured. For example, the incidence of plasmids in marine bacteria and of genetic exchange amongst bacteria of non-marine and marine origin provides useful information from which one can extrapolate.

PLASMIDS IN ESTUARINE BACTERIA

In 1975, Guerry characterized plasmids in Chesapeake Bay bacteria (Guerry, 1975). Of the bacteria tested, 40.7% contained plasmids. Furthermore, *in vitro* transfer of plasmids from *Escherichia coli* to *Vibrio parahaemolyticus*, an autochthonous Chesapeake Bay vibrio, was demonstrated. The transfer rate was found to be low for laboratory matings, about 10^{-5}. However, demonstration of inter-generic transfer was concluded to be significant. Because the number of heterotrophic bacteria in Chesapeake Bay water has been shown to be about $10^5 \, \text{ml}^{-1}$, it is highly probable that genetic transfer occurs between bacteria in this major estuary.

In a series of studies over a 10 year period, the following was observed. Sewage effluent and outfall confluence samples at the Barceloneta Regional Treatment Plant in Barceloneta, Puerto Rico, and outfall confluence samples at Ocean City, Maryland, were collected. Samples from uncontaminated open ocean areas served as clean-water controls. Bacteria were enriched in marine broth 2216 amended with 1 μg of one of a set of chemicals selected for study per millilitre: nitrobenzene, dibutyl phthalate, *m*-cresol, *o*-cresol, 4-nitroaniline, bis(tributyltin) oxide and quinone. Minimum Inhibitory Concentrations (MICs) of the chemicals were determined individually for all isolates. The bacterial isolates were evaluated for resistance to nine different antibiotics and for the presence of plasmid DNA. Treated sewage was found to contain large numbers of bacteria simultaneously possessing antibiotic resistance, chemical resistance, and multiple bands of plasmid DNA. Bacteria resistant to a combination of antibiotics, including kanamycin, chloramphenicol, gentamicin, and tetracycline, were isolated only from sewage effluent samples. Thus, isolates derived from toxic chemical wastes were observed more frequently to contain plasmid DNA and to demonstrate antimicrobial resistance than bacterial isolates from domestic sewage-impacted waters or from uncontaminated open ocean sites (Baya *et al.*, 1986); the incidence of plasmids in these isolates is shown in Table 1.

Site	Number of plasmid bands													Mean number of bands per strain
	0	1	2	3	4	5	6	7	8	9	10	11	12	
Barceloneta effluent	18	27	11	7	5	4	3	6	4	0	0	1	1	2.6
Barceloneta confluence[b]	21	9	4	2	0	1	1	2	0	0	0	0	0	1.2
Ocean City confluence[b]	52	5	1	0	0	0	0	0	0	0	0	1	0	0.4
Clean water (control)	35	2	4	0	1	0	0	0	0	0	0	0	0	0.3

[a]Data summarized from Baya *et al.* (1986).
[b]Sewage–seawater confluence.

Table 1 Number of bacterial strains exhibiting plasmid bands detected with agarose gel electrophoresis[a]

PLASMIDS IN MARINE BACTERIA

In 1975, Sizemore studied the incidence of R plasmids in marine bacteria collected from the Atlantic Ocean (Sizemore, 1975). Table 2 shows that the largest number of R plasmid-bearing strains were isolated from harbours, coastal waters, and other areas under anthropogenic influence. Similar observations subsequently have been reported by other investigators. Devanas *et al.* (1980) showed the presence of cadmium and antibiotic resistance among bacteria isolated from New York Bight

Sample	Resistant colonies/ml			TVC[c]
	Tc[b]	Cm[b]	Sm[b]	
Chesapeake Bay, USA				
Colgate Bay	22	22	25	4.0×10^4
Kent Island	5	6	3	9.8×10^4
Atlantic Ocean				
Charleston Harbour	TNTC	TNTC	TNTC	TNTC
4 miles off shore	TNTC	TNTC	TNTC	TNTC
100 miles off shore	0.09	0.24	0.24	1.5×10^2
Puerto Rico				
San Juan Harbour	13	5	TNTC	8.1×10^2
60 miles off shore	0.04	0.01	0.14	2.0 ± 10^0
Research vessel bilge	1.9×10^4	7.7×10^3	2.1×10^5	2.4×10^7

[a]Compiled from data gathered by Sizemore (1975).
[b]Tc = tetracycline, Cm = chloramphenicol, Sm = streptomycin (all at $15 \, \mu g \, ml^{-1}$.
[c]TVC = total viable count (colony-forming units per millilitre).
TNTC × too numerous to count.

Table 2   Incidence of antibiotic-resistant bacteria in water samples[a]

Apex sediment. Interestingly, the evidence suggests that, once they enter the marine environment, multiply resistant bacteria persist, even in the apparent absence of selective pressure (Goyal and Adams, 1984). As shown by Guerry (1975), Sizemore (1975) also succeed in transferring resistance markers and recorded *in vitro* marker transfer from naturally occurring marine isolates to *E. coli* recipients (Table 3).

GENE TRANSFER IN THE NATURAL ENVIRONMENT

Anthropogenic materials have been associated with alterations in aquatic bacteria and, presumably, their genetic material. Several studies have been done, yielding results that are in support of this hypothesis. Results obtained from a variety of studies, although predominantly circumstantial, support the hypothesis that bacteria in the marine environment exhange genetic information *in situ*.

The data for fresh water systems are convincing. For example, Grabow *et al.* (1974) detected antibiotic resistance transfer between strains of *E. coli* suspended in cellulose dialysis bags in a South African river at 20 °C. The frequency of transfer was observed to be significant for *in situ* transfer, e.g. 3×10^{-1}. Actual transfer of R plasmids was not proven, only phenotypic transfer, and it was assumed that conjugation was the mechanism of transfer.

Transduction was shown to occur *in situ* by Morrison *et al.* (1978) who inoculated membrane chambers with either streptomycin-sensitive (Str[s]) *Pseudomonas aeruginosa* or Str[r] *P. aeruginosa* lysogenic for phage F116. The chambers were placed in a Tennessee reservoir for 10 days at 25 °C. High rates of transfer were observed, e.g. 5×10^{-6}, 1.2×10^{-2}, and 9.5×10^{-1} transductants per recipient at

Donor	Recipient	Marker selected[b]	Result
WR26	Ec1485	Cm, N	0
E8	Ec1485	Cm, N	0
E23	Ec185	Cm, N	0
E108	Ec185	Cm, N	0
E118	Ec185	Tc, N	+
E126	Ec185	Tc, N	0
E129	Ec185	Tc, N	0
E262	Ec185	Tc, N	+
E263	Ec185	Tc, N	+
E270	Ec185	Sm, N	0
E292	Ec185	Tc, N	+·
E342	Ec185	Sm, N	0
B390	Ec185	Tc, N	+
B391	Ec185	Sm, N	+
B394	Ec185	Tc, N	+
B398	Ec185	Sm, N	+
B410	Ec185	Sm, N	0
B411	Ec185	Sm, N	0

[a]Taken from Sizemore (1975).
[b]Cm = chloramphenicol ($30\,\mu g\,ml^{-1}$); N = nalidixic acid ($50\,\mu g\,ml^{-1}$); Tc = tetracycline ($30\,\mu g\,ml^{-1}$; and Sm = streptomycin ($250\,\mu g\,ml^{-1}$).

Table 3   Results of mating experiments using strains of marine bacteria as donors and *E. coli* as recipient[a]

1 h, 4 days, and 10 days respectively. It was concluded by Morrison *et al.* (1978) that such rates were capable of supporting high levels of genetic reassortment in nature.

Circumstantial evidence for *in situ* plasmid transfer among bacteria contained in the gut of deep sea amphipods was recently obtained by Wortman and Colwell (1987). Identical plasmids were identified in a *Vibrio* sp. and a *Pseudomonas* sp. isolated from the gut of an amphipod collected from the Bay of Biscay at 4300 m. Identification was based on agarose mobility, restriction similarity, and DNA–DNA hybridization. Interestingly, several bacteria, including *E. coli*, have been found to show increased longevity at increased hydrostatic pressure (500 atm) (Baross *et al.*, 1975).

RISKS

Major considerations regarding release of genetically engineered organisms are fate, movement, dispersal and survival in the environment (Stotzky and Babich, 1984).

Survival of bacteria in the marine environment has been studied for decades, with many examples in the literature of the decline or persistence of bacteria exposed to the natural environment, including seawater. Factors affecting survival in the ocean include temperature, pH, absorption, water type, nutrient concentration, ionic strength and composition. Indeed, a variety of factors contribute to lack of survival, including toxic concentrations of pollutants in seawater in estuaries, solar irradiation, acidity, starvation and predators, to mention a few. Factors influencing growth of microorganisms in the natural environment are not so clearly understood, but community structure and species interactions play an important role.

Various viewpoints have been expressed concerning potential risks of releasing genetically modified microorganisms, ranging from the suggestion that the outcome of introducing a new species is not predictable, since there is, at the present time, no systematic understanding of the natural factors influencing success or failure of microorganisms in the environment, to the suggestion that the probabilities of survival and establishment are small, but that the potential consequences may be significant (Stotzky and Babich, 1984; Gillett et al., 1984; Brill, 1985; Colwell and Grime, 1986).

Considerations associated with risks are primarily genetic, e.g. characteristics of the parental strain and molecular genetic constitution of the engineered strain, and environmental factors, as discussed above, which include habitat and geographical factors affecting survival, reproduction and dispersal, and biological and/or biogeochemical interactions.

That microorganisms in the environment can exchange genetic material has been well documented (Colwell and Grimes, 1985; Shaw, 1985). In fact processes of genetic exchange and uptake of DNA within and between species are widespread in nature and have been reported. New nucleotide sequences or genes, indeed, may occur. Theoretically, therefore, any gene combination can potentially be found in any single organism and integration of introduced DNA can occur by both homologous and heterologous recombination. In general, recombinants contain DNA derived from organisms within a single, natural exchanger group. Homologous transformation–transfection, transduction, plasmid and conjugative transposon transfer during conjugation, and mobilization of non-conjugative plasmids are known to occur in the natural environment.

Drawing the line between what is risky and what is not with respect to genetically engineered organisms is not an easy task. Plasmids may be a special problem because, quite frequently, desired genes are introduced via plasmids, which have the potential of transferring to other cells. The possibility of pre-selection, because the microorganisms are designed to be selected under the conditions by which they are released, and pre-emption, because the conditions of release are also designed to favour the genetically engineered organism, is not unique to genetically engineered organisms. This occurs in many enrichment processes carried out using non-recombinant DNA methods. It may be that the real problem is the introduction of organisms into new environments. The problem then is how to determine whether the ecosystem will be affected in a harmful or undesirable way, if a genetically engineered organism is released into a given environment that is new to the organism. The vast majority of introductions of naturally occurring organisms into a new environment have been harmless, although some have not, e.g. the gypsy moth, English sparrow, and Japanese beetle. Some introductions have been very successful, as, for example, biological control of insect pests.

It is impossible to maximize benefit and minimize risk by selecting the parental strain for genetic engineering appropriately. For example, a strain can be selected for limited ability to transfer genes or to disperse. On the other hand, an organism can be selected that is already in widespread use and has been employed safely, e.g. the Vaccina virus as a vector for construction of new vaccines.

Finally, in dealing with assessment of risk associated with genetically engineered organisms, the assumption is that the risk is conjectural, e.g. risk of an event that has not occurred before and, therefore, is not an analysis based on the statistics of experience, such as living to a given age or being run over by a train, as is the case with actuarial tables. The small genetic changes generally involved in organisms engineered by recombinant DNA methods are usually neutral or confer loss of fitness in most environments. Nevertheless, the possibility exists that a few such organisms may have an increased fitness.

It is not possible to predict whether members of well-established marine microbial communities in seawater or sediment will be so well adapted to their environment that they will counteract entry of new organisms to their community, so that a new organism will not be able to compete effectively. Such communities may, in fact, be invadable, with concomitant changes in biogeochemistry that could have serious consequences for the resident, i.e. native (autochthonous), species. It may well be that disturbed marine ecosystems may be particularly vulnerable. In any case, the deliberately released organism must grow, if it is to be any risk at all. Whether GEMs grow only temporarily and then die back or persist indefinitely is not yet known and cannot be accurately predicted at the present time.

Gene transfer among bacteria, including marine bacteria, is a highly evolved, natural process, not an odd event occurring only in the laboratory. Furthermore, some plasmids and transposons clearly are highly mobile and, because of that property, are used to introduce new genes. Control of the spread of these genes to resident organisms depends on many factors in a very complex environment, such as seawater or marine sediment, making prediction of ensuing behaviour very difficult indeed, unless special care is taken to construct organisms with non-conjugating, non-transposable vectors.

In summary, there are many benefits accruing to aquaculture, fishery stock assessment, biodegradation of wastes in the sea, and other specific marine or estuarine systems from marine biotechnology. The evidence for substantial interaction at the molecular genetic level of marine microbial communities is convincing, although sparse, relative to what is known for fresh water bacteria and bacteria of medical significance. From known genetic interactions amongst marine bacteria and with the new information now being gathered, a more realistic assessment of risks and benefits will be possible.

REFERENCES

Ahmed, S.I. and Attaway, D.H. 1987. *Marine biotechnology: status and prospects*. National Sea Grant Program, Rockville, MD.

Baross, J.A., Hanus, F.J. and Morita, R.Y. 1975. Survival of human enteric and other sewage microorganisms under simulated deep-sea conditions. *Applied Microbiology*, **30**: 309–318.

Baya, A.M., Brayton, P.R., Brown, V.L., Grimes, D.J., Russek-Cohen, E. and Colwell, R.R. 1986. Coincident plasmids and antimicrobial resistance in marine bacteria isolated from polluted and unpolluted Atlantic Ocean samples. *Applied and Environmental Microbiology*, **51**: 1285–1292.

Brill, W.J. 1985. Safety concerns and genetic engineering in agriculture. *Science*, **227**: 381–384.

Colwell, R.R. 1983. Biotechnology in the marine sciences. *Science*, **222**: 19–24.

Colwell, R.R. and Grimes, D.J. 1986. Evidence for genetic modification of microorganisms occurring in natural aquatic environments. In *ASTM Special Technical Publication 921*, Vol. 9, *Aquatic toxicology and environmental fate* (eds. T.M. Poston and R. Purdy) pp. 222–230. ASTM, Philadelphia, PA.

Devanas, M.A., Litchfield, C.D., McClean, C. and Gianni, J. 1980. Coincidence of cadmium and antibiotic resistance in New York Bight Apex benthic microorganisms. *Marine Pollution Bulletin*, **11**: 264–269.

Gillett, J.W., Levin, S.A., Harwell, S.A., Andow, D.A., Alexander, M. and Stern, A.M. 1984. *Potential impacts of environmental release of biotechnology products: assessment regulation and research needs*. Ecosystems Research Center, Cornell University, Ithaca, NY.

Goyal, S.M. and Adams, W.N. 1984. Drug-resistant bacteria in continental shelf sediments. *Applied and Environmental Microbiology*, **48**: 861–862.

Grabow, W.O.K., Prozesky, O.W. and Smith, L.S. 1974. Drug resistant coliforms call for review of water quality standards. *Water Research*, **8**: 1–9.

Guerry, P. 1975. The ecology of bacterial plasmids in Chesapeake Bay. *PhD Thesis*. University of Maryland.

Morrison, W.D., Miller, R.V. and Sayler, G.S. 1978. Frequency of F116-mediated transduction of *Pseudomonas aeruginosa* in a freshwater environment. *Applied and Environmental Microbiology*, **36**: 724–730.

Shaw, P.D. 1985. Plasmid ecology. In *Plant microbe interactions*. Vol. 2, (eds. E. Nester and T. Kosgue) pp. 3–39. Macmillan, New York.

Sizemore, R.K. 1975. A study of plasmids isolated from antibiotic-resistant marine bacteria. *PhD Thesis*. University of Maryland.

Stotzky, G. and Babich, H. 1984. Fate of genetically-engineered microbes in natural environments. *National Institutes of Health Recombinant DNA Technical Bulletin*, **7**: 163–188.

Tamplin, M.L., Colwell, R.R., Hall, S., Kogure, K. and Strichartz, G.R. 1987. Sodium channel inhibitors produced by enteropathogenic *Vibrio cholerae* and *Aeromonas hydrophila*. Lancet, April 25, 975.

Wortman, A.T. and Colwell, R.R. 1987. Evidence of plasmid transfer among bacteria isolated from the deep sea. In preparation.

9

Sulphate-reducing bacteria and the mechanism of corrosion in the marine environment

W. Allan Hamilton, Department of Genetics and Microbiology, Marischal College, University of Aberdeen, Aberdeen AB9 1AS, UK

Renewed interest in the practical consequences of microbial corrosion in the marine environment has arisen at a time when our knowledge of both microbial consortia and, more specifically, the physiology of the sulphate-reducing bacteria is also advancing rapidly. This conjunction has resulted in an increased awareness of the environmental conditions likely to give rise to corrosion effects. It has also brought a greater understanding of and interest in the mechanism of anaerobic corrosion and the interplay between microbiology and metallurgy.

INTRODUCTION

Corrosion is an electrochemical process where both anodic and cathodic reactions occur (Hamilton, 1985):

$$M \rightleftharpoons M^{2+} + 2e^- \qquad\qquad \text{anode}$$

$$\left. \begin{array}{l} \tfrac{1}{2}O_2 + H_2O + 2e^- \rightleftharpoons 2OH^- \\[2ex] 2H^+ + 2e^- \rightleftharpoons 2H \rightleftharpoons H_2 \end{array} \right\} \quad \text{cathode}$$

At the anode dissolution of the metal produces electrons, and it is therefore necessary to have the cathodic reaction where these electrons are absorbed. The most familiar cathodic reaction occurs in the presence of oxygen and gives rise to rust, with the alternative reaction with protons as electron acceptors being possibly important under anaerobic conditions.

In microbially induced corrosion the role of the microorganisms can be direct, creating such an electrochemical cell, or it can be indirect in that it maintains a pre-existing electrochemical cell by stimulating either the cathodic or the anodic reaction. There are a number of microorganisms and a number of mechanisms which are considered to be involved in these roles. For example, the growth of microbial colonies or slimes involves a whole range of possible organisms, including Pseudomonads and other aerobic species. These can create a concentration cell. Most commonly a differential aeration cell is created with a low concentration of oxygen shielded underneath the slime or colony growth as compared with the high concentration externally in the bulk environment. Under these conditions the surface in the low concentration area becomes anodic with the dissolution of metal, while the electrons react at the cathodic region with the high concentration of oxygen giving rise to hydroxide, with the ultimate formation of the metal oxides and hydroxides characteristic of the aerobic corrosion, or rusting process.

This is the commonest corrosion mechanism occurring aerobically. It is also involved in the loss of protection against corrosion in stainless steels where the so-called passivation is dependent on the formation and maintenance of an oxide film in the presence of air. Therefore the absence of oxygen beneath microbial growth can have this additional effect. The corrosion effect itself can be exacerbated where there are present, either as the sole or as a component organism, the iron-oxidizing bacteria which cause the formation of a tubercule at the outer edges of the bacterial growth where the oxidation of ferrous to ferric iron causes the precipitation of ferric hydroxide. This last reaction constitutes an anodic stimulation of the corrosion process.

Very often as a further expression of low oxygen concentrations anaerobic conditions develop within the colony or slime growth, and under these conditions sulphate-reducing bacteria may be present. The sulphate-reducing bacteria are the principal causative organism when it comes to microbially induced corrosion, although not the only culprit as is clear from the instances cited above. While the sulphate-reducing bacteria cause significant economic damage where the total environment is anaerobic (for example with buried pipes in clay soils), in fact the more usual situation is one in which there is the potential both for aerobic corrosion mechanisms involving oxygen concentration cells and for anaerobic corrosion from sulphate reducers growing within the anaerobic anodic region. That is to say, the normal situation with microbially induced corrosion is complex where there is likely to be more than one mechanism of corrosion operating. Even where the process causing the major damage is indeed due to the sulphate reducers, they are generally dependent upon other organisms to create the physicochemical and nutrient conditions necessary for their growth and activity. This is an important point which will be dealt with more fully later.

MECHANISM OF ANAEROBIC CORROSION

Hypotheses
There are a number of hypotheses and a considerable degree of argument over the mechanism of anaerobic corrosion. The classical hypothesis (von Wolzogen Kühr and van der Vlugt, 1934) proposes an anodic reaction with metal dissolution coupled to a cathodic reaction in which it is suggested that protons are the electron acceptor giving rise to hydrogen. Sulphate reducers then oxidize that hydrogen to produce

sulphide which reacts with the iron to give ferrous sulphide as the principal corrosion product. This is an over-simplification because in fact there are two reactions at the cathode; the first forms atomic hydrogen from the proton and electron, followed by the second combination reaction which gives molecular hydrogen. It is the molecular hydrogen which sulphate reducers can oxidize with the resulting production of sulphide. King and Miller (1971) proposed a modification of this classical hypothesis in which the ferrous sulphide corrosion product exists as a film at the metal surface. In purely electrochemical terms this ferrous sulphide film is cathodic with reference to unreacted iron so that it increases corrosion by stimulating the cathodic reaction. There is also the point that the ferrous sulphide film effectively increases the surface area of the cathode to a very large extent and so facilitates the reaction of the sulphate reducers with cathodic hydrogen. In these two ways therefore, King and Miller (1971) have modified the classical hypothesis and have stressed the importance of sulphide as stimulating the overall reaction, but the fundamentals of the mechanism are not greatly altered. Costello (1974), in contrast, was highly critical of some of the analysis and experimentation behind these ideas and suggested that it was a great deal more likely that hydrogen sulphide itself was the electron acceptor at the cathode according to the equation

$$2H_2S + 2e^- = 2HS^- + H_2$$

It is noteworthy that it is molecular hydrogen which is formed directly by this reaction, and which again requires to be oxidized by the sulphate-reducing bacteria.

Two other hypotheses have proposed altogether different mechanisms. Schaschl (1980) has suggested that elemental sulphur is the corrodent. Elemental sulphur could be produced from sulphide by autoxidation and here the presence of oxygen is an important component of the reaction mechanism. Schaschl suggests that a sulphur concentration cell, analogous to an oxygen concentration cell, creates an anodic region underneath the microbial growth shielded from the higher concentration of the sulphur generated by autoxidation in the aerobic zone surrounding the cellular mass. Work done by Hardy (1983) might relate directly to such a mechanism. Corrosion resulting from the growth of sulphate-reducing bacteria was measured by electrical resistance probe. When the cells were growing anaerobically under a nitrogen gas phase there was very little corrosion, but when the reaction vessel was sparged with air the rate of corrosion increased very sharply. Under these conditions, Hardy noted a stimulation of corrosion from about 30.5 μm per year up to 2.25 mm per year; that is, approximately a 100-fold increase in corrosion resulting from air sparging. No mechanism was proposed directly, but it is possible that the sparging with air gives rise to the autoxidation of biogenic sulphide to sulphur. Whatever the mechanism, these data stress again that in anaerobic corrosion oxygen appears to play a critical role.

There is a fifth hypothesis due to Iverson (1984). Originally one of the champions of the classical hypothesis, Iverson has recently suggested that the corrodent is in fact a volatile phosphorus compound, not yet identified but produced by the sulphate-reducing bacteria. In indirect support of this concept, various phosphorous compounds have been identified by a number of workers as corrosion products, notably vivianite, although it is generally considered to be protective.

These five broad hypotheses have points of overlap and points of difference. What they do have in common is the necessity for the growth and active metabolism of the sulphate reducers, and in most cases they stress the importance of hydrogen oxidation and of sulphide production although there is argument over which is qualitatively the more important of these two.

Experimental studies: pure culture

Recent experimental studies therefore have looked further at the corrosion mechanism from the point of view of the growth of sulphate reducers, hydrogen oxidation and sulphide production. Hardy and Bown (1984) looked at the production of $[^{35}S]$-labelled sulphide from $[^{35}S]$-labelled sulphate in their experimental system which contained a marine sulphate reducer in the presence and the absence of a steel working electrode. In the absence of this electrode there was effectively no sulphate reduction. Although the organisms are grown on hydrogen:CO_2 with acetate as carbon source, in the experimental assay they are being incubated under nitrogen:CO_2. These organisms cannot oxidize the acetate; they can only use it as a source of carbon and require an oxidant, e.g. hydrogen, in order to reduce sulphate and to supply the energy for carbon assimilation. When the steel electrode was introduced, however, it provided an oxidant and sulphate reduction was recorded. The theory is that the steel electrode is corroding and therefore producing cathodic hydrogen which is then being oxidized by the sulphate-reducing bacteria. This is a clear demonstration that cathodic hydrogen produced electrochemically can be oxidized and act as a source of metabolic energy for the sulphate reducers. Hardy and Bown suggested, however, that it is still likely that sulphide production is the more important aspect of corrosion.

Recent studies in my own laboratory have substantiated these findings (Pankhania *et al.*, 1986). The experimental system (Fig. 1) consists of a reaction vessel sparged with nitrogen:CO_2 and containing acetate medium with hydrogen:CO_2 plus acetate-grown cells. There are reference electrode, working electrode, pH electrode and a range of electrical control devices so that the potential of the working electrode can be subject to experimental control.

The experimental findings are demonstrated in Fig. 2. Initially, in the absence of the working electrode, there is no growth of sulphate reducers in acetate medium with the nitrogen:CO_2 gas phase. Neither is there significant increase in hydrogenase levels, nor decrease in acetate or sulphate concentrations. If the mild steel working electrode (diameter 1 cm, polished to a 600 grit finish and degreased in acetone) is then introduced, but the potential is held at a value (-100 mV) such that corrosion would not be expected to take place nor hydrogen to be produced, there is still no growth, nor increase in hydrogenase or sulphate utilization. If, however, the potential of the working electrode is lowered to a corrosive value (-1400 mV), where cathodic hydrogen is being produced at the electrode, there is growth of the population, and an increase in hydrogenase activity and sulphate utilization. Nonetheless, this is a transient phenomenon and growth, hydrogenase production and sulphate utilization all cease after a period of some hours. If the working electrode is removed and at this point the gas phase is altered to hydrogen:CO_2, there is a rapid re-establishment of growth and increased hydrogenase, with a further decrease in sulphate. The cells have not been killed therefore by the presence of the

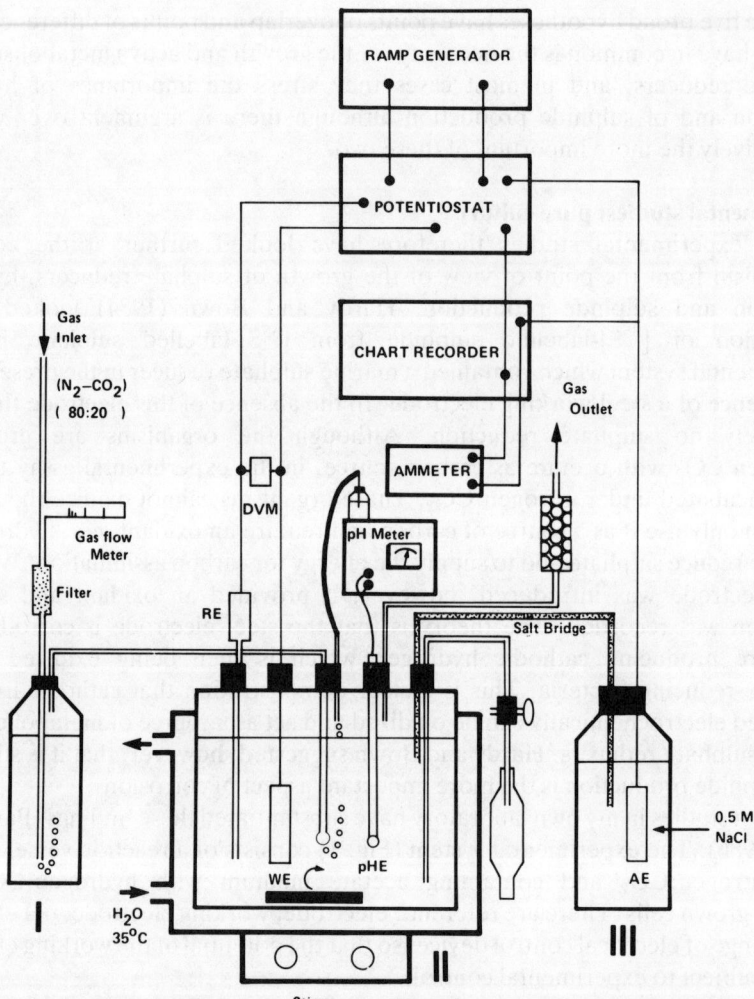

Fig. 1—Electrochemical cell for polarization experiments. WE, working electrode; pHE, pH electrode; AE, auxiliary electrode; RE, reference electrode; DVM, digital voltmeter. (Reproduced from Pankhania *et al*. (1986) with permission of *Journal of General Microbiology*.)

corroding electrode; it is simply that the reaction has stopped, and when hydrogen is supplied in the gas phase then metabolic activity is readily re-established.

A similar transitory effect was reported by Hardy and Bown (1984) and both sets of authors have suggested that the sulphide produced during sulphate reduction may have poisoned the combination reaction which converts atomic to molecular hydrogen. These experiments have therefore shown that cathodic hydrogen can indeed support the growth of sulphate reducers, as required both by the classical hypothesis and in the modifications due to King and Miller and to Costello.

These experimental systems, and the whole range of studies carried out by earlier workers in this field, have had this much in common: they have involved a pure culture of a sulphate-reducing bacterium in homogeneous suspension under

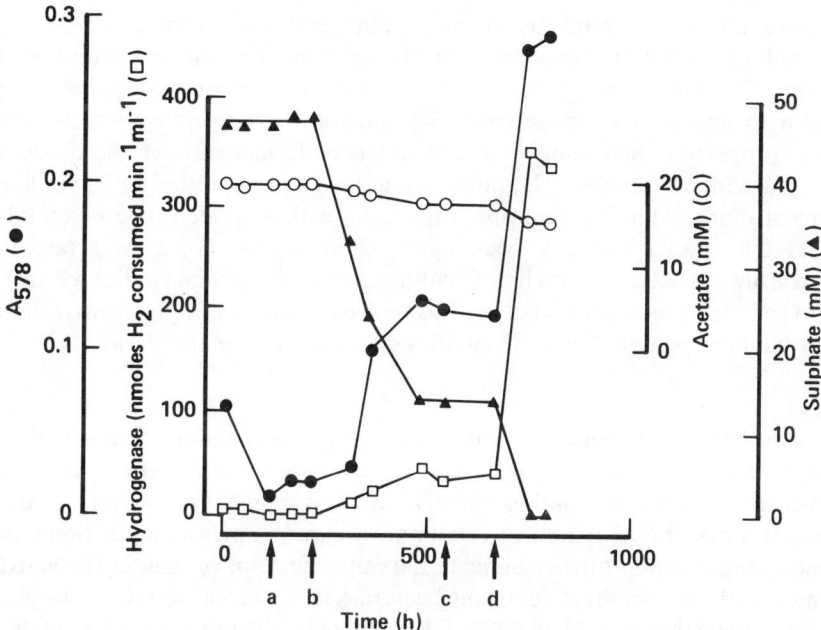

Fig. 2—Growth of *D. vulgaris* in acetate medium under $N_2:CO_2$. ●, population density (light absorption at 578 nm); □, hydrogenase; ▲, sulphate; ○, acetate. a, working electrode introduced, potential held at −100 mV; b, potential lowered to −1400 mV; c, working electrode removed; d, gas phase changed to $H_2:CO_2$. (Reproduced from Pankhania *et al.* (1986) with permission of *Journal of General Microbiology*.)

regulated anaerobic conditions in a fairly simple defined medium. In the cases cited above acetate was the carbon source, but most often it has been a lactate medium that has been employed. This reflects not so much what happens in nature, but rather what we thought we knew about the sulphate-reducing bacteria; strict anaerobes using a restrictive number of carbon sources such as lactate and metabolically independent of other microorganisms. That this is no longer seen as an adequate description of the sulphate-reducing bacteria and their role in microbial ecosystems is of major importance both to our understanding of anaerobic microbial corrosion and to the experimental systems we develop for its study.

SULPHATE-REDUCING BACTERIA
Physiology

The sulphate reducers were previously thought of as being a restricted range of organisms, perhaps just two genera and with a very limited range of nutrients (Postgate, 1984). Careful and exhaustive studies, particularly by Widdel working in Pfennig's group (Pfennig *et al.*, 1981), and since then by others, have demonstrated within the last ten years or so that there are at least nine different genera of sulphate reducers. They can be either Gram positive or Gram negative; most of them are rod shaped or vibrios, but some are ovoid and others filamentous. Nutritionally they can be divided into two groups: those capable of only partial oxidation of a limited range of carbon sources with acetate as an end product, and those capable of oxidation of short and long chain fatty acids (including acetate) and organic compounds such as benzoate. The latter group can be sub-divided into those giving only partial oxidation to acetate and those oxidizing substrates completely to CO_2; a very common property is that many sulphate reducers do indeed have the capacity to oxidize molecular hydrogen. The point is that the sulphate-reducing bacteria are a very much more extensive group of organisms with a wider range of metabolic activities than had previously been appreciated. If, therefore, a lactate-based medium is always used for enrichment culture and experimental studies we are very liable to get a false picture of what organisms are present in any given environment, and of what their potential or real activities might be (Taylor and Parkes, 1985).

Ecology

These data have been obtained from laboratory experimental studies with pure cultures, but there are also ecological studies (Sørensen *et al.*, 1981) which indicate that hydrogen and acetate, and to a lesser extent propionate and butyrate, are the principal sources of energy and carbon in most natural environments. Some other ecological studies again further indicate the importance of oxygen in the activities associated with the sulphate-reducing bacteria. In a marine sediment Jørgensen (1977) has shown that more than 60% of the sulphate reduction takes place in the top 10 cm of the sediment. The significance is that the sulphide produced will then be able to diffuse into the oxidized sediment or into the oxidized sea water above, and to be reoxidized by microorganisms such as the *Beggiatoa*. In this way there is only limited build-up of sulphide which would otherwise poison the system. At the same time, the sulphate required in the anaerobic zone comes by diffusion from the aerobic zones. Thus the maximal activity of the sulphate reducers, although they are obligate anaerobes, occurs close to the aerobic interface. This involvement of oxygen and the role of the aerobic–anaerobic interface is very important to the rapid turn-over in such an ecological situation, and these conditions can profitably be carried forward in considering the mechanism of anaerobic microbial corrosion.

BIOFILMS

Model for anaerobic microbial corrosion

Fig. 3 shows the biofilm model which we have put forward (Hamilton, 1985) to describe corrosion by sulphate reducers, incorporating the nutritional, physiological and ecological concepts which have been briefly described above. The model depicts a metal surface with an adherent microbial biofilm. At the outer surface of the biofilm the bulk phase may be a totally aerobic environment; for example, with marine fouling on the outer surface of an offshore oil platform, while with a water-or oil-carrying pipeline the flowing liquid may or may not be aerobic depending on a number of factors. Irrespective of the conditions of aerobiosis however, there will be present in the biofilm a wide range of organisms; in the case of marine fouling it will include higher organisms of marine flora and fauna, while with growth in internal pipework the biofilm will generally be exclusively bacterial.

The organisms present may be phototrophic or heterotrophic, aerobic or facultative, and capable of metabolizing whatever material is present in the bulk

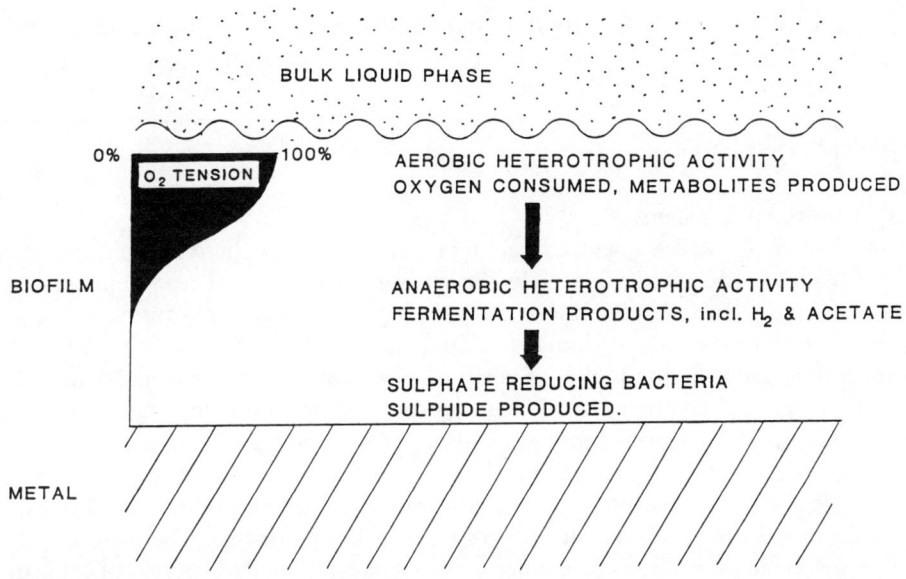

Fig. 3—Biofilm model for anaerobic corrosion. (Reproduced, with permission, from the *Annual Review of Microbiology* **39**. © 1985 by Annual Reviews Inc.)

phase; for example, hydrocarbons or general organic detritus. As a result of the activities of the organisms present in the biofilm, end-products will be produced; breakdown of hydrocarbons giving rise to fatty acids, of polysaccharide materials to sugars, and so forth. These degradative activities generally require oxygen and are carried out by aerobic and facultative species. At a certain thickness of biofilm the rate of diffusion of oxygen into the film will be less than the rate at which it is utilized by microbial activity so that within the biofilm anaerobic conditions will develop. Facultative and anaerobic organisms will also be present therefore, and they will further degrade the fatty acids and sugars to a variety of fermentation products of which hydrogen and acetate will be key examples. Thus ideal conditions will have been created for sulphate reducers. From material that they cannot metabolize such as hydrocarbons, suitable nutrients will have been produced; in what would otherwise have been an aerobic environment, anaerobic conditions will have been developed. Thus the necessary nutrient and physicochemical conditions will have been evolved to enable sulphate-reducing bacteria to live and grow, but to do so totally dependent upon the activities of other biological species present in the same biofilm, and ultimately even dependent upon the aerobic conditions which allowed these organisms to carry out their initial degradative metabolism.

There is also the potential for sulphide to diffuse outwards to the aerobic regions, either through the film or where there is a patchiness in the film itself, and by chemical oxidation to give rise to sulphur, if indeed this is an important component of the mechanism. There is the potential for hydrogen produced electrochemically at the surface to act as a source of metabolic energy. The complete biofilm therefore is necessary for the metabolic activities and physicochemical conditions required for the development of the maximum potential for anaerobic corrosion by the sulphate-reducing bacteria. It therefore becomes vitally important to develop a methodology which allows us to test corrosion mechanisms under the conditions defined by the biofilm model.

Experimental test systems

In my own group we have been examining the role of the sulphate-reducing bacteria in corrosion processes experienced by the offshore oil industry. The systems we have been looking at include concrete and steel jacket production platforms with their associated water injection systems, oil risers, platform-to-shore pipelines, discarded drill cuttings, and marine fouling. Specifically, we have been concerned with water injection systems where microbiological problems are very important from the points of view of corrosion of the system itself, and the potential dangers of reservoir souring. Other areas of particular interest to us are the enclosed environments within drilling legs on concrete production platforms, and marine fouling and discarded drill cuttings on the external surfaces of steel jacket platforms. The technique we employ is to put in position corrosion coupons which after a fixed period of exposure can be retrieved and subjected to corrosion and microbiological analyses. The coupons can either be unprotected or cathodically protected at $-950\,mV$ by attachment of a sacrificial zinc anode. Coupons are then sited in the environment of choice for periods of a few weeks (water injection system) or up to two years (marine fouling or sea-bed sediments). An essential feature of this technique is that the critical first steps in microbiological analysis are carried out immediately the coupons are retrieved on the platform. With this system we have analysed weight loss, corrosion products, and the numbers and types of microorganisms present, as determined by standard counting techniques. The relevance of the numbers of bacteria that may be grown as pure cultures in the laboratory on selective media to their *in situ* environmental activity is always open to doubt however. If the biofilm model has any validity at all, then what is required is an assay of the activity of the complete undisturbed biofilm.

^{35}S-labelled sulphate reduction assay

Fortunately there is an activity unique to the sulphate reducers that can be assayed directly by adding ^{35}S-labelled sulphate and observing its conversion to ^{35}S-labelled sulphide (Fig. 4). Not only does this assay measure directly the critical activity of the sulphate-reducing bacteria, but it does so without perturbation of the physical and physiological integrity of the biofilm which is essential for any true measure of the metabolic activity of these organisms. The assay has been developed for field analysis using metal corrosion coupons, and the necessary time and control parameters established (Maxwell and Hamilton, 1986a).

Field data

We have applied this technology in a number of offshore environments. For example, coupons were suspended for 72 days in the flooded drill leg of a concrete production platform, and the numbers and activities of organisms in the bulk phase compared with those on the metal surface of the corrosion coupons. Whereas the

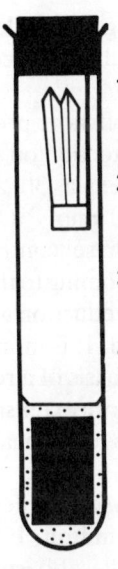

1. Metal coupon placed into 4 ml of anaerobic, filtered, sterile <u>sea water</u> containing <u>10 µCi</u> ^{35}S—sulphate

2. Bung seated securely and 0.5 ml of oxygen-free 2 N zinc acetate <u>immediately</u> injected onto the filter paper wick

3. Metal coupon incubated at desired temperature for <u>optimum incubation time</u> (in this case 5 h)

4. 0.5 ml of oxygen-free 6 N hydrochloric acid injected past the wick into the solution

5. Acid volatile sulphides, including any [H_2 ^{35}S] formed, trapped during a <u>2 h equilibration</u> period at 35°, 100 oscillations min^{-1} in a shaking water bath

    (All manipulations carried out under oxygen-free nitrogen)

Fig. 4—Modified radiorespirometric method for use with metal coupons. (Reproduced from Maxwell and Hamilton (1986a) with permission of *Journal of Microbiological Methods.*)

numbers of sulphate reducers in the bulk sea water phase were 10^3 1^{-1}, on the coupons there were 10^8 m^{-2}. More importantly, a ^{35}S-sulphate reduction rate of 500 µmol m^{-2} day^{-1} was recorded for the coupons as compared with non-detectable levels in the bulk sea water. Although these data give no information on any corrosion effects resulting from microbial activity, they do stress the concentration of bacterial numbers at surfaces and the likely requirement for biofilm development to create the necessary conditions for sulphate reduction activity.

Other studies have looked at the possible effects of cathodic protection on the microbial activity associated with metal test coupons (Maxwell and Hamilton, 1986b). After some 160 days exposure in a mildly polluted harbour environment, it was noted that the sulphate reduction rates on protected and unprotected coupons were 40 nmol coupon^{-1} day^{-1} and 230 nmol coupon^{-1} day^{-1} respectively. The addition of glucose to the assay system raised these activities to 380 nmol coupon^{-1} day^{-1} and 490 nmol coupon^{-1} day^{-1}. It appears therefore that in addition to minimizing corrosion *per se* cathodic protection may also be reducing microbial activity, at least as regards the sulphate-reducing bacteria. The very considerable stimulation given to both protected and unprotected steels by the addition of glucose, however, suggests that the rate-determining factor may be substrate limitation. In this regard it is important to point out that glucose is not itself a substrate for sulphate-reducing bacteria and that this finding reinforces the concept that these bacteria are critically dependent upon other species of microorganisms within the biofilm. Our view of cathodic protection and sulphate reduction activity is that any effect is likely to be indirect in that the formation of corrosion products and sulphide films helps to generate at the metal surface the anaerobic conditions necessary for the growth and biological activity of the sulphate-reducing bacteria.

This therefore adds a further component to the suggestion of King and Miller (1971) that the sulphide film is itself directly involved in the electrochemical corrosion reaction.

Longer exposures of up to two years under marine fouling on offshore production platforms in general do not show a similar effect of cathodic protection on sulphate reduction. The range of activities recorded with unprotected steel was 30–200 nmol $(10\,cm)^{-2}$ day^{-1} and with protected samples, 0–150 nmol $(10\,cm)^{-2}$ day^{-1} (Sanders and Hamilton, 1986). One explanation of these contradictory findings might be that over the longer time periods the anodes are coming to the end of their useful life, and that therefore at the time when the sulphate reduction activity is being assayed the effective cathodic protection is greatly reduced. It is more likely, however, that we are trying to establish an overall picture on the basis of a restricted number of single snapshot pictures and that only when we have more extensive data, acquired regularly at a single site over an extended period, will a less equivocal picture emerge.

Such an extended study is now under way, with coupons buried in sea bed deposits of discarded drill cuttings. We have already noted (Sanders and Hamilton, 1986) high rates of sulphate reduction associated with cuttings from diesel-based muds $(100–5000\,nmol\ g^{-1}\ day^{-1})$ as compared with water-based muds $(100–200\,nmol\ g^{-1}\ day^{-1})$ or natural sediment from a polluted estuary $(100–400\,nmol\,g^{-1}\ day^{-1})$. These activities are matched by the numbers of sulphate-reducing and hydrocarbon-oxidizing bacteria that can be cultured from the various deposits. When corrosion coupons are introduced into cutting deposits similar patterns of activity and numbers are noted. With diesel-based muds the figures are 500 nmol $coupon^{-1}$ day^{-1}, 10^4 sulphate reducers $coupon^{-1}$ and 10^2 hydrocarbon degraders $coupon^{-1}$; with low toxicity muds the figures are 400, 10^3 and 10^1, and with water-based muds, 300, 10^2 and 10^1. The corrosion rates in these three environments are respectively 60, 50 and 30 mg dm^{-2} day^{-1}. Where cathodic protection is applied these figures reduce to 2, 0.5 and 0.5 mg dm^{-2} day^{-1}. The most interesting statistic however relates to the anode life. In each case the anodes had a design life of eight years. Although the anodes remained active in low toxicity and water-based drill cuttings for longer than three years (duration of study, so far), those in the diesel cuttings were exhausted in less than two years. That is to say, although cathodic protection reduces corrosion even in the presence of high sulphate reduction activity, under these conditions there is excessive pressure on the anode, leading to reduced life and ultimately to increase corrosion risk.

Although such studies in the offshore environment have direct relevance to the problems experienced by the oil industry, they are inevitably limited in terms of their scientific rigour. We are therefore mounting a parallel series of laboratory experiments designed to give less equivocal data on the actual mechanisms of anaerobic microbial corrosion.

ACKNOWLEDGEMENTS

The experimental findings discussed in this paper have been the achievement of my colleagues Stephen Maxwell, Peter Sanders, Ish Pankhania and Ali Moosavi. My laboratory has received support from the Science and Engineering Research Council (Biological Sciences Sub-Committee and Marine Technology Directorate), British Petroleum, Shell and Britoil.

REFERENCES

Costello, J.A. 1974. Cathodic depolarisation by sulphate-reducing bacteria, *South African Journal of Science*, **70**: 202–240.

Hamilton, W.A. 1985. Sulphate-reducing bacteria and anaerobic corrosion, *Annual Reviews of Microbiology*, **39**: 195–217.

Hardy, J.A. 1983. Utilisation of cathodic hydrogen by sulphate-reducing bacteria. *British Corrosion Journal*, **18**: 190–193.

Hardy, J.A. and Brown, J.L. 1984. Sulphate-reducing bacteria: their contribution to the corrosion process. *Corrosion*, **40**(12): 650–654.

Iverson, W.P. 1984. Mechanism of anaerobic corrosion of steel by sulfate reducing bacteria. *Materials Performance*, **23**(3): 28–30.

Jørgensen, B.B. 1977. The sulphur cycle of a coastal marine sediment (Limfjorden, Denmark). *Limnology and Oceanography*, **22**: 814–832.

King, R.A. and Miller, J.D.A. 1971. Corrosion by the sulphate-reducing bacteria. *Nature*, **233**: 491–492.

Maxwell, S. and Hamilton, W.A. 1986a. Modified radiorespirometric assay for determining the sulfate reduction activity of biofilms on metal surfaces. *Journal of Microbiological Methods*, **5**: 83–91.

Maxwell, S. and Hamilton, W.A. 1986b. Activity of sulphate-reducing bacteria on metal surfaces in an oilfield situation. In *Biologically induced corrosion* (ed. S.C. Dexter), pp. 284–290. National Association of Corrosion Engineers, Houston.

Pankhania, I.P., Moosavi, A.N. and Hamilton, W.A. 1986. Utilization of cathodic hydrogen by *Desulfovibrio vulgaris* (Hildenborough). *Journal of General Microbiology*, **132**: 3357–3365.

Pfennig, N., Widdel, F. and Trüper, H.G. 1981. The dissimilatory sulfate-reducing bacteria. In *The prokaryotes* (eds. M.P. Starr, H. Stolp, H.G. Trüper, A. Balows and H.G. Schlegel), pp. 926–940. Springer-Verlag, Berlin.

Postgate, J.R. 1984. *The sulphate-reducing bacteria*, 2nd edn. Cambridge University Press, Cambridge.

Sanders, P.F. and Hamilton, W.A. 1986. Biological and corrosion activities of sulphate-reducing bacteria in industrial process plant. In *Biologically induced corrosion* (ed. S.C. Dexter), pp. 47–68. National Association of Corrosion Engineers, Houston.

Schaschl, E. 1980. Elemental sulphur as a corrodent in deaerated, neutral aqueous solutions. *Materials Performance*, **19**(7): 9–12.

Sørensen, J., Christensen, D. and Jørgensen, B.B. 1981. Volatile fatty acids and hydrogen as substrates for sulfate-reducing bacteria in anaerobic marine sediment. *Applied and Environmental Microbiology*, **42**: 5–11.

Taylor, J. and Parkes, R.J. 1985. Identifying different populations of sulphate-reducing bacteria within marine sediment systems using fatty acid biomarkers. *Journal of General Microbiology*, **131**: 631–642.

von Wolzogen Kühr, C.A.H. and van der Vlugt, L.S. 1934. The graphitization of cast iron as an electrobiochemical process in anaerobic soils. *Water (den Haag)*, **18**: 147–165.

10

Short term responses to energy fluctuation by marine heterotrophic bacteria

Staffan Kjelleberg and Malte Hermansson, Department of Marine Microbiology, University of Göteborg, Carl Skottsbergs Gata 22, S-413 19 Göteborg, Sweden

The notion that marine waters are heterogeneous with respect to microenvironments, e.g. a wide range of interfaces, and levels, composition and activities of substrates, has formed the basis for physiological and molecular studies of responses by some marine heterotrophic bacteria to energy fluctuations. Several regulatory mechanisms and pathways allow the cell to adapt from growing to non-growing conditions. Some alterations involved therein are discussed. Emphasis is given to the changes in cell surface characteristics and adhesion, the pronounced cellular reorganization during the initial period of non-growth, the starvation-specific protein turnover and modulation, including *de novo* synthesis during non-growth, and some characteristics of amino acid uptake and incorporation during short as well as long term starvation. It is concluded that small, energy-limited, non-growing bacterial cells are not dormant, but are active cells that differ significantly from their growing counterparts, and that they have the capacity to capture and incorporate organic substrates at very low concentrations and to do so during periods of a decrease in biomass and no nucleic acid synthesis.

INTRODUCTION

The net result of marine bacterial activity is cell division and growth. Ultimately, each species must reproduce to avoid extinction. However, short term measurements of the bacterial activity of heterogeneous populations will reflect a large variation of physiological types and metabolic activities, including species or groups of species in a non-growing state. During the relatively long mean bacterial generation times commonly found in natural waters, intermittent growth, with periods of unbalanced growth as well as non-growth between division events, is likely to be induced by fluctuations in substrate concentration and composition. Microenvironments that are quite distinct from each other and from the bulk water mass are found, for example, at various interfaces and in the vicinity of excreting living organisms. It also appears that some marine bacteria are not able to show continuous growth at the low ambient nutrient concentration in the bulk water. When bacteria are transported between different microenvironments, the cells have to adjust to changes in nutrient and energy availability. For example, the detachment of cells from a particle or a faecal pellet may represent a drastic downshift from an environment of plenty to starvation conditions, from a growing to a non-growing, survival, phase.

The basis for the present discussion is a series of experiments in which attempts have been made to study the physiological and molecular aspects of downshift and starvation of some bacterial isolates that may be representative of a significant group of the marine bacterial population. One may ask whether physiological and molecular research, by necessity performed using a few strains only, can improve our understanding of bacteria in the natural ecosystem. We suggest that because of the complexity of the marine system, marine microbiological problems should be examined using the broadest possible approach. We hope that the ideas and results presented in this paper will contribute towards a somewhat less static concept of the marine microbial ecosystem. An understanding of the starvation survival traits may prove to be of some importance, not only in the area of microbial ecology, but also in the areas of public health in relation to survival of pathogens, risk assessment of released engineered organisms and biotechnology.

ECOLOGICAL CONSIDERATIONS

Before the presentation of some cellular pathways of downshift and starvation survival, a few aspects of the ecological background related to these events will be discussed.

The nutritional diversity

It is known that there is a great nutritional diversity among marine bacteria. The population is often divided into distinct groups based on their growth patterns and nutritional requirements. The bacteria which are able to grow at the low concentrations of nutrients found in the bulk water have been defined as oligotrophs (e.g. Poindexter, 1981a). The oligotrophs resemble the K-strategists (Jannasch, 1974). The copiotrophs (Poindexter, 1981b), by contrast, require higher nutrient concentrations for growth. These may be called r-strategists (Jannasch, 1974). Carlucci et al. (1986) showed that bacteria growing on relatively rich media outnumbered the ones that were able to grow on unsupplemented seawater agar in most seawater samples tested, including those from the open ocean. In fact, in 24 out of 29 samples, less than 15% of the numbers growing on rich media grew on unsupplemented agar. On the other hand, many cases are reported where marine bacterial isolates are not able to grow in unsupplemented seawater (e.g. Novitsky and Morita, 1978; Carlucci et al., 1986). On the other hand, Ammerman et al. (1984) were able to show an increase in bacterial density in filter-sterilized seawater cultures inoculated with seawater samples. The results obtained in these experiments indicate that a part of, but not the entire, free-living bacterial population is able to grow on unsupplemented seawater. The percentage of the original population that showed active growth in unsupplemented seawater is difficult to assess, however.

Field studies

Field investigations of bacterial production, using for example incorporation of tritiated thymidine into DNA, provide useful information about precursor utilization by the fraction of the population that is actively growing. The natural diversity, both in bacterial types and in metabolic states, is often overlooked, however, in these studies. The frequent changes in the nutrition state in any environment will be expressed as physiological shifts in certain fractions of the population, and the existence of starving cells or cells in a transient phase between different states of growth may be expected. Microautoradiographic studies of natural samples clearly show that there is considerable variation in the proportion of metabolically active bacteria in different habitats (e.g. Meyer-Reil, 1978; Douglas *et al.*, 1987). Furthermore, the interesting finding that greater numbers of cells incorporated ^{3}H-glutamate than ^{3}H-thymidine led Douglas *et al.* (1987) to discuss the possibility that some populations are in a state of unbalanced growth, i.e. that cells incorporate carbon and energy substrates at times when they are not synthesizing nucleic acids. This situation is clearly supported by laboratory experiments in which periods of growth and non-growth are defined. Cells that have been exposed to starvation conditions during a period of 48 h starvation showed very low rates of incorporation of thymidine, while uptake and incorporation of amino acids were appreciable (Nyström *et al.*, 1986; Mårdén, Hermansson and Kjelleberg, to be published; Mårdén, Nyström and Kjelleberg, to be published).

Size of bacterial cells

A significant decrease in cell size appears to be a general phenomenon when growing bacteria are exposed to starvation conditions (e.g. Novitsky and Morita, 1977; Kjelleberg *et al.*, 1983). Five marine bacterial isolates frequently used in starvation-survival studies in our laboratory decreased to a size characteristic of natural marine bacteria *in situ* (Mårdén, Hermansson and Kjelleberg, to be published). The outgrowth of 'normal' sized cells of *Vibrio, Pseudomonas, Aeromonas* and *Alcaligenes* spp. on dilute agar from an inoculum of $0.22\,\mu m$ filtered seawater indicated that the very small cells found in the sea are 'normal' bacteria which have adapted to the nutrient poor environment (MacDonell and Hood, 1982).

Bacteria found on particles are usually larger than the free-living ones (Ferguson and Rublee, 1976; Hodson *et al.*, 1981; Caron *et al.*, 1982). Starvation experiments of marine bacteria in unsupplemented filter-sterilized seawater demonstrated that the cell volumes were approximately twice as large in the presence of micro-aggregates. The marine *Pseudomonas* sp. S9 showed a cell size of $0.1\,\mu m^3$ without particles, while the mean cell volume of cells in the presence of inanimate micro-aggregates was $0.2\,\mu m^3$ (Mårdén, Hermansson and Kjelleberg, to be published). This suggests that the microenvironment of the particle influences the metabolic state of the attached cells.

Grazing of bacteria seems at least in part to be determined by the size of the bacteria. Ammerman *et al.* (1984) suggested that the large sized cells that were observed during growth of bacteria in unsupplemented seawater cultures were not seen under natural conditions because of the removal of such cells by size-selective grazing. Size-selective grazing by microflagellates of a mixed seawater population was shown by Andersson *et al.* (1986). Wright (1984) suggested that grazing would control the upper end and that survival mechanisms during periods of non-growth would set the lower limit of bacterial numbers in natural waters. Small starved bacteria may not be under the same grazing pressure as larger cells. In fact, the decrease in cell size during starvation may be a way to protect them from predation (Pomeroy, 1984; Morita, 1985).

Influence of bacterial attachment

Many reports lend support to the notion that marine waters are highly heterogeneous with respect to both physical and substrate discontinuities (e.g. Fletcher and Marshall, 1982; Mopper and Lindroth, 1982). The microenvironmental conditions are different at an interface compared with the bulk phase (e.g. Marshall, 1976). Surface active organic molecules are accumulated at surfaces (e.g. Hunter, 1980; Norkrans, 1980). Such surface-localized nutrients can be utilized by both irreversibly and reversibly attached bacteria (Kefford *et al.*, 1982; Hermansson and Dahlbäck, 1983; Hermansson and Marshall, 1985). It was therefore argued that particles may support a loosely associated population that may be difficult to distinguish from the free-living bacteria by conventional methods of size fractionation and sampling (Hermansson and Marshall, 1985). Macroscopic particles or 'marine snow' are difficult to sample with conventional sampling equipment (Silver *et al.*, 1978; Fellows *et al.*, 1981; Knauer *et al.*, 1982). Such fragile aggregate particles are found in oceanic and coastal waters (Shanks and Trent, 1979; Fellows *et al.*, 1981; Caron *et al.*, 1982; Knauer, *et al.*, 1982) and form zones of enriched organic matter and nutrients (Alldredge, 1979; Shanks and Trent, 1979). These particles have been reported to be inhabited by bacterial densities two to five orders of magnitude higher than the free-living population in the surrounding water (Silver and Alldredge, 1981; Caron *et al.*, 1982; Davoll and Silver, 1986). Attached cells have often been found to be more biochemically active than those that are free-living (e.g. Hodson *et al.*, 1981; Kirchman and Mitchell, 1982; Laanbroek and Verplanke, 1986), although equal or lower activity of the surface associated population has also been noted (e.g. Alldredge *et al.*, 1986).

A possible life cycle

Based on the discussion and the information presented above, a life cycle for a part of the bacterial population may be proposed. Small non-growing cells are transported from a nutrient-deficient bulk phase to a microzone where growth is possible. Such a zone may be a nutrient-laden surface or may surround an excreting organism. At or in the vicinity of the interface, the cell utilizes the accumulated nutrients and increases in size. The large cells may eventually detach, either by a change in the cell surface character so as to affect the strength of adhesion, or by the detachment of daughter cells. The latter possibility is in agreement with the suggestion that faecal pellets may act as 'baby machines' which produce free-living cells (Jacobsen and Azam, 1984). The detachment results in a drastic change in nutrient availability and composition. The adaptation during the downshift to the new situation will be described in some detail subsequently. This possible life-cycle is clearly in agreement with the spinning wheel aggregate hypothesis of Goldman (1984). Azam and Ammerman (1984) demonstrated that the motility of bacterial cells allowed them to migrate to and stay in spheres of diffusing organic matter around phytoplankton cells.

PHYSIOLOGICAL AND MOLECULAR PROCESSES OF SUBSTRATE DOWNSHIFT AND STARVATION

During the course of the relatively long generation times displayed by bacteria in the marine environment, a range of survival mechanisms is likely to be utilized. Although several of these mechanisms should preferably be discussed, this presentation will deal mainly with bacterial adhesion and alterations in surface characteristics and the synthesis and degradation of proteins by non-growing marine bacteria. The presentation does not represent a review of the entire research area. Such a review was recently published by Kjelleberg *et al.* (1987).

Bacterial surface characteristics and adhesion

The alteration in bacterial surface characteristics and adhesion during substrate fluctuations may prove to be of ecological significance. There are several such observations reported in the literature. It appears that the most common alteration may be that of an increased liability for adhesion and degree of cell surface hydrophobicity as a consequence of the energy downshift (Dawson *et al.*, 1981; Kjelleberg and Hermansson, 1984; Conway *et al.*, 1986; Hermansson, Jones and Kjelleberg, to be published), although reports of no apparent alteration also exist (Kjelleberg *et al.*, 1985). The former situation has been obserbed for both bacterial isolates and mixed natural bacterial assemblages. It is therefore of interest to examine the specific changes in bacterial surface components that take place during non-growth.

Such changes were first reported by Dawson *et al.* (1981). Formation of fibrillar structures on the surface of the marine *Vibrio* DW1 was seen during the initial phase of starvation. These structures may be directly involved in adhesion as seen by electron microscopy preparations and the fact that increased adhesion to glass surfaces (Dawson *et al.*, 1981) and increased degree of hydrophobicity (Kjelleberg and Hermansson, 1984) correlated with the detection of the structures. This organism showed an apparent reduction in the amount of mucopeptide in the wall without a corresponding reduction in the outer membrane (Dawson *et al.*, 1981). Other microscopic observations of the bacterial surface during non-growth refer to an apparent folding of the outer membrane of several strains that increased their degree of adhesion and hydrophobicity (Kjelleberg and Hermansson, 1984). The formation of outer membrane vesicles has been observed for both *Vibrio cholerae* (Hood *et al.*, 1986) and the marine *Pseudomonas* sp. S9 (Mårdén *et al.*, 1985). The role in adhesion of these structures is unclear, however.

The marine *Pseudomonas* S9 displays yet another pattern of surface alteration that significantly alters the organism's capacity to bind to surfaces. The production and release of a neutral exopolysaccharide are observed during starvation (Wrangstadh *et al.*, 1986). The presence of the non-ionic polysaccharide is inversely correlated to the cell's ability to adhere to a surface, particularly those with a relatively low surface free energy. Using antibodies that were produced against the polymer, these workers demonstrated that the polymer was not present on the cell surface until after 3–4 h of starvation, after which time the number of polymer-coated cells increased until 6–9 h. The polymer was subsequently released from the cells into the starvation medium, resulting in a low percentage of polymer-coated cells at 24 h. The bacteria were relatively more adhesive before and after the polymer was apparent on the surface. It should be stressed that bacterial exopolysaccharides forming a glycocalyx-like structure around the cell are generally suggested to be adhesive in character (e.g. Costerton *et al.*, 1985), partly because of gel-forming characteristics and exposure of charged groups.

With respect to polysaccharides, another possible starvation-induced alteration has been suggested. Based on specific carbohydrate degradation and compositional shifts of starved *V. cholerae*, it is possible that the more hydrophilic molecules of the O-side chains may be more readily utilized under starvation conditions (Hood *et al.*, 1986). These workers showed that three- and five-carbon sugars were decreased, while seven- and six-carbon sugars, probably making up the oligosaccharides of the R-core, were relatively conserved. The possible role of changes in the lipopolysaccharides (LPS) was also examined during starvation of the marine isolate S14 (Albertson *et al.*, 1987). We could not find any differences in the gel electrophoresis patterns of LPS digests from growing and starved cells. Neither could we detect any antibody–antigen reaction by immunoblotting of the LPS. Antibodies specific for starved cells of S14 had previously been produced. The fact that the LPS molecule is constant during starvation conditions in some cells was utilized to detect *V. cholerae* in the viviform stage using fluorescent antibodies against the LPS of the bacteria (e.g. Grimes *et al.*, 1986). Although this strain of *V. cholerae* undergoes major morphological and physiological changes during starvation, the LPS molecule appears to remain the same. Long periods of starvation do not appear to alter the presence or the number of flagella and fimbriae on several bacterial strains (Amy and Morita, 1983a; L. Rüdén, personal communication).

General starvation-induced traits

There are many reports of starvation-induced changes, particularly in non-marine bacteria (for reviews, see Dawes, 1976; Morita, 1982; Mason *et al.*, 1986; Kjelleberg *et al.*, 1987). Most of these refer to endogenous metabolism and constituents and attempts are made to elicidate the possible correlation between survivability and a specific starvation-induced trait. Factors that have been investigated include the number and nature of uptake and transport mechanisms, the stringent response, the proton-motive force, the release of catabolite repression which may improve the cells ability to make use of diverse substrates, and the type of degradation of storage polymers and endogenous macromolecules. Recent studies on protein pattern shifts during starvation of *E. coli* strongly support the notion of a preferential synthesis of starvation and stress proteins (Groat and Matin, 1986). The study of DNA turnover during starvation has also revealed interesting information (reviewed by Kjelleberg *et al.*, 1987). It has recently been found that starving cells of *S. typhimurium* preferentially degrade supercoiled plasmid DNA while they retain copies of the plasmid in relaxed open circular form (Robert, 1986). It is possible that the same process occurs also during starvation of *V. cholerae* (M. Hood, personal communication). A general conclusion pertinent to factors correlating to loss of viability is that the latter is related to the rate at which constituents are metabolized rather than the absolute levels of a particular constituent at the time of starvation. A low rate of endogenous metabolism that nearly matches the provision of the maintenance energy requirements will improve the chances of prolonged viability.

The persistence of copiotrophic cells in nutrient-limited systems such as the marine environment is probably defined, among other things, by the capacity to undergo a number of adaptions, such as the miniaturization of cells (Poindexter, 1981a), the induction of low endogenous respiration (Novitsky and Morita, 1977), the use of previously stored reserves (Jones and Rhodes-Roberts, 1981), and the regulation of protein turnover (Nyström *et al.*, 1986), including the synthesis of new, starvation-specific, peptides (Jouper-Jaan *et al.*, 1986; Albertson *et al.*, 1987; Albertson, Nyström and Kjelleberg, to be published). Furthermore, the capacity to scavenge and transport nutrients seems to be of utmost importance for substrate-limited cells (Morita, 1984). As mentioned above, it is a significant feature of marine bacteria to form small cells. The downshift and miniaturization process occurs in two distinct phases with a demarcation line dividing the initial period of pronounced cellular activities and the subsequent long term survival when very low endogenous metabolism and small alterations in morphology, activity and levels of endogenous constituents are seen. During the initial transient phase of starvation, varying in length of time depending on the organism, the following alterations have been observed.

(i)   A temporary increase in endogenous respiration for both free-living and surface-bound cells (Amy *et al.*, 1983; Kjelleberg *et al.*, 1983; Mårdén *et al.*, 1985).

(ii) A temporary increase in amino acid uptake and incorporation (Nyström *et al.*, 1986; Mårdén, Nyström and Kjelleberg, to be published).

(iii) A non-gradual alteration in the membrane fatty acid composition (Malmcrona-Friberg *et al.*, 1986). An increased ratio of mono-unsaturated to saturated fatty acids and an increased fraction of short chain fatty acids may partly explain the improved uptake capacity of starved cells. The fatty acid changes may cause an increased fluidity of the membranes.

(iv) A rapid decline of total carbohydrates (Hood *et al.*, 1986) and lipids (Oliver and Stringer, 1984; Hood *et al.*, 1986) including polyhydroxybutyrate (PHB) (Mårdén *et al.*, 1985; Malmcrona-Friberg *et al.*, 1986), which may reflect the energy source for the first rather active reorganization phase of starvation.

The increased rate of leucine uptake, that occurred after 3 and 5 h of starvation of S14 cells, appears to correlate with the increase in protein synthesis and the increase in respiratory activity previously shown for this strain. These increases also coincide with the disappearance of the storage polymer PHB (Malmcrona-Friberg *et al.*, 1986). The utilization of PHB might permit a relatively high metabolic and uptake activity in S14 cells during the initial phase of non-growth.

Changes in protein pattern
Assuming that the environment around the bacterial cell is not identical during its life cycle and thereby fails to provide for balanced growth with a continuous increase in biomass of the organism, a series of regulatory responses will be established. We have

restricted the following discussion to some observations of changes in the protein pattern during short term energy deprivation of a few marine bacterial isolates.

With respect to shifts in protein synthesis, Reeve *et al.*, (1984a, b) showed, by using peptidase deficient mutants of *E. coli* with decreased stability during starvation, that protein degradation is important in the survival of starving cells in order to provide amino acids for protein synthesis. These mutant cells lacked free amino acids for protein production. The *E. coli* carbon starvation work suggests that the proteins synthesized initially during the first few hours of starvation are the most critical for survival. This may not be applicable to other strains, however. Starvation of *S. typhimurium* cells for a period of 20 days showed that inhibitors of protein synthesis became more effective and resulted in greater loss of viability as the period of starvation increased (Kjelleberg, Conway and Stenström, to be published). Furthermore, using two-dimensional gel electrophoresis, the marine psychrophilic *Vibrio* Ant 300 was shown not to synthesize starvation specific proteins prior to 7 days of starvation and the production increased in intensity over a period of 30 days (Amy and Morita, 1983b). In fact, dramatic changes in the protein fingerprint pattern were not detected until after 30 days of starvation. The importance of protein synthesis during relatively long term starvation is also anticipated for mesophilic marine bacteria. Subsequent to a temporary increase in protein synthesis during the first hours of starvation of the marine bacterium S14, a decreased but clearly measurable synthesis was followed throughout a starvation period of 120 h (Nyström *et al.*, 1986).

It is reasonable to assume that protein profiles may be a powerful tool for elucidating some of the regulatory networks for survival. Dynamic starvation-induced protein rearrangements occur in two marine bacterial isolates which we have studied in some detail. This is evidenced by the disappearance of some proteins and a new production of others, as well as the shift in the protein size distribution, during periods of non-growth. The rate of protein degradation increases significantly in starving as compared with growing cells (cf. Mandelstam, 1958, 1960; Reeve *et al.*, 1984b).

We initiated these studies by examining whole cell homogenates of growing and non-growing cells with two-dimensional gel electrophoresis (Jouper-Jaan *et al.*, 1986). The two isolates exhibited different pathways of the starvation survival process as judged by the shifts in the two-dimensional protein pattern and measurements of protein concentrations at the onset of and after 24 h of starvation. The *Vibrio* DW1 showed a relatively large decrease in cell volume and total protein content that was not reflected by the loss of only a few proteins. This possibly represents a general decrease, but not a complete loss of, several proteins. The loss of two proteins was observed and two new protein molecules appeared to be synthesized. This included large amounts of a 30 kD protein. A different pattern was observed for the isolate S14. A highly specific change of the protein pattern was demonstrated. A total of 17–20 proteins had disappeared and two novel or previously undetected proteins had accumulated by 24 h of starvation. S14 showed a large decrease in cell volume but not in cellular protein concentration during starvation. It is noteworthy that there may be an initial increase in the total protein per biomass that is due to degradation of other constituents such as PHB.

It was subsequently shown that some of the proteins observed in higher amounts are starvation-specific proteins. This was demonstrated by both producing antibodies against starved cells and pulse labelling with tritiated amino acids after various times of starvation (Albertson *et al.*, 1987; Albertson, Nyström and Kjelleberg, to be published). To determine at what time and to what extent the starvation antigens appeared, cells starved for various periods of time were titrated against serum that had been adsorbed with growing cells. Washed S14 cells produced starvation antigens after less than 1 h in the starvation regime. The titre increased and reached a plateau after 14 h of starvation. No antibody–antigen reaction was detected with either early or late stationary phase cells. The titre for DW1 cells decreased with time of starvation and reached a minimum after 4 h. This decrease was observed using serum that was incompletely adsorbed with growing cells in order to demonstrate the decrease in surface antigens during starvation. Total protein digests of growing and 24 h starved cells were separated on a gel, transferred to nitrocellulose paper and immunoassayed. While no starvation-specific antigens were detected for DW1, several such proteins were detected for S14. One small, 14 kD, and several larger, 100–200 kD, proteins were seen. The location of these were determined by fractionation of the cell, which revealed that the smaller protein found in the whole cell digests was located in the outer membrane (OM) and that the larger protein bands corresponded to periplasmic proteins.

It was mentioned earlier that DW1 increases its cell surface hydrophobicity during starvation. The decrease in the antiserum titre with starvation time implies that growing and starved cells expose the same antigens but that starved cells reveal fewer of these surface structures. Apparently, the surface-exposed adhesion-mediating structures have a strong affinity for binding. DW1 may have a preferential degradation of surface proteins to provide for the synthesis of other starvation proteins not located at the cell surface. As mentioned, protein degradation is important in the survival of starving cells.

Using *in vivo* pulse labelling, it was found that the starvation-specific antigens of S14 cells are not products of degradation (Albertson, Nyström and Kjelleberg, to be published). These studies were carried out by pulse labelling with ^{14}C-leucine at various times of starvation up to 1 week. The radioactivity that was incorporated in the various proteins was measured in slices from SDS gels of the various subfractions of the cell. In this way, a more detailed resolution of the time of starvation-dependent modulation of the protein pattern was seen. For example, of five OM proteins that were found on the SDS gel by 3 h but not 0 h of starvation, at least three were demonstrated to be the result of *de novo* synthesis. With respect to the OM, changes in the protein profile upon starvation were also found in *Klebsiella* (Sterkenburg *et al.*, 1984). SDS gels revealed that two periplasmic proteins were synthesized during starvation (Albertson, Nyström and Kjelleberg, to be published). Pulse-labelling experiments showed that one of these polypeptides was synthesized after 3 h of starvation and constituted the dominating protein by 1 week. The other was detected by pulse labelling after 24 h of starvation and gave rise to starvation-specific antibodies at that time. While most of the new proteins appear after 3 h of starvation, coinciding with other temporary alterations as previously discussed, new peptides were evidently also synthesized after 24 h and 1 week of starvation. The radioactivity that was incorporated into proteins during growth was partly degraded and converted to other protein components during the starvation period studied. It was demonstrated, for example, that the periplasmic space displayed an increased fraction of smaller size proteins during starvation.

It is relevant to ask whether the energy downshift of marine bacteria is similar to that of a stress situation. Groat and Matin (1986) recently analysed more precisely the changes in protein synthesis of carbon-starved *E. coli*. Two-dimensional gel electrophoresis of *in vivo* pulse-labelled proteins resolved at least 30 polypeptides that were either new or synthesized at higher levels during the first 3–4 h. It is interesting to note that there were several heat-shock proteins among those that were synthesized, although the majority were considered to be starvation specific. The overlap in proteins that are induced in *E. coli* by various shock treatments is not clear. Conditions that resulted in the stringent response also yielded at least five of 17 heat-shock proteins (Grossman *et al.*, 1985) and some SOS (emergency repair) inducing agents can induce the heat-shock response (Walker, 1984). A less extensive overlap was recently found in *S. typhimurium* (Spector *et al.*, 1986). So far, it appears that there is a quantitative difference between the starvation induced proteins that are synthesized by *E. coli* and *S. typhimurium* on the one hand and the marine bacteria on the other hand. The energy costs involved in the stress type response to starvation in *E. coli* would not be an ecologically suitable survival tactic for marine bacteria. By the synthesis of fewer proteins, these bacteria may be better adjusted to nutrient deprivation and fluctuation in substrate levels and thereby avoid depletion of limited intracellular reserves.

Functional aspects of the protein modulation

Some functional aspects of the starvation-induced protein modulation are seen by studying substrate capture during periods of non-growth. Both oligotrophic and copiotrophic bacteria have both low and high affinity uptake systems (e.g. Hodson *et al.*, 1979; Glick, 1981; Faquin and Oliver, 1984; Morita, 1984; Davis, 1985; Carlucci *et al.*, 1986). Growth of oligotrophic bacteria on unsupplemented seawater may demand an affinity for organic substrates at concentrations less than $10 \mu g \, C \, l^{-1}$ (Carlucci *et al.*, 1986). It is interesting that the substrate levels corresponding to the K_m values of the high affinity uptake system of a series of copiotrophic bacteria are in the same order of magnitude as the concentration considered as the upper limit for oligotrophic growth. This indicates that copiotrophic bacteria under non-growing conditions are capable of competing for and obtaining solutes at concentration levels found in the marine environment. It is worthwhile to reiterate the fact that isolates on enriched agar media significantly outnumber those on unsupplemented seawater media in samples from a range of sampling stations in marine waters (Carlucci *et al.*, 1986).

Several studies on substrate uptake by starved bacteria have revealed that a high affinity uptake is switched on or that a lower K_m of the high affinity system is induced during starvation (Glick, 1981; Faquin and Oliver, 1984; Davis, 1985). This has been noted also after several weeks of non-growth. A high affinity leucine uptake was induced or derepressed by starvation of S14 cells (Mårdén, Nyström and Kjelleberg, to be published). Non-growing cells of S14 increased their leucine uptake via the high affinity system. The increase that was caused by an increase in V_{max} was most pronounced at 3 and 24 h of starvation. V_{max} continued to increase, as expressed per cellular protein concentration, during 72 h, and remained at this level also after 1 week of non-growth. V_{max} of the low affinity uptake system, however, decreased gradually during this starvation period. This was also seen by the relative leucine binding activity of the periplasmic binding proteins, released by osmotic shock, of the high and low affinity uptake system. The proteins were separated by FPLC gel filtration. A 44 and a 37 kD sized protein were shown to represent the high and low affinity binding proteins respectively. While the high affinity system was shown to be repressed during growth in amino acid supplemented medium, the relative binding activity of the low affinity system decreased markedly during starvation. It appears that the high affinity uptake is favoured during periods of non-growth.

Protein synthesis during starvation was also demonstrated subsequent to the osmotic shock release of periplasmic proteins (Mårdén, Nyström and Kjelleberg, to be published). While all cells remained viable, the osmotic shock resulted in a complete loss of substrate uptake by binding proteins. Independent of the time of this shock release, the continued starvation included a gradual resynthesis of the periplasmic binding proteins until almost complete uptake capacity was regained after 72 h. It is noteworthy that this resynthesis took place during a period of decreasing total protein concentration in the cells.

Another example of protein synthesis during starvation has been demontrated by the recovery from Cd^{2+} damage of S14 cells during starvation. Cadmium is reported to induce single strand breakage in DNA (Mitra and Bernstein, 1978) and to inactivate DNA polymerase I (Springgate *et al.*, 1973). Cd^{2+}-exposed starved cells lose their ability to grow on agar plates as a result of this damage, although they are still viable as seen by respiratory measurements (Nyström and Kjelleberg, 1987). Five new cytoplasmic proteins were synthesized during starvation of cadmium-exposed S14 cells. The transfer of these cells to Cd^{2+}-free starvation regimes allows the gradual and complete recovery in the colony-forming capacity on agar plates.

CONCLUDING REMARKS

Although there are many conclusions to be drawn from various experiments in this research area, we will close this discussion reiterating a few points.

(I)   There are several regulatory pathways or mechanisms involved as heterotrophic marine bacteria are exposed to short term energy depletion. Some of these might be understood by using starvation-specific antibodies to select for mutants which are unable to survive starvation conditions or differ in their response to starvation.

(II)   Starvation survival is an active process and non-growing cells are not dormant but capable of maintaining an efficient substrate capture. The possession of different uptake systems for the same substrate might reflect the fact that copiotrophic cells are adapted to a life at different nutrient levels. It is obvious that the terms dormancy and dormant cells, often used for small marine bacteria, are misleading. In fact, it may be suggested that these cells take part in the carbon flow in the environment. The non-growing stage as experienced by copiotrophic bacteria is thus merely an intermittent period of maintenance until favourable conditions for regrowth occur.

(III) The protein modulation observed as a result of starvation may not resemble that of a stress response. Relatively limited changes are observed for marine strains, particularly with respect to the synthesis of new proteins. The function of these is largely unknown. They may aid in, for example, improved substrate capture. It is also known that derepression of catabolic enzymes takes place in some starving heterotrophic bacteria (Matin, 1979). Whether the changes in surface characteristics that allow for increased adhesion specifically can be related to the synthesis of new, starvation-induced protein is not clear. No such correlation was seen by using antibodies against starved DW1 cells, There was a slight increase in surface hydrophobicity and degree of irreversible adhesion at inanimate surfaces after 5 h of starvation of S14 cells (Kjelleberg and Hermansson, 1984) and this is concomitant with the most pronounced change and increase in proteins.

(IV) In terms of ecology it may be reinforced that starvation survival prevents the bacterial number from declining: it sets the lower limit, while predation controls the upper limit of bacterial numbers (Wright, 1984). This explains the apparently small variations in bacterial concentrations in natural waters. In the area of public health it is apparent that survival rather than growth may be expected for allochthonous bacteria in the marine environment. Starvation-survival research in this area will most likely also provide the knowledge for developing specific probes for rapid and specific detection of a given species in the marine environment.

ACKNOWLEDGEMENTS

We thank N. Albertson, M. Hood, P. Mårdén and T. Nyström for contributing unpublished results included in this paper. We are grateful to P. Conway for language corrections. Work from the authors' laboratory was supported in part by grants from the Swedish Natural Science Research Council.

REFERENCES

Albertson, N.H., Jones, G.W. and Kjelleberg, S. 1987. The detection of starvation specific antigens in two marine bacteria. *Journal of General Microbiology*, in press.

Alldredge, A.L. 1979. The chemical composition of macroscopic aggregates in two neritic seas. *Limnology and Oceanography*, **24**: 855–866.

Alldredge, A.L., Cole, J.J. and Caron, D.A. 1986. Production of heterotrophic bacteria inhabiting macroscopic organic aggregates (marine snow) from surface waters. *Limnology and Oceanography*, **31**: 68–78.

Ammerman, J.W., Fuhrman, J.A., Hagström, Å. and Azam, F. 1984. Bacterio-plankton growth in seawater: I. Growth kinetics and cellular characteristics in seawater cultures. *Marine Ecology — Progress Series*, **18**: 31–39.

Amy, P.S. and Morita, R.Y. 1983a. Starvation-survival patterns of sixteen freshly isolated open-ocean bacteria. *Applied and Environmental Microbiology*, **45**: 1109–1115.

Amy, P.S. and Morita, R.Y. 1983b. Protein patterns of growing and starved cells of a marine *Vibrio* sp. *Applied and Environmental Microbiology*, **45**: 1748–1752.

Amy, P.S., Pauling, C. and Morita, R.Y. 1983. Starvation-survival processes of a marine vibrio. *Applied and Environmental Microbiology*, **45**: 1041–1048.

Andersson, A., Larsson, U. and Hagström, Å. 1986. Size-selective grazing by a microflagellate on pelagic bacteria. *Marine Ecology — Progress Series*, **33**: 51–57.

Azam, F. and Ammerman, J.W. 1984. Cycling of organic matter by bacterioplankton in pelagic marine ecosystems: microenvironmental considerations. In *Flows of energy and materials in marine ecosystems: theory and practice* (ed. M.J.R. Fasham), pp. 345–360. Plenum Press, New York.

Carlucci, A.F., Shimp, S.L. and Craven, D.B. 1986. Growth characteristics of low-nutrient bacteria from the north-east and central Pacific Ocean. *FEMS Microbiology Ecology*, **38**: 1–10.

Caron, D.A., Davies, P.G., Madin, L.P. and Sieburth, J.McN. 1982. Heterotrophic bacteria and bacterivorous protozoa in oceanic macroaggregates. *Science*, **218**: 795–797.

Conway, P.L., Maki, J., Mitchell, R. and Kjelleberg, S. 1986. Starvation of marine flounder, squid and laboratory mice and its effect on the intestinal microbiota. *FEMS Microbiology Ecology*, **38**: 187–195.

Costerton, J.W., Marrie, T.J. and Cheng, K.-J. 1985. Phenomena of bacterial adhesion. In *Bacterial adhesion* (eds. D.C. Savage and M. Fletcher), pp. 3–43. Plenum Press, New York.

Davis, C.L. 1985. Physiological and ecological studies of mannitol utilizing marine bacteria. *PhD Thesis*. University of Cape Town, South Africa.

Davoll, P.J. and Silver, M.W. 1986. Marine snow aggregates: life history sequence and microbial community of abandoned larvacean houses from Monterey Bay, California. *Marine Ecology — Progress Series*, **33**: 111–120.

Dawes, E.A. 1976. Endogenous metabolism and the survival of starved prokaryotes. In *The survival of vegetative microbes* (eds. T.R.G. Gray and J.R. Postgate), pp. 19–53. Cambridge University Press, Cambridge.

Dawson, M.P., Humphrey, B.A. and Marshall, K.C. 1981. Adhesion: a tactic in the survival strategy of a marine vibrio during starvation. *Current Microbiology*, **6**: 195–199.

Douglas, D.J., Novitsky, J.A. and Fournier, R.O. 1987. Microautoradiography-based enumeration of bacteria with estimates of thymidine-specific growth and production rates. *Marine Ecology — Progress Series*, **36**: 91–99.

Faquin, W.C. and Oliver, J.D. 1984. Arginine uptake by a psychrophilic marine *Vibrio* sp. during starvation-induced morphogenesis. *Journal of General Microbiology*, **130**: 1331–1335.

Fellows, D.A., Karl, D.M. and Knauer, G.A. 1981. Large particle fluxes and the vertical transport of living carbon in the upper 1500 m of the northeast Pacific Ocean. *Deep-Sea Research*, **28**A: 921–936.

Ferguson, R.L. and Rublee, P. 1976. Contribution of bacteria to standing crop of coastal plankton. *Limnology and Oceanography*, **21**: 141–145.

Fletcher, M. and Marshall, K.C. 1982. Are solid surfaces of ecological significance to aquatic bacteria? In *Advances in microbial ecology* (ed. K.C. Marshall), Vol. 6, pp. 199–236. Plenum Press, New York.

Glick, M.A. 1981. Substrate capture, uptake, and utilization of some amino acids by starved cells of a psychrophilic marine Vibrio. *MS Thesis*. Oregon State University, USA.

Goldman, J.C. 1984. Conceptual role for microaggregates in pelagic waters. *Bulletin of Marine Science*, **35**: 462–476.

Grimes, D.J., Attwell, R.W., Brayton, P.R., Palmer, L.M., Rollins, D.M., Roszak, D.B., Singleton, F.L., Tamplin, M.L. and Colwell, R.R. 1986. Fate of enteric pathogenic bacteria in estuarine and marine environments. *Microbiological Sciences*, **3**: 324–329.

Groat, R.G. and Matin, A. 1986. Synthesis of unique proteins at the onset of carbon starvation in *Escherichia coli*. *Journal of Industrial Microbiology*, **1**: 69–73.

Grossman, A.D., Taylor, W.E., Burton, Z.F., Burgess, R.R. and Gross, C.A. 1985. Stringent response in *Escherichia coli* induces expression of heat shock proteins. *Journal of Molecular Biology*, **186**: 357–365.

Hermansson, M. and Dahlbäck, B. 1983. Bacterial activity at the air/water interface. *Microbial Ecology*, **9**: 317–328.

Hermansson, M. and Marshall, K.C. 1985. Utilization of surface localized substrate by non-adhesive marine bacteria. *Microbial Ecology*, **11**: 91–105.

Hodson, R.E., Carlucci, A.F. and Azam, F. 1979. Glucose transport in a low nutrient marine bacterium. *Abstract Annual Meeting. American Society of Microbiology, N-59*, p. 189.

Hodson, R.E., Maccubbin, A.E. and Pomeroy, L.R. 1981. Dissolved adenosine triphosphate utilization by free-living and attached bacterioplankton. *Marine Biology*, **64**: 43–51.

Hood, M.A., Guckert, J.B., White, D.C. and Deck, F. 1986. Effect of nutrient deprivation on lipid, carbohydrate, DNA, RNA, and protein levels in *Vibrio cholerae*. *Applied and Environmental Microbiology*, **52**: 788–793.

Hunter, K.A. 1980. Microelectrophoretic properties of natural surface-active organic matter in coastal seawater. *Limnology and Oceanography*, **25**: 807–822.

Jacobsen, T.R. and Azam, F. 1984. Role of bacteria in copepod fecal pellet decomposition: colonization, growth rates and mineralization. *Bulletin of Marine Sciences*, **35**: 495–503.

Jannasch, H.W. 1974. Steady state and the chemostat in ecology. Comment. *Limnology and Oceanography*, **19**: 716–720.

Jones, K.L. and Rhodes-Roberts, M.E. 1981. The survival of marine bacteria under starvation conditions. *Journal of Applied Bacteriology*, **50**: 247–258.

Jouper-Jaan, Å., Dahlöf, B. and Kjelleberg, S. 1986. Changes in the protein composition of three bacterial isolates from marine waters during short term energy and nutrient deprivation. *Applied and Environmental Microbiology*, **42**: 1419–1421.

Kefford, B., Kjelleberg, S. and Marshall, K.C. 1982. Bacterial scavenging: utilization of fatty acids localized at a solid-liquid interface. *Archives of Microbiology*, **133**: 257–260.

Kirchman, D. and Mitchell, R. 1982. Contribution of particle-bound bacteria to total microheterotrophic activity in five ponds and two marshes. *Applied and Envirnomental Microbiology*, **43**: 200–209.

Kjelleberg, S. and Hermansson, M. 1984. Starvation induced effects on bacterial surface characteristics. *Applied and Environmental Microbiology*, **48**: 497–503.

Kjelleberg, S., Humphrey, B.A. and Marshall, K.C. 1983. Initial phases of starvation and activity of bacteria at surfaces. *Applied and Environmental Microbiology*, **46**: 978–984.

Kjelleberg, S., Marshall, K.C. and Hermansson, M. 1985. Oligotrophic and copiotrophic marine bacteria — observations related to attachment. *FEMS Microbiology Ecology*, **31**: 89–96.

Kjelleberg, S., Hermansson, M., Mårdén, P. and Jones, G.W. 1987. The transient phase between growth and nongrowth of heterotrophic bacteria, with emphasis on the marine environment. *Annual Review of Microbiology*, **41**: 25–49.

Knauer, G.A., Hebel, D. and Cipriano, F. 1982. Marine snow; major site of primary production in coastal waters. *Nature*, **300**: 630–631.

Laanbroek, H.J. and Verplanke, J.C. 1986. Seasonal changes in percentages of attached bacteria enumerated in a tidal and a stagnant coastal basin: relation to bacterioplankton productivity. *FEMS Microbiology Letters*, **38**: 87–98.

MacDonell, M.T. and Hood, M.A. 1982. Isolation and characterization of ultramicrobacteria from a Gulf Coast estuary. *Applied and Environmental Microbiology*, **43**: 566–571.

Malmcrona-Friberg, K., Tunlid, A., Mårdén, P., Kjelleberg, S. and Odham, G. 1986. Chemical changes in cell envelope and poly-β-hydroxybutyrate during short term starvation of a marine bacterial isolate. *Archives of Microbiology*, **144**: 340–345.

Mandelstam, J. 1958. Turnover of protein in growing and non-growing populations of *Escherichia coli*. *Biochemical Journal*, **69**: 110–119.

Mandelstam, J. 1960. The intracellular turnover of protein and nucleic acids and its role in biochemical differentiation. *Bacteriological Reviews*, **24**: 289–308.

Mårdén, P., Tunlid, A., Malmcrona-Friberg, K., Odham, G. and Kjelleberg, S. 1985. Physiological and morphological changes during short term starvation of marine bacterial isolates. *Archives of Microbiology*, **142**: 326–332.

Marshall, K.C. 1976. *Interfaces in microbial ecology*. Harvard University Press, Cambridge, MA, USA.

Mason, C.A., Hamer, G. and Bryers, J.D. 1986. The death and lysis of microorganisms in environmental processes. *FEMS Microbiology Reviews*, **39**: 373–401.

Matin, A. 1979. Microbial regulatory mechanisms at low nutrient concentrations as studied in chemostat. In *Strategies of microbial life in extreme environments* (ed. M. Shilo), pp. 323–339. Berlin: Dahlem Konferenzen, Verlag Chemie, Weinheim.

Meyer-Reil, L.-A. 1978. Autoradiography and epifluorescence microscopy combined for the determination of number and spectrum of actively metabolizing bacteria in natural waters. *Applied and Environmental Microbiology*, **36**: 506–512.

Mitra, R.S. and Bernstein, I.A. 1978. Single-strand breakage in DNA of *Escherichia coli* exposed to Cd^{2+}. *Journal of Bacteriology*, **133**: 75–80.

Mopper, K. and Lindroth, P. 1982. Diel and depth variations in dissolved free amino acids and ammonium in the Baltic Sea determined by shipboard HPLC analysis. *Limnology and Oceanographya, 27*: 336–347.

Morita, R.Y. 1982. Starvation-survival of heterotrophs in the marine environment. In *Advances in microbial ecology* (ed. K.C. Marshall), Vol. 6, pp. 171–198. Plenum Press, New York.

Morita, R.Y. 1984. Substrate capture by marine heterotrophic bacteria in low nutrient waters. In *Heterotrophic activity in the sea* (eds. J.E. Hobbie and P.J.leB. Williams), pp. 83–100. Plenum Press, New York.

Morita, R.Y. 1985. Starvation and miniaturisation of heterotrophs, with special emphasis on maintenance of the starved viable state. In *Bacteria in their natural environments* (eds. M.M. Fletcher and G.D. Floodgate), pp. 111–130. Academic Press, London.

Norkrans, B. 1980. Surface microlayers in aquatic environments. In *Advances in microbial ecology* (ed. M. Alexander), Vol. 4, pp. 51–85. Plenum Press, New York.

Novitsky, J.A. and Morita, R.Y. 1977. Survival of a psychrophilic marine vibrio under long-term nutrient starvation. *Applied and Environmental Microbiology*, **33**: 635–641.

Novitsky, J.A. and Morita, R.Y. 1978. Possible strategy for the survival of marine bacteria under starvation conditions. *Marine Biology*, **48**: 289–295.

Nyström, T. and Kjelleberg, S. 1987. The effect of cadmium on starved heterotrophic bacteria isolated from marine waters. *FEMS Microbiology Ecology*, in press.

Nyström, T., Mårdén, P. and Kjelleberg, S. 1986. Relative changes in incorporation rates of leucine and methionine during starvation survival of two bacteria isolated from marine waters. *FEMS Microbiology Ecology*, **38**: 285–292.

Oliver, J.D. and Stringer, W.F. 1984. Lipid composition of a psychrophilic marine *Vibrio* sp. during starvation-induced morphogenesis. *Applied and Environmental Microbiology*, **47**: 461–466.

Poindexter, J.S. 1981a. Oligotrophy: fast and famine existence. In *Advances in microbial ecology* (ed. M. Alexander), Vol. 5, pp. 63–89. Plenum Press, New York.

Poindexter, J.S. 1981b. The caulobacters: ubiquitous unusual bacteria. *Microbiological Reviews*, **45**: 123–179.

Pomeroy, L.R. 1984. Microbial processes in the sea: diversity in nature and science. In *Heterotrophic activity in the sea* (eds. J.E. Hobbie and P.J.leB. Williams), pp. 1–23. Plenum Press, New York.

Reeve, C.A., Amy, P.S. and Matin, A. 1984a. Role of protein synthesis in the survival of carbon-starved *Escherichia coli* K-12. *Journal of Bacteriology*, **160**: 1041–1046.

Reeve, C.A., Bockman, A.T. and Matin, A. 1984b. Role of protein degradation in the survival of carbon-starved *Escherichia coli* and *Salmonella typhimurium*. *Journal of Bacteriology*, **157**: 758–763.

Robert, D.K. 1986. Studies on the 63 megadalton cryptic plasmid of *Salmonella typhimurium*. *PhD Thesis*. University of Michigan, Ann Arbor, USA.

Shanks, A.L. and Trent, J.D. 1979. Marine snow: microscale nutrient patches. *Limnology and Oceanography*, **24**: 850–854.

Silver, M.W. and Alldredge, A.L. 1981. Bathy-pelagic marine snow: deep sea algal and detrital community. *Journal of Marine Research*, **39**: 501–530.

Silver, M.W., Shanks, A.L. and Trent, J.D. 1978. Marine snow: microplankton habitat and source of small-scale patchiness in pelagic populations. *Science*, **201**: 371–373.

Spector, M.P., Aliabadi, Z., Gonzalez, T. and Foster, J.W. 1986. Global control in *Salmonella typhimurium*: two-dimensional electrophoretic analysis of starvation, anaerobiosis-, and heat shock-inducible proteins. *Journal of Bacteriology*, **168**: 420–424.

Springgate, C.F., Mildran, A.S., Abramson, R., Engle, J.L. and Loeb, L.A. 1973. *Escherichia coli* deoxyribonucleic acid polymerase I, a zinc metallo enzyme; nuclear quadrupolar relaxation studies of the role of bound zinc. *Journal of Biological Chemistry*, **248**: 5987–5993.

Sterkenburg, A., Vlegels, E. and Wouters, J.T.M. 1984. Influence of nutrient limitation and growth rate on the outer membrane proteins of *Klebsiella aerogenes NCTC 418*. *Journal of General Microbiology*, **130**: 2347–2355.

Walker, G.C. 1984. Mutagenesis and inducible responses to deoxyribonucleic acid damage in *Escherichia coli*. *Microbiological Reviews*, **48**: 60–93.

Wrangstadh, M., Conway, P.L. and Kjelleberg, S. 1986. The production and release of an extracellular polysaccharide during starvation of a marine *Pseudomonas* sp. and the effect thereof on adhesion. *Archives of Microbiology*, **145**: 220–227.

Wright, R.T. 1984. Dynamics of pools of dissolved organic carbon. In *Heterotrophic activity in the sea* (eds. J.E. Hobbie and P.J.leB. Williams), pp. 121–154. Plenum Press, New York.

11

Marine microbial adhesion and its consequences

B. Egan, School of Ocean Sciences, University College of North Wales, Menai Bridge, Gwynedd LL59 5EH, UK

The paper reviews the current status of our understanding of the basic forces acting upon microbial cells during the attachment process to submerged substrata, and the importance of the surface characteristics of the cells and the substrata on this process. The effects of attachment to various surfaces on the activity of the microbial population are also discussed as are the problems of assessing such effects. The structure of attached microbial communities is examined and the steps in the development of microbial biofilms and their relevance to man-made structures are outlined with particular reference to the problems of heat transfer surfaces.

Microbial adhesion may be crudely defined as an association of a microbial cell with the surface of a substratum, involving an interaction between the two which requires an input of energy to separate them. The sequence of events that occur in the adhesion of microbial cells to substrata has been the centre of much research interest in the last 15 years.

MECHANISMS OF ADHESION

The processes involved in microbial adhesion to surfaces are complex and involve a range of forces which can be divided into long range (London–Van der Waals, electrical double layer), short range (dipole–dipole, ion–dipole, hydrogen bonding, dipole-induced dipole etc.) and hydrodynamic (shear etc.); the contributions of these to the net attractive or repulsive force acting on a cell approaching a surface are dependent upon the nature of the cell, the surface and the bulk fluid in which the attachment is to occur.

To understand the mechanisms operating it is necessary to look at a simplified model of the event. This can best be done by treating the microbial cell as if it were a colloidal particle of approximately the same dimensions. By applying the theory of Derjaquin, Landau, Verwey and Overbeek (DLVO) (Derjaquin and Landau, 1941; Verwey and Overbeek, 1948), which explains the stability of colloidal suspensions, to the behaviour of such a particle as it approaches a surface, we can assess the forces acting upon it and their relative importance. In quiescent fluid conditions the main forces acting upon the cell at a distance from the surface are Van der Waals forces and the effect of the overlap of the electrical double layers of the cell and the surface. In marine systems where the pH ranges from 6.5 to 8.5 most bacteria and substratum surfaces have a net negative charge (Harden and Harris, 1953; Neihof and Loeb, 1972, 1974); therefore the forces mentioned will be acting antagonistically to one another. The resultant force can be schematically represented as in Fig. 1 (Rutter and Vincent, 1984). The effect of increasing electrolyte concentration is to reduce the ζ potential, or net charge, of the bodies involved and so to reduce the effect of the electrical double layer.

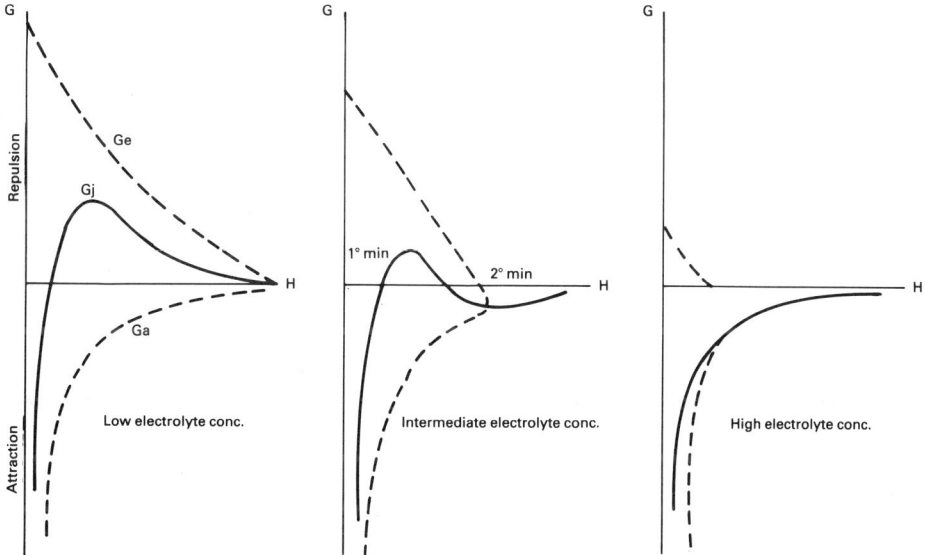

Fig. 1—Interaction free energy (G) versus distance (H) between two surfaces at differing electrolyte concentrations. G_e = electrical double layer force, G_a = London–Van der Waals force, G_j = resultant net force. (After Rutter and Vincent, 1984.) Reproduced with permission from Springer-Verlag, Berlin.

In the marine environment electrolyte concentrations would normally fall into the medium concentration bracket shown in Fig. 1. In this situation a cell approaching with sufficient kinetic energy from the bulk fluid will achieve the primary minimum close to the surface and be held by the strong attractive force whilst a cell with less energy may well only attain the secondary minimum and be unable to approach closer owing to the net repulsive force acting between it and the substratum. In the latter case the relative weakness of the attractive force at the secondary minimum would suggest that a relatively small input of energy would be required to remove the cell and to re-entrain it into the bulk fluid and that the stability of this equilibrium would be poor.

As the cell approaches very close to the surface a further range of forces become increasingly important. These occur at a molecular level and include dipole–dipole, dipole-induced dipole, ion–dipole and hydrogen bonding. All of these interactions may be repulsive or attractive in nature depending upon the characteristics of the surfaces involved. Amongst the factors to be considered of importance to adhesion is the wettability of a surface, which may be measured either as hydrophobicity (Pringle *et al.*, 1983) or surface free energy (Zisman, 1964; Wu, 1980; Neumann *et al.*, 1974). In general terms, in aqueous systems, if both of the surfaces involved are hydrophilic a short range repulsive force may exist, whilst if both surfaces are hydrophobic there will be a net attractive force. If this were true we would expect hydrophilic bacteria to adsorb to hydrophobic surfaces (e.g. Teflon), more strongly than to hydrophilic surfaces (e.g. glass) and this has generally been found to be the case (Fletcher and Loeb, 1979; Mahachanchawalit *et al.*, 1986; Busscher *et al.*, 1986).

This very simple model assumes that the two surfaces involved are of a regular nature and that the surface charges of the surfaces are evenly distributed. In practice this is not the case and our model is greatly complicated by the presence of macromolecules on both surfaces. The immersion of a substratum into a marine environment has been shown to be rapidly followed by the development of a 'conditioning film' of macromolecules which is mainly proteinaceous in nature, although polysaccharides and other macromolecules may also be present (Baier, 1972; Goupil *et al.*, 1980). The surface of a microbial cell presents a wide range of protein, lipopolysaccharide, polysaccharide and lipid molecules which vary not only between species but also with nutritional status of the organism (Malmcrona-Friberg *et al.*, 1986). The presence of these macromolecules will affect many of the factors in the simple model. Their release into the bulk fluid will cause an increase in viscosity and may reduce the adsorption rate constant for the cells; if produced in sufficient quantities they may bind into loose flocs, as has been shown in colloidal suspensions (Vrij, 1976), and may entrap cells in the fluid phase. The presence of cell and surface bound macromolecules may well affect both the electrical double layer and the Van der Waals forces.

It has been shown that polymers can adsorb to solid surfaces in an apparently irreversible manner by means of quite weak polymer surface bonds, the strength of the adhesion being determined by the very large number of such bonds which may occur along the length of the polymer molecule (Barnett *et al.*, 1981). Such molecules also change their shape from their tertiary structure in the fluid phase to a much more two-dimensional structure when in contact with the surface and may exhibit different surface characteristics at the fluid–polymer interface as a result (Goupil *et al.*, 1980), the degree of disruption of the molecules being dependent on the surface characteristics of the substratum.

The optimal molecular size for these extracellular polymers is less than 10^6 Daltons as larger molecules become very sensitive to shear stress (Robb, 1984). The radius of gyration of a polyelectrolyte of molecular weight of approximately 5×10^5 is between 15 and 100 nm (Miyamoto *et al.*, 1981); this is comparable with the distance at which the electrical double layer may exert an effect, which in an electrolyte concentration of 10^{-3} mol dm^{-3} is around 10 nm (Robb, 1984). It is therefore quite possible that the effect of the surface charge of the surfaces may be restricted to the polymer film and that the net surface charge present may depend solely on the charged groups exposed on the outer surface of the polymers, this being particularly true where the ζ potentials of the surfaces are not very large. In such a case, it would be possible for like-charged surfaces to approach close enough for polymer bridging to occur. Experimental observation has shown that many cells which impact on a surface remain for a period of time, but show no specific adhesive processes and may subsequently be re-entrained into the bulk fluid. The formation of polymer bridges seems to be of critical importance in the adhesive process and can be correlated with the development of the irreversible phase of adhesion. Polymer bridging has been demonstrated clearly, both in cultures (Fletcher and Floodgate, 1973) and in environmental samples (Pearl, 1984).

To summarize, in a model colloidal system we should expect the surface charge and hydrophobicity of both surfaces, the bulk fluid electrolyte concentration, the flow conditions and the presence and composition of polymers to affect the adhesive process.

COLONIZATION OF SURFACES BY MICROORGANISMS

In natural systems in which microbial adhesion occurs the subtle combination of these factors can lead to significant variation in the colonization of surfaces (McEldowney and Fletcher, 1986). The role of polymers, for example, is not as clear-cut as our model supposes. The surface of cells is covered by a heterogeneous layer of polymer molecules and a wide range of interactions must be occurring between them and the polymers coating the substratum.

Polysaccharides have long been held responsible for the bridging between cells and substrata during the irreversible phase of adhesion, although their role in the initial events has been unclear (Fletcher and Floodgate, 1973). However, a number of workers have shown that the production of extracellular polysaccharides is not directly related to the strength of adhesion of a cell (Pringle et al., 1983), and that although they are a major component of slime films, the presence of hydrophilic polymers such as polysaccharides may actually inhibit initial adhesion (Rosenberg et al., 1983; Pringle and Fletcher, 1986), and may be produced to aid desorbtion from surfaces in adverse environmental conditions (M. Wrangstadh, P.L. Conway and S. Kjelleberg, personal communication). Thus we should perhaps look to the more hydrophobic polymers such as lipopolysaccharides (Wood, 1980), proteins (Imam et al., 1984: Bagg and Silverwood, 1986) and lipoteichoic acids (Sherman and Savage, 1986) to explain the positive role of polymers in the initial stages of microbial adhesion.

The metabolic nature of microbes and their ability to alter their glycocalyx in response to environmental conditions (Costerton et al., 1985; Hood et al., 1986), as well as the presence of a wide range of polymers and structures such as fimbriae, pili and flagella, may influence the selectivity of some microbe–surface interactions and may also allow cells a choice of mechanisms of attachment for use on different substrata (Paul and Jeffrey, 1985). This complicates greatly our colloidal model, as does the realization that the free surface energy of bacterial surfaces varies considerably from species to species (Magnusson et al., 1977; Dahlback et al., 1981). It is also likely that upon a single cell the distribution of hydrophobic sites may not be even and may play a part in determining the orientation of cells on surfaces (Marshall and Cruickshank, 1973).

A consequence of the adhesion of cells to substrata is the development of attached microbial communities. In examining these communities it is first necessary to examine the process involved in the transport of cells to a substratum from the bulk fluid. The relative importance of the various factors affecting transport will alter in different flow conditions, but it is possible to identify a number of the processes involved. In a turbulent flow regime convective firces will dominate the transport processes; in such a system eddy diffusion maintains a dispersion of the microbial cells in the bulk fluid, and where the direction of motion places them near the substratum, and travelling faster than the surrounding fluid, a force acts upon the cell towards the surface (Characklis, 1981a). In close proximity to the substratum the cell enters an area known as the boundary or viscous sub-layer in which flow does not conform to that of the bulk fluid but is relatively stationary. Upon entering this region the cell will suffer considerable frictional drag from the water through which it is travelling and also a force exerted by the draining fluid film between itself and the substratum surface (Characklis, 1981a) which will slow the cell's progress; it may also be affected by turbulent bursts which penetrate the boundary layer (Merry and Wright, 1982). The depth of the boundary layer is dependent upon both the surface and the flow conditions in the bulk fluid but for a shear stress of $6.5-7.9\,\mathrm{N\,m}^{-2}$ the viscous sublayer is about $40\,\mu\mathrm{m}$ thick (Characklis, 1981b).

In quiescent flow conditions the significance of sedimentation increases, although this only affects the larger sized particles, such as aggregates and flocs (Leech and Hefford, 1980), while most individual bacteria are close to neutrally buoyant in seawater systems and form a stable suspension. Other factors which contribute to mass transfer include Brownian motion (Characklis, 1981a) and the effects of cell hydrophobicity (Marshall and Cruichshank, 1973), although these are only likely to be significant in flowing systems in the crossing of the boundary layer. The chemotactic responses of motile bacteria are well documented (Characklis, 1981b) and may be ecologically significant, particularly in low flow environments.

The effect of attachment on the metabolism of microbial cells is hard to assess, particularly in natural environments. The problem is partly due to the difficulty in finding a measure of the overall activity of attached and free-living microbial cells. A wide range of activity measurements have been carried out, including changes in attached and free living cell numbers (Harwood and Pirt, 1972; Hattori, 1972), changes in cell size (Kjelleberg et al., 1982); substrate uptake (Bright and Fletcher 1983b; Fletcher, 1986), electron transport activity (Bright and Fletcher. 1983a), respiration rate measured as carbon dioxide production or oxygen uptake (Kefford et al., 1983), substrate breakdown (Estermann et al., 1959), product formation (Hattori and Hattori, 1963), thymidine incorporation (Jeffrey and Paul, 1985) and heat production (Gordon et al., 1983; Lock and Ford, 1985). The problem of interpretation of this wealth of data is further complicated by the fact that the population structures of the adherent and free-living populations are often different in natural environments (Mills and Maubrey, 1981; Marsh et al., 1983), and differences in activity may be attributable to this fact rather than to an actual increase in the cellular activity of comparable attached and free living cells.

In considering suspended particles in the water column, particle concentration seems to show a general relationship with the proportion of attached cells (Goulder, 1977; Harvey and Young, 1980; Bell and Albright, 1982; Kirchman and Mitchell, 1982), although some studies have related the degree of microbial attachment to particles with the status of the microphytoplankton population in the water column (Lovell and Konopka, 1985; Albright et al., 1986). In sediments there are large populations of bacteria (Meyer-Reil, 1984), most of which are attached to the sediment particles (Meyer-Reil et al., 1978); however, individual particles normally have relatively few microcolonies on their surface and frequently do not develop continuous microbial films, although a few particles may be quite densely populated. The size of microcolonies on marine sedimentary particles tends to be small with 5–20-celled colonies common on sand grains, whilst larger colonies of up to 100 cells are rare (Weise and Rheinheimer, 1978).

The effects of attachment to particles on the activity of cells are complex. A number of studies have suggested increased activity in attached organisms (Harvey and Young, 1980; Kirchman and Mitchell, 1982; Jeffrey and Paul, 1986; Fletcher, 1986), but on closer examination a more complicated picture emerges. In a study by Bell and Albright (1982) of a large number of marine, estuarine and freshwater samples, particle-bound cells showed an increase in the uptake of amino acids over free-living cells, but the reverse proved true for glucose assimilation, whilst Fletcher (1986) found increases in both attached and recently detached cells over free-living cells for a range of substrata. However, under some environmental conditions no stimulation and even inhibition of microbial activity have been seen (Williams, 1970; Fletcher and Marshall, 1982; Ducklow and Kirchman, 1983). The use of uptake rates related to cell numbers can also lead to discrepancies since the cell volumes of attached cells are often greater than those of the free-living population so that uptake per unit volume may remain constant although uptake per cell may change (Hodson et al., 1981). It would appear that, in oligotrophic systems in particular, attachment to particles does increase the activity of microbial populations, although this may not be reflected throughout the whole spectrum of substrates available, and may depend upon a wide range of environmental factors such as substrate concentrations (Bell and Albright, 1982), salinity (Bell and Albright, 1982), size range and loading of suspended solids (Kirchman and Mitchell, 1982) and cell size (Hodson et al., 1981).

Mathematical modelling of the size distribution of microcolonies in deep sea sediment samples has shown that the desorption rate of cells has to exceed the growth rate significantly to give the distributions found in these samples (Davidson and Fry, 1987), which suggests that cells are highly mobile within the sediment and undergo regular desorption and re-attachment. This supports work which has shown that the alteration of the salinity of marine sediments can drastically reduce the adhesion of bacteria to the particles of the sediment when the electrolyte concentration approaches the colloidal dispersion value (Roper and Marshall, 1974), thus suggesting that the cells are remaining reversibly bound to the sediment particles.

On macroscopic submerged surfaces the situation is somewhat different and leads to the development of irreversibly bound microbial biofilms, which are of great importance to man-made structures in the marine environment owing to their effects in altering the physical characteristics of materials and in their role in corrosive reactions. The corrosive consequences of microbial colonization of marine surfaces will be covered elsewhere in this Symposium. The alteration of material characteristics is particularly important where they impinge upon the design criteria of the material. This can often involve significant reductions in operational efficiency and increased costs such as in pipelines, heat exchangers, optical systems, ships hulls etc., where the effects include increased heat transfer resistance, increased frictional resistance and reduced optical transmission (Characklis, 1973a, b; Norrman, 1976).

DEVELOPMENT OF MICROBIAL COMMUNITIES ON SURFACES

The developmental history of primary microbial slime films in seawater is now generally known. Initially, within hours of immersion, a conditioning film of glycoproteins and polysaccharides adheres to the surface to a depth of 10–80 nm (Baier *al et.*, 1983). There is evidence that structural alteration of the molecules occurs as they attach; this may relate to studies which have shown that the free surface energy of the substratum affects the strength of adhesion of this layer (Baier, 1981). The relationship is complex, but in general high surface energy surfaces produce a tightly bonded conditioning film, whilst low surface energy surfaces (such as those with exposed methyl or hydroxyl groups) produce a less tightly bonded film which requires substantially less energy to remove.

This apparently direct relationship of free surface energy to strength of film attachment unfortunately breaks down on very low surface energy substrata such as PTFE upon which very tightly bound films form. The reasons for this effect are explained by Pringle and Fletcher in thermodynamic terms as the work required for adhesion to occur. The work described is that needed to remove the water molecules from the two surfaces involved, and the greater this work is then the less likely it is that adhesion will occur. The practical implications are that the surface energies of materials for use in coating or sheathing of marine surfaces to reduce fouling should lie within certain critical values, i.e. 20–$30\,\mathrm{N\,m^2}$ (Baier, 1972; Fletcher, 1979; Fletcher and Loeb, 1979).

The second phase of fouling development is the attachment to the conditioned surface of a wide range of periphytic bacteria, which usually occurs within several days of immersion. The organisms which initially settle in marine systems are Gram-negative rods, including *Pseudomonas* spp. and *Vibrio* spp., although the exact composition of the primary colonizing population has been shown to be site specific (Zambon *et al.*, 1984). During this preliminary phase the removal of the film is not difficult and the surface is easily cleaned. The bacteria rapidly begin to synthesize extracellular polymers which form bridging fibres crossing the spaces between the bacterial cells and the surface. Once these are in position the cell is irreversibly bound to the surface and requires stringent cleaning techniques to remove it. The nature of the extracellular polymers produced seems to vary but both polysaccharides and amino sugars have been identified (Zaidi *et al.*, 1984; Christensen *et al.*, 1985).

During the next few days of exposure the attached organisms multiply and form microcolonies enclosed in a slime of their extracellular polymers which probably serves to insulate them from any direct shear forces and from the effects of toxins

(Mittelman and Geesey, 1985). These colonies soon form a continuous layer of slime film on the surface, composed mainly of extracellular polymers and water, and local environmental conditions generated within the film lead to shifts in population structure within the microbial community.

These early stages of the development of the film seem to be remarkably constant irrespective of surface composition or water treatment (Baier *et al.*, 1983). There follows a period of rapid succession of species within the community of the microfilm away from the initial species, which seem to be mainly Gram-negative rods that produce copious quantities of extracellular bridging materials, and towards filamentous and stalked species (Corpe, 1976; Gerchakov *et al.*, 1976).

The film continues to thicken by means of multiplication of cells, adsorption of particulates and new cells at the surface and slime production. The relative importance of these sources of mass input to the film is not clear and is dependent on a range of environmental factors, but it has been shown in test systems of aluminium tubes that growth within the film was the dominant source of mass input after the initial period of colonization (Bott and Miller, 1983). The film continues to thicken until a point is reached when the rates of diffusion of nutrients and/or oxygen limit further growth of the organisms in the basal layers of the film close to the substratum surface. Subsequent growth of the film is accompanied by a weakening of the basal layers of the film, possibly as a result of nutrient deficiency or anaerobiosis, which leads to the sloughing off of portions of the film. In high flow conditions the overall growth curve of biofilms tends to be sinusoidal with a slow growth or 'lag' phase, a rapid 'growth' phase and an optimal 'stationary' phase, the level of which is determined by the local environmental conditions (Fig. 2). In low flow or quiescent conditions the stationary phase may be replaced by dramatic decreases owing to the sloughing off of large portions of biofilm. The depth of microbial films can range from less than 100 μm in high flow or oligotrophic environments to 3000 μm or more in low flow, eutrophic situations. The rate-limiting steps for the process are probably the rates of the development of the conditioning film and the initial rate of microbial attachment, although the latter can occur very rapidly when sloughing off exposes a new surface for colonization.

BIOFOULING OF HEAT EXCHANGERS

Microbial fouling of heat exchanger surfaces can have significant repercussions, a slime layer of 60–100 μm can lead to fouling resistance (Rf) values of greater than $5 \times 10^4 \,°C \, m^2 \, W^{-1}$ at heat exchanger surfaces, which will impair the function of heat exchange units (Little and Lavoie 1979).

Field studies on the development of biofilms on heat exchangers have been carried out at many sites including the Gulf of Mexico (Little and Lavoie, 1979), the Virgin islands (Aftring *et al.*, 1978), and Hawaii (Berger *et al.*, 1979). The results of these tests show that heat transfer resistance remains constant for several weeks after the initial immersion of the clean surface and then begins to increase rapidly. This compares well with the 30 day induction period found in experimental rigs before fouling films begin to influence heat transfer (Baier *et al.*, 1983). The heat transfer resistance declines on cleaning, either chemical or mechanical, but subsequently the

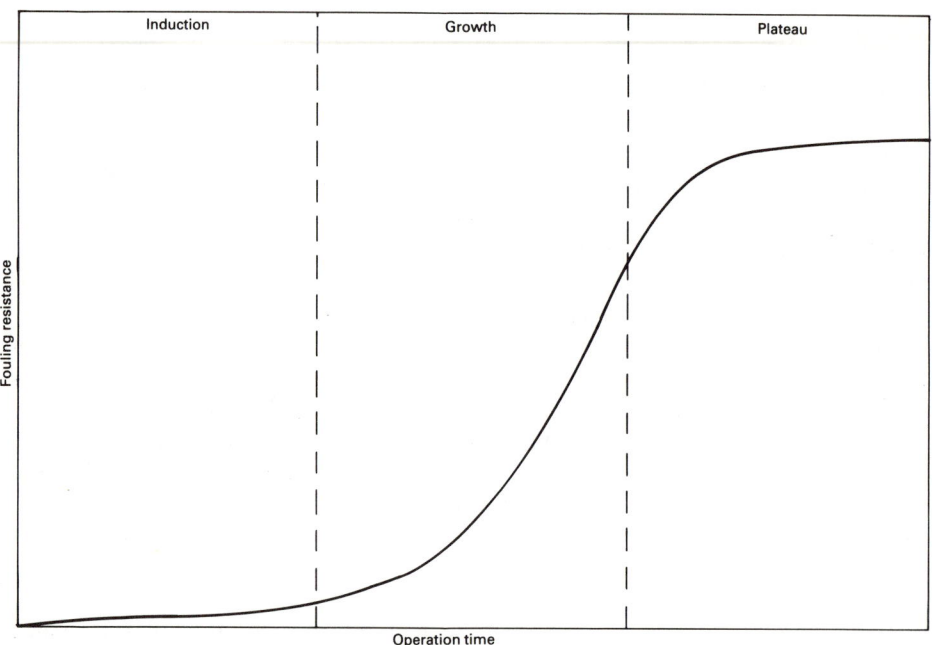

Fig. 2—Typical biofilm development curve.

resistance increases again rapidly, there being little or no lag phase. The slope of the growth curve and the length of the initial and post-cleaning lag phases seem to be both site and substratum specific. In contrast with the above data the information currently available from the US test facility in Hawaii suggests that with deep-sea cold water there is little or no fouling on the smooth tube samples tested (Panchal *et al.*, 1984; Berger and Berger, 1986) (see Fig. 3). These tests have been carried out over a 3 year period. It is likely that fouling is occurring but at a very low rate because of the temperature of the seawater, and so the film is still in its lag phase relative to *Rf* values.

It would appear that the lag phase represents a period of film growth which has no apparent effect on heat resistance but that at a critical thickness the film begins to cause increased heat resistance. The decrease in lag phase after cleaning may be due to the persistence on the surface of organic debris acting as a nutrient source to stimulate regrowth of the film, and also to the formation of a plaque on the surface made up of filamentous organisms which are extremely resistant to cleaning that develops after repeated washings (Nickels *et al.*, 1981).

FACTORS IN BIOFILM DEVELOPMENT

The factors involved in the control of slime production in natural populations of film-forming bacteria are as yet poorly understood, as are the relationships between biofilm thickness, density and polymer production, and thermal heat resistance.

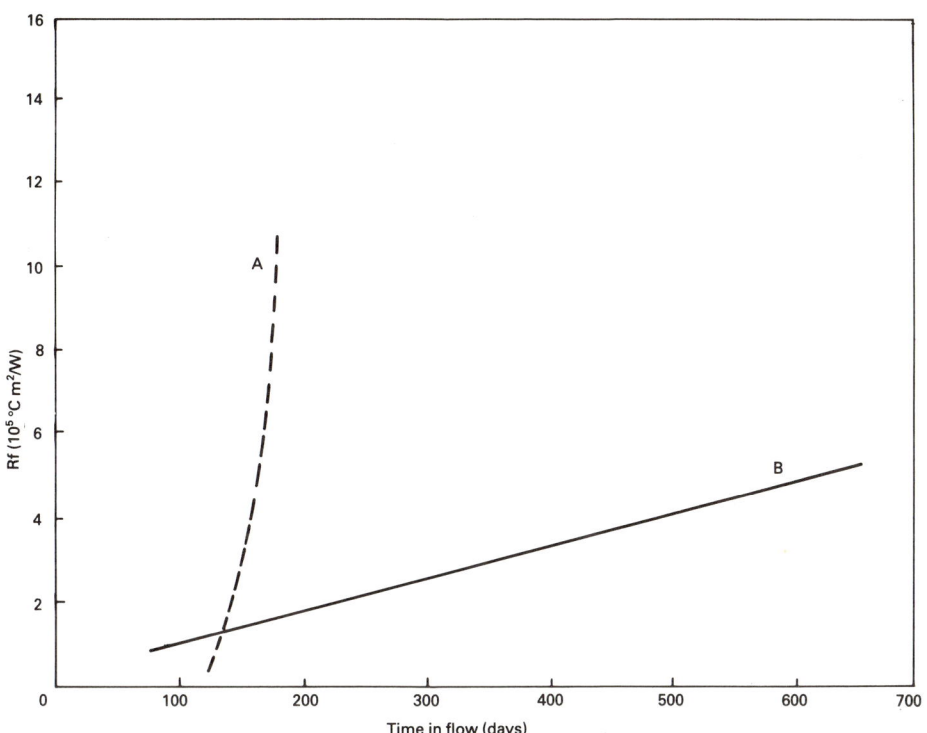

Fig. 3—Heat transfer resistance increase versus time for aluminium 5052. Curve A, with warm surface seawater; curve B, with cold, deep seawater. (After Berger and Berger, 1986).

Bobbie *et al.* (1979) have shown that total organic carbon (TOC) correlates well with *Rf* value after the initial fouling period, and this can be used in working heat exchangers to assess the degree of fouling without having to stop the fluid flow whilst samples are removed.

As we have previously noted, the rate of microbial attachment to a surface is affected by a range of factors and this is also true for the subsequent development of biofilms. The relative importance of the factors involved can best be examined by breaking down the development of the film into a series of fundamental processes which can be explained in terms of mass and energy balances. The effects of environmental factors on each process can then be examined and their contribution to the overall development of the biofilm assessed. Where research has been directed towards this end (Bryers *et al.*, 1979; Bryers and Characklis, 1981) a much deeper insight into the relationship of these underlying processes has been obtained.

The effects of nutrient concentration in the bulk fluid on the development of biofilms have been the subject of a number of studies (Bott and Gunatillaka, 1982; Bott and Miller, 1983), in which it has been shown that nutrient enrichment of the bulk fluid passing over a biofilm increases both the thickness and the cellular density of the film. Increases in the C:N ratio of the bulk fluid also led to thicker films. The C:N ratio of the water is of importance since nitrogen is essential for the production of proteins and biomass. A high ratio tends to lead to the production of large quantities of extracellular polysaccharides which may have several effects. It may enhance the adhesion of the biofilm to its substratum; it may also increase the efficiency of scavenging, both by increasing the adhesion of impacting particles and by adsorption of dissolved organic compounds which in turn increase the nutrient loading of the film upon which the subsequent growth depends. Even in nutrient-depleted waters the polymer matrix, once developed, can scavenge nutrients, and the nutritional status of the film may be far higher than that of the surrounding waters. Hence, although water chemistry has a major influence in the early stages, the fully developed film is substantially buffered against short term variations (Bott and Miller, 1983).

The relationship between biofilm development and flow rate is a complex one. Increases in the flow rate increase the flow of nutrients and oxygen past the film surface and maintain higher concentrations in this critical area, although the effect is not linear. At low fluid velocities a diffusion gradient may develop in proximity to the surface and so reduce the available nutrients at the film surface; however, once the flow rate is sufficient to overcome this effect, subsequent increases do not increase the nutrient uptake rate (Characklis, 1981b). The maximum film mass has been shown in a test system to develop at flow rates of about $1.8\,\mathrm{m}\,\mathrm{s}^{-1}$ using river water on an aluminium test section (Duddridge et al., 1982). In addition to the direct effects of increased shear on reducing initial attachment and increasing the mass flow of nutrients to the biofilm, increases in flow rate also increase turbulence, and the forces involved both in turbulent eddies in the main body of the fluid and in turbulent bursts within the boundary layers close to the substratum surface are increased (Merry and Wright, 1982). Consequently, the probability of re-entrainment of particles and cells increases, and indeed, above a certain critical value, re-entrainment becomes the dominant feature and net loss of biomass from the film may occur. It has also been reported that the topographical variability of the film is much reduced at high flow velocities.

The rate of biofilm growth increases with increased temperature up to a temperature of around 40 °C, above which it tails off (see Fig. 4).

The choice of materials may affect the rate of biofouling or the density, depth or population structure of the mature film. Tests carried out on aluminium, titanium, copper and stainless steel, all in both a rough and a polished finish, showed that the surface texture was not important but that significant differences occurred between the different metals (Berk et al., 1981). Titanium was shown to foul more rapidly than aluminium during the initial developmental period, for example. The role of inorganic salts and particles in biofilm development is unclear. The addition of kaolin to a test system reduced the attachment of microbial cells significantly (Lowe et al., 1984), although once the film developed the dry weight of the film was significantly higher with kaolin present; this was probably due to the impaction and subsequent incorporation of particles into the polymer matrix. The inorganic composition of biofilms varies with the composition of the bulk fluid passing over it. Calcium, magnesium and iron are known to affect the bonding of the polymers in the matrix and may have an effect on the structural integrity of the film. The role of corrosion products in the development and adhesion of biofilms on metal surfaces is still unclear.

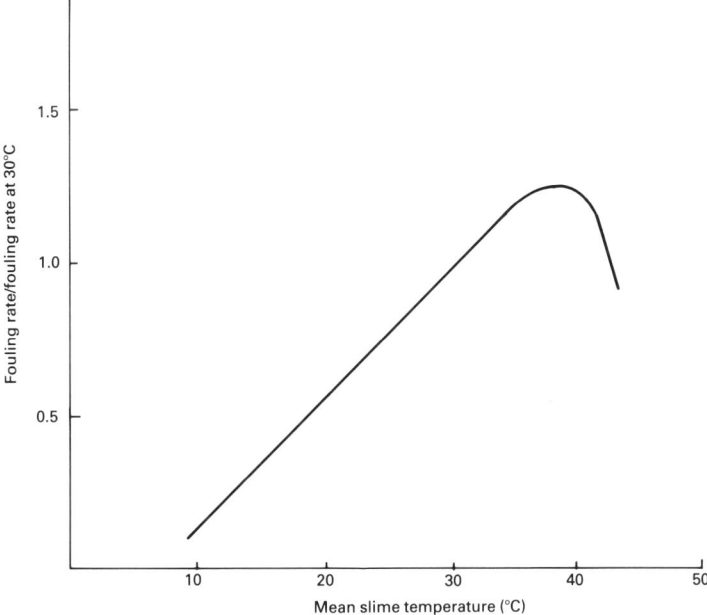

Fig. 4—Temperature effect on the development of a biofouling film. (From Merry and Wright, 1982.)

ASSESSMENT OF THE EFFECTS OF BIOFOULING

A range of methods have been developed for the assessment of biofouling films. These include methods for measuring the effects of biofilms on thermal resistivity and frictional resistance, whilst others have been developed which are more concerned with measuring the events occurring within the biofilm for a greater understanding of its underlying processes. Biological investigations of films have been carried out using a number of techniques. These include scanning electron microscopy, the estimation of adenosine triphosphate (ATP) and (TOC) (Berger *et al.*, 1979). The information obtained from electron microscopy is largely qualitative in nature, and whilst giving insights into the structure and attachment mechanisms of the organisms involved does little to determine the underlying biochemical processes involved in them. The ATP method provides data on the stored energy pool of the total film. This method was modified later to include AMP and ADP to allow calculation of the 'adenylate charge' ratio, which is a measure of the physiological state of the population, and is more sensitive for use in the early stages of fouling. ATP measurements were found to be too insensitive to be of general use, especially in the later stages of film life when senescent portions of the biofilm contain little ATP. The TOC method was found to correlate closely with *Rf* values and gives a measure of both biomass and organic extracellular products such as slimes (Berger *et al.*, 1979). TOC is therefore a very useful tool for examining the relationship between biofouling and increased heat resistance.

A number of methods have been used to examine the biochemistry of films and the relationship with heat resistance. It has been shown that total biofilm nitrogen and protein correlate well with Rf, suggesting that either changes are directly related to biomass or to a proteinaceous extracellular product (Berger *et al.*, 1979). Capillary gas chromatography has been used to detect fatty acid methyl esters to give insights into the biochemical structure of the microbial population (Bobbie *et al.*, 1980). A combination of this technique with Fourier transform infrared analysis has provided a very powerful tool for analysing the transformations occurring within films (White, 1987). The same group have also developed a series of sensitive, quantitative analyses for the polymers found in films.

Our work in the School of Ocean Sciences currently centres upon the fouling films that develop in pipe sections exposed to high flow rates of seawater. We have designed and constructed a rig to assess the films which develop under these conditions, and we are currently comparing the fouling behaviour of a range of materials including both plastics and metals. We are also interested in the specificity of the polymer matrix–substratum interactions occurring in naturally developed biofilms and are attempting to quantify these. The results of this work will be discussed during the oral presentation.

REFERENCES

Aftring, R.P., Capone, D.C., Duguay, L., Fell, J., Master, I.M. and Taylor, B.F. 1978. Biofouling and site characterisation studies in the Ocean Thermal Energy Conversion (O.T.E.C.) experiment at St. Croix, U.S. Virgin Islands. In *Proceedings of the Fifth Ocean Thermal Energy Conversion Conference, Florida, USA* (eds. A. Lavi and T.N. Veziroglu), pp. VIII-45–VIII-73. Department of Energy, Washington, USA.

Albright, L.J., McCrae, S.K. and May, B.E. 1986. Attached and free floating bacterioplankton in Howe Sound, British Columbia, a coastal marine fjord–embayment. *Applied and Environmental Microbiology*, **51**: 614–621.

Bagg, J. and Silverwood, R.W. 1986. Coagglutination reactions between *Candida albicans* and oral bacteria. *Journal of Medical Microbiology*, **22**: 165–169.

Baier, R.E. 1972. Influence of the initial surface condition of materials on bioadhesion. In *Proceedings of the Third International Congress on Marine Corrosion and Fouling* (eds. R.F. Acker, B.F. Brown, J.R. DePalma and W.P. Iverson, pp. 633–639. North Western University Press, Evanston, USA.

Baier, R.E. 1981. Early events of microfouling of all heat transfer equipment. In *Fouling of heat transfer equipment* (eds. E.F.C. Somerscales and J.G. Knudsen). Hemisphere, Washington, USA.

Baier, R.E., Meyer, A.E., Depalma, V.A., King, R.W. and Fornalik, M.S. 1983. Surface microfouling during the induction period. *Journal of Heat Transfer*, **105**: 619–624.

Barnett, K.G., Cosgrove, T., Vincent, B., Sissons, D.S. and Cohen-Stuart, M. 1981. Measurement of the polymer bound fraction at the solid–liquid interface by pulsed nuclear magnetic resonance. *Macromolecules*, **14**: 1018–1020.

Bell, C.R. and Albright, L.J. 1982. Attached and free-floating bacteria in a diverse selection of water bodies. *Applied and Environmental Microbiology*, **43**: 1227–1237.

Berger, L.R. and Berger, J.A. 1986. Countermeasures to microfouling in simulated ocean thermal energy conversion heat exchangers with surface and deep ocean waters in Hawaii. *Applied and Environmental Microbiology*, **51**: 1186–1198.

Berger, L.R., McCoy, W.F. and Berger, J.A. 1979. Biofouling assay for O.T.E.C. pipes. In *Proceedings of the Ocean Thermal Energy Conversion (O.T.E.C.) Biofouling, Corrosion and Materials Workshop, Rosslyn, Virginia, USA*, pp. 38–56. Argonne National Laboratory, Argonne, Illinois, USA, ANL/OTEC-BCM-002.

Berk, S.G., Mitchell, R., Bobbie, R.J., Nickels, J.S. and White, D.C. 1981. Microfouling on metal surfaces exposed to seawater. *International Biodeterioration Bulletin*, **17**: 29–37.

Bobbie, R.J., Nickels, J.S., Davis, W.M., White, D.C., Lott, D.F., Dyjak, R. and Hollowell, J. 1979. Measurement of microfouling mass and community structure during succession in O.T.E.C. simulators — a preliminary report. In *Proceedings of the Ocean Thermal Energy Conversion (O.T.E.C.) Biofouling, Corrosion and Materials Workshop, Rosslyn, Virginia, USA*, pp. 101–121. Argonne National Laboratory, Argonne, Illinois, USA, ANL/OTEC-BCM-002.

Bobbie, R.J., White, D.C. and Benson, P.H. 1980. Biochemical analysis of the response of the marine microfouling community structure to cleaning procedures designed to increase heat transfer efficiency. In *Proceedings of the Fifth International Congress on Marine Corrosion and Fouling, Barcelona, Spain*, pp. 391–401.

Bott, T.R. and Gunatillaka, M. 1982. Nutrient composition and biofilm thickness. In *Proceedings of Fouling of Heat Exchange Surfaces Conference, Whitehaven, Pennsylvania, USA*, pp. 724–734.

Bott, T.R. and Miller, P.C. 1983. Mechanisms of biofilm formation on aluminium tubes. *Journal of Chemical Technology and Biotechnology*, **33b**: 177–184.

Bright, J.J. and Fletcher, M. 1983a. Amino acid assimilation and electron transport system activity in attached and free living marine bacteria. *Applied and Environmental Microbiology*, **45**: 818–825.

Bright, J.J. and Fletcher, M. 1983b. Amino acid assimilation and respiration by attached and free living populations of a marine *Pseudomonas* sp. *Microbial Ecology*, **9**: 215–226.

Bryers, J.D. and Characklis, W.G. 1981. Early fouling biofilm formation in a turbulent flow system: overall kinetics. *Water Research*, **15**: 483–491.

Bryers, J.D., Characklis, W.G., Zelver, N. and Nimmons, M.G. 1979. Microbial film development and associated energy losses. In *Proceedings of the Sixth O.T.E.C. Conference, Washington, USA*, (ed. G.L. Dugger), pp. 12–15. Department of Energy, USA.

Busscher, H.J., Uyen, M.H.W.J.C., Van Pelt, A.W.J., Weerkamp, A.H. and Arends, J. 1986. Kinetics of the adhesion of the oral bacterium *Streptococcus sanguis* CH3 to polymers with different surface free energies. *Applied and Environmental Microbiology*, **51**: 910–914.

Characklis, W.G. 1973a. Attached microbial growths — I. Attachment and growth. *Water Research*, **7**: 1113–1127.

Characklis, W.G. 1973b. Attached microbial growths — II. Frictional resistance due to microbial slimes. *Water Research*, **7**: 1249–1258.

Characklis, W.G. 1981a. Bioengineering report: fouling biofilm development: a process analysis. *Biotechnology and Bioengineering*, **23**: 1923–1960.

Characklis, W.G. 1981b. Microbial fouling: a process analysis. In *Fouling of Heat Transfer Equipment* (eds. E.F.C. Somerscales and J.G. Knudsen), pp. 251–291. Hemisphere, Washington, DC.

Christensen, B.E., Kjosbakken, J. and Smidsrod, O. 1985. Partial chemical and physical characterisation of two extracellular polysaccharides produced by marine, periphytic *Pseudomonas* strain NCMB2021. *Applied and Environmental Microbiology*, **50**: 837–845.

Corpe, W.A. 1976. Primary bacterial films and marine microfouling. In *Proceedings of the Fourth International Congress of Marine Corrosion and Fouling, Juan des Pins, France*, pp. 105–108.

Costerton, J.W., Marrier, T.J. and Cheng, K.J. 1985. Phenomenon of bacterial adhesion. In *Bacterial adhesion; mechanisms and physiological significance* (eds. D.C. Savage and M. Fletcher), pp. 3–43. Plenum Press, New York.

Dahlback, B., Hermansson, M., Kjelleberg, S. and Nordkrans, B. 1981. The hydrophobicity of bacteria, an important factor in their initial adhesion at the air/water interface. *Archives of Microbiology*, **128**: 267–270.

Davidson, A.M. and Fry, J.C. 1987. A mathematical model for the growth of bacterial microcolonies on marine sediment. *Microbial Ecology*, **13**: 31–45.

Derjaquin, B.V. and Landau, L. 1941. Theory of the stability of strongly charged lyophobic solutions and of the adhesion of strongly charged particles in solutions of electrolytes. *Acta Physicochimica, USSR*, **14**: 633–662.

Ducklow, H.W. and Kirchman, D.L. 1983. Bacterial dynamics and distribution during a spring diatom bloom in the Hudson River plume, USA. *Journal of Plankton Research*, **5**: 333–355.

Dudderidge, J.E., Kent, C.A., Miller, C., Laws, J.F., Pritchard, A.M. and Bott, T.R. 1982. Effects of flow on biofilm development. In *Proceedings of a Conference: Fouling of Heat Exchange Surfaces, Whitehaven, Pennsylvania, USA*, pp. 717–725.

Estermann, E.F., Peterson, G.H. and McLaren, A.D. 1959. Digestion of clay protein, lignin protein and silica protein complexes by enzymes and bacteria. *Proceedings of the Amercian Society of Soil Science*, **23**: 31–36.

Fletcher, M. 1979. A microautoradiographic study of the activity of attached and free living bacteria. *Archives of Microbiology*, **122**: 271–274.

Fletcher, M. 1986. Measurement of glucose utilisation by *Pseudomonas fluorescens* that are free-living and that are attached to surfaces. *Applied and Environmental Microbiology*, **52**: 672–676.

Fletcher, M. and Floodgate, G.D. 1973. An electron microscope demonstration of an acidic polysaccharide involved in adhesion of a marine bacterium to solid surfaces. *Journal of General Microbiology*, **74**: 325–334.

Fletcher, M. and Loeb, G.I. 1979. Influences of substratum characteristics on the attachment of a marine Pseudomonad to solid surfaces. *Applied and Environmental Microbiology*, **37**: 67–72.

Fletcher, M. and Marshall, K.C. 1982. Are solid surfaces of ecological significance to aquatic bacteria? In *Advances in microbial ecology*, Vol. 6 (ed. K.C. Marshall), pp. 199–236. Plenum Press, New York.

Gerchakov, S.M., Marszalek, D.S., Roth, F.J. and Udey, L.R. 1976. Succession of periphytic microorganisms on metal and glass surfaces in natural seawater. In *Proceedings of the Fourth International Congress of Marine Corrosion and Fouling, Jaun des Pins, France*, pp. 203–211.

Gordon, A.S., Gerchakov, S.M. and Millero, F.J. 1983. Effects of inorganic particles on metabolism by a periphytic marine bacterium. *Applied and Environmental Microbiology*, **45**: 411–417.

Goulder, R. 1977. Attached and free bacteria in an estuary with abundant suspended solids. *Journal of Applied Bacteriology*, **43**: 399–405.

Goupil, D.W., DePalma, V.A. and Baier, R.E. 1980. Physical/chemical characteristics of the macromolecular conditioning film in biological fouling. In *Proceedings of the Fifth International Congress on Marine Corrosion and Fouling, Madrid, Spain*, pp. 401–410.

Harden, V.P. and Harris, J.O. 1953. The iso-electric point of bacterial cells. *Journal of Bacteriology*, **65**: 198–202.

Harvey, R.W. and Young, L.Y. 1980. Enumeration of particle-bound and unattached respiring bacteria in the salt-marsh environment. *Applied and Environmental Microbiology*, **40**: 156–160.

Harwood, J.H. and Pirt, S.J. 1972. Quantitative aspects of growth of the methane oxidising bacterium *Methylococcus capsulatus* on methane in shake flask and continuous chemostat culture. *Journal of Applied Bacteriology*, **35**: 597–607.

Hattori, R. 1972. Growth of *E. coli* on the surface of an anion-exchange resin in continuous flow system. *Journal of General and Applied Microbiology*, **18**: 319–330.

Hattori, R. and Hattori, T. 1963. Effect of a liquid/solid interface on the life of micro-organisms. *Ecological Reviews*, **16**: 64–70.

Hodson, R.E., MacCubbin, A.E. and Pomeroy, L.R. 1981. Dissolved adenosine triphosphate utilisation by free-living and attached bacterioplankton. *Marine Biology*, **64**: 43–51.

Hood, M.A., Guckert, J.B., White, D.C. and Deck, F. 1986. Effect of nutrient deprivation on lipid, carbohydrate, DNA, RNA and protein levels in *Vibrio cholerae*. *Applied and Environmental Microbiology*, **52**: 788–793.

Imam, S.H., Bard, R.F. and Tosteson, T.R. 1984. Specificity of marine microbial surface interactions. *Applied and Environmental Microbiology*, **48**: 833–839.

Jeffrey, W.H. and Paul, J.H. 1985. Activity of an attached and free living *Vibrio* sp. as measured by thymidine incorporation, *p*-iodonitrotetrazolium reduction and ATP/DNA ratios. *Applied and Environmental Microbiology*, **51**: 150–156.

Jeffrey, W.H. and Paul, J.H. 1986. Activity measurements of planktonic microbial and microfouling communities in a eutrophic estuary. *Applied and Environmental Microbiology*, **51**: 157–162.

Kefford, B., Kjelleberg, S. and Marshall, K.C. 1983. Bacterial scavenging: utilisation of fatty acids localised at a solid/liquid interface. *Archives of Microbiology*, **133**: 257–260.

Kirchman, D. and Mitchell, R. 1982. Contribution of particle bound bacteria to total microheterotrophic activity in five ponds and two marshes. *Applied and*

Environmental Microbiology, **43**: 200–209.

Kjelleberg, S., Humphrey, B.A. and Marshall, K.C. 1982. Effect of interfaces on small, starved marine bacteria. *Applied and Environmental Microbiology*, **43**: 1166–1172.

Leech, R. and Hefford, R.J.W. 1980. The observation of bacterial deposition from a flowing suspension. In *Microbial adhesion to surfaces* (eds. R.C.W. Berkeley, J.M. Lynch, J. Melling, P.R. Rutter and B. Vincent), pp. 544–545. Ellis Horwood, Chichester.

Little, B. and Lavoie, D. 1979. Gulf of Mexico ocean thermal energy conversion (O.T.E.C.) biofouling and corrosion experiment. In *Proceedings of the Ocean Thermal Energy Conversion (O.T.E.C.) Biofouling, Corrosion and Materials Workshop, Rosslyn, Virginia, USA*, pp. 60–101. Argonne National Laboratory, Argonne, Illinois, USA, ANL/OTEC-BCM-002.

Lock, M.A. and Ford, T.E. 1985. Microcalorimetric approach to determine relationships between energy supply and metabolism in river epilithon. *Applied and Environmental Microbiology*, **49**: 408–412.

Lovell, C.R. and Konopka, A. 1985. Thymidine incorporation by free-living and particle-bound bacteria in a eutrophic dimictic lake. *Applied and Environmental Microbiology*, **49**: 501–504.

Lowe, M.J., Duddridge, J.E., Pritchard, A.M. and Bott, T.R. 1984. Biological–particulate fouling interactions: effects of suspended particles on biofilm development. In *First UK Conference on Heat Transfer, Leeds, UK*, pp. 391–400. Institute of Chemical Engineering.

Magnusson, K.E., Stendahl, O., Tagesson, C., Edebo, L. and Johansson, G. 1977. The tendency of smooth and rough *Salmonella typhimurium* bacteria and lipopolysaccharide to hydrophobic and ionic interactions as studied in an aqueous polymer two-phase system. *Acta Pathologica et Microbiologica Scandinavica, Section B*, **85**: 212–218.

Mahachanchawalit, P., Maeda, M. and Simidu, U. 1986. Succession of ciliated protozoa on the solid surfaces of substrata in seawater. *Nihon Biseibutsu Seitai Gakkaiho*, **1**: 9–17.

Malmcrona-Friberg, K., Tunlid, A., Marden, P., Kjelleberg, S. and Odham, G. 1986. Chemical changes in cell envelope and poly-β-hydroxybutyrate during short term starvation of a marine bacterial isolate. *Archives of Microbiology*, **144**: 340–345.

Marsh, P.D., Hunter, J.R., Bowden, G.H., Hamilton, I.R., McKee, A.S., Hardie, J.M. and Ellwood, D.C. 1983. The influence of growth rate and nutrient limitation on the microbial composition and biochemical properties of a mixed culture of oral bacteria grown in a chemostat. *Journal of General Microbiology*, **129**: 755–770.

Marshall, K.C. and Cruickshank, R.H. 1973. Cell surface hydrophobicity and the orientation of certain bacteria at interfaces. *Archives of Microbiology*, **91**: 29–40.

McEldowney, S. and Fletcher, M. 1986. Variability of the influence of physico-chemical factors affecting bacterial adhesion to polystyrene substrata. *Applied and Environmental Microbiology*, **52**: 460–465.

Merry, H. and Wright, K.H.R. 1982. Cooling water studies at NEL. In *Fouling and Corrosion of Metals in Seawater, Proceedings of a Meeting, Oban, UK*, (ed. J. Mauchline), pp. 263–290.

Meyer-Reil, L.A. 1984. Bacterial biomass and heterotrophic activity in sediments and overlying water. In *Heterotrophic activity in the sea* (eds. J.E. Hobbie and P.J.LeB. Williams), pp. 523–546. Plenum Press, New York.

Meyer-Reil, L.A., Dawson, R., Liebezeit, G. and Tiedje, H. 1978. Fluctuations and interactions of bacterial activity in sandy beach sediments and overlying waters. *Marine Biology*, **48**: 161–171.

Mills, A.L. and Maubrey, R. 1981. Effect of mineral composition on bacterial attachment to submerged rock surfaces. *Microbial Ecology*, **7**: 315–322.

Mittelman, M.W. and Geesey, G.G. 1985. Copper binding characteristics of exopolymers from a freshwater sediment bacterium. *Applied and Environmental Microbiology*, **49**: 846–851.

Miyamoto, S., Ishii, Y. and Ohnuma, H. 1981. Intrinsic viscosity of polyelectrolytes having hydrophobic side residues. *Macromolecular Chemistry*, **182**: 483–500.

Neihof, R.A. and Loeb, G.I. 1972. The surface charge of particulate matter in seawater. *Limnology and Oceanography*, **17**: 7–16.

Neihof, R.A. and Loeb, G.I. 1974. Dissolved organic matter in seawater and the electric charge of immersed surfaces. *Journal of Marine Research*, **32**: 5–12.

Neumann, A.W., Good, R.J., Hope, C.J. and Sejpal, M. 1974. An equation of state approach to determine surface tensions of low-energy solids from contact angles. *Journal of Colloids and Interface Science*, **49**: 291–304.

Nickels, J.S., Bobbie, R.J., Lott, R.F., Martz, R.F., Benson, P.H. and White, D.C. 1981. Effect of manual brush cleaning on biomass and community structure of microfouling film formed on aluminium and titanium surfaces exposed to rapidly flowing seawater. *Applied and Environmental Microbiology*, **41**: 1442–1453.

Norrman, G. 1976. Control of microbial fouling in circular tubes with chlorine. *MSc. Thesis*. Rice University, USA.

Panchal, P.B., Stevens, H.S., Genens, L.E., Hillis, D.L., Larsen-Basse, J., Zaidi, S. and Daniel, T. 1984. Biofouling and corrosion studies at the seacoast test facility in Hawaii. In *Proceedings of Oceans '84, A Conference Record: Industry, Government, Education: Designs for the Future*, pp. 364–369. Marine Technology Society–IEEE, New York, USA.

Paul, J.H. and Jeffrey, W.H. 1985. Evidence for separate adhesion mechanisms for hydrophilic and hydrophobic surfaces in *Vibrio proteolytica*. *Applied and Environmental Microbiology*, **50**: 431–437.

Pearl, H.W. 1984. Influence of attachment on microbial metabolism and growth in aquatic ecosystems. In *Bacterial adhesion: mechanisms and physiological significance* (eds. D.C. Savage and M. Fletcher), pp. 363–395. Plenum Press, New York.

Pringle, J.H. and Fletcher, M. 1986. Influence of substratum hydration and adsorbed macromolecules on bacterial attachment to surfaces. *Applied and Environmental Microbiology*, **51**: 1321–1325.

Pringle, J.H., Fletcher, M. and Ellwood, D.C. 1983. Selection of attachment mutants during the continuous culture of *Ps. fluorescens* and relationship between attachment ability and surface composition. *Journal of General Microbiology*, **129**: 2557–2569.

Robb, I.D. 1984. Stereo-chemistry and function of polymers. In *Microbial adhesion and aggregation* (ed. K.C. Marshall), pp. 39–49. Springer-Verlag, Berlin.

Roper, M.M. and Marshall, K.C. 1974. Modification of the interaction between *E. coli* and bacteriophage in saline sediment. *Microbial Ecology*, **1**: 1–13.

Rosenberg, E., Kaplan, N., Pines, O., Rosenberg, M. and Gutnick, D. 1983. Capsular polysaccharides interfere with adherence of *Acinetobacter calcoaceticus* to hydrocarbons. *FEMS Microbiology Letters*, **17**: 157–160.

Rutter, P.R. and Vincent, B. 1984. Physicochemical interactions of the substratum, microorganisms, and the fluid phase. In *Microbial adhesion and aggregation* (ed. K.C. Marshall), pp. 21–38. Springer-Verlag, Berlin.

Sherman, L.A. and Savage, D.C. 1986. Lipoteichoic acids in *Lactobacillus* strains that colonise the mouse gastric epithelium. *Applied and Experimental Microbiology*, **52**: 302–304.

Verwey, E.J.W. and Overbeek, J.T.G. 1984. *Theory of the stability of lyphobic colloids*. Elsevier, Amsterdam.

Vrij, A. 1976. Polymers at interfaces and the interactions in colloidal dispersion. *Pure and Applied Chemistry*, **48**: 471–483.

Weise, W. and Rheinheimer, G. 1978. Scanning electron microscopy and epifluorescence investigation of bacterial colonisation of marine sand sediments. *Microbial Ecology*, **4**: 175–188.

White, D.C. 1987. Environmental effects testing with quantitative microbial analysis: *in situ* biofilm analysis by FT/IR correlated to chemical signatures. In *Toxicology assessment* (eds. D. Lui and B.J. Dutka), Wiley, New York, in press.

Williams, P.J. 1970. Heterotrophic utilisation of dissolved organic compounds in the sea. I. Size distribution of population and relationship between respiration and incorporation of growth substrates. *Journal of the Marine Biological Association of UK*, **50**: 859–870.

Wood, J.M. 1980. The incubation of micro-organisms with ion exchange resins. In *Microbial adhesion to surfaces* (eds. R.C.W. Berkeley, J.M. Lynch, J. Melling, P.R. Rutter and B. Vincent), pp. 163–185. Ellis Horwood, Chichester.

Wu, S. 1980. Surface tension of solids: generalisation and reinterpretation of critical surface tension. In *Adhesion and adsorption of polymers* (ed. L.-H. Lee), pp. 53–65. Plenum Press, New York.

Zaidi, B.R., Bard, R.F. and Tosteson, T.R. 1984. Microbial specificity of metallic surfaces exposed to ambient seawater. *Applied and Environmental Microbiology*, **48**: 519–524.

Zambon, J.J., Huber, P.S., Mayer, A.E., Slots, J., Fornalik, M.S. and Baier, R.E. 1984. An *in situ* identification of marine bacterial species in marine microfouling films by using an immunofluorescence technique. *Applied and Environmental Microbiology*, **48**: 1214–1220.

Zisman, W.A. 1964. Relation of equilibrium contact angles to liquid and solid constitution. In: *Contact angle, wettability and adhesion* (ed. R.F. Gould). American Chemical Society, Washington, DC.

Index